D0908181

From Darwin to behaviourism

From Darwin to behaviourism

Psychology and the minds of animals

ROBERT BOAKES

Laboratory of Experimental Psychology,
University of Sussex

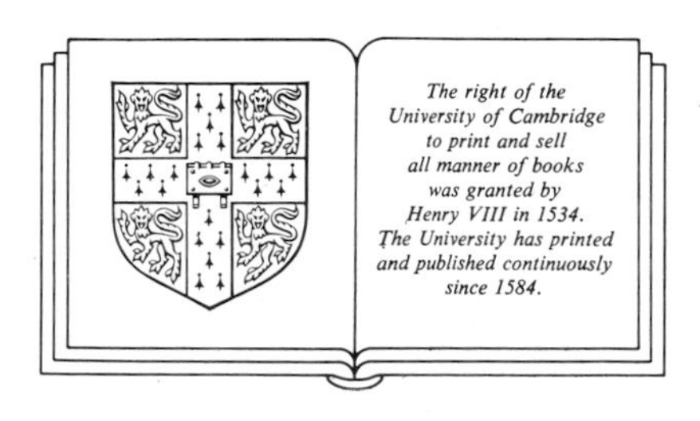
The right of the
University of Cambridge
to print and sell
all manner of books
was granted by
Henry VIII in 1534.
The University has printed
and published continuously
since 1584.

CAMBRIDGE UNIVERSITY PRESS
Cambridge
London New York New Rochelle
Melbourne Sydney

Published by the Press Syndicate of the University of Cambridge
The Pitt Building, Trumpington Street, Cambridge CB2 1RP
32 East 57th Street, New York, NY 10022, USA
296 Beaconsfield Parade, Middle Park, Melbourne 3206, Australia

First published 1984

Printed in Great Britain at the University Press, Cambridge

Library of Congress catalogue card number: 83–10091

British Library cataloguing in publication data

Boakes, Robert Alan
From Darwin to behaviourism: psychology and the
minds of animals.

1. Animal behavior 2. Animal psychology – History
I. Title
591.51'092'2 QL751

ISBN 0 521 23512 X hard covers
ISBN 0 521 28012 5 paperback

To Mary, Stephen, Christopher and Leila

Contents

Illustrations

Acknowledgements

The opportunity to follow up a latent interest in the history of psychology first arose during a year spent at the Department of Psychology of Princeton University. I am grateful to the members of this department, particularly Leon Kamin and Sam Glucksberg, for providing this opportunity and to Julian Jaynes who game me the initial encouragement to make the interest a serious one. This turned out to involve far more work than I had anticipated and persisting with it has needed the combination of specific criticisms and general enthusiasm that has been offered over the years by Vin LoLordo, Fergus Lowe and David Murray; the advice on what a non-psychologist might well find obscure or tedious given by John Mogg; the invaluable assistance and professional reassurance contributed by Philip Pauly from the early days of the project; and the devastatingly accurate evaluation of what was wrong with the early drafts of each chapter ever cheerfully and promptly delivered by Mic Burton.

It is also a pleasure to acknowledge the useful, lengthy and informative notes and evaluations of individual chapters, provided by Fred Westbrook for Chs. 1 & 2; by Bob Bolles for Ch. 3; by Bernard Singer for Ch. 4; by Jeffrey Gray and Stephen Walker for Ch. 5; by Lexa Logue and Cedric Larson for Ch. 6; by the late Hans-Lukas Teuber, Mary Henle and John Last for Ch. 7; and by Bob Bolles, Steve Cross and Franz Samelson for Ch. 8. One of the many things I learned in preparing this book was how lengthy a job it can be to obtain suitable illustrations. This task would have been even more arduous without the help of a number of people including Eliot Hearst, Dexter Gormley, Bernard Singer, John Mollon, Marion Zunz and William Schupbach. I also very much appreciated the care and interest shown by Colin Atherton in the photographic work he did for this book.

Finally, I would like to express my appreciation for the help and support provided by various fellow members of the Laboratory of Experimental Psychology at Sussex. I am very grateful to Valerie Leroy, Stella Frost, Anne Doidge and Susanne Westgate for typing various chapters. Writing a history of some subject is, for me, of interest only if the subject is fascinating in its own right. In this respect the book reflects the inspiration of the many friends who have worked in animal psychology at Sussex, including Seb Halliday, who was to have been a co-author at one early stage, Tony Dickinson, Geoff Hall, Euan Macphail and Nick Mackintosh.

Some of the research for this book was supported by a grant from the Royal Society. Permission was kindly given by the Adolf Meyer Archive (The Alan Mason Chesney Medical Archives of the Johns Hopkins Medical Institutions) to quote excerpts from the Watson/Meyer correspondence and to reproduce Fig. 6.11; by Yale University Library to quote excerpts from the Watson/Yerkes correspondence and reproduce Figs. 6.4, 6.5., 7.9, 7.10 and 7.11 from its Robert M. Yerkes Papers; by the Wellcome Institute Library, London to reproduce Figs. 1.9, 2.3, 2.4, 3.1, 4.4, 4.5, 4.8, 4.9 and 4.11; by the National Portrait Gallery, London to reproduce Figs. 1.3, 1.6 and 3.5; by the Osler Library, McGill University to reproduce Figs. 5.1, 5.2 and 5.6 from its Babkin Collection; by the Syndics of Cambridge University Library to reproduce Fig. 3.2; and by the other institutions indicated in the List of Illustrations for single reproductions. This list also gives the names of various individuals who very generously lent and allowed me to reproduce items from their own private collections; I am particularly grateful for help in this respect to members of the Morgan family, especially Mary Denniston.

Preface

My aim in this book has been to provide an account of the study of animal behaviour and of various ideas during the period from around 1870 to 1930 as to what kind of minds animals possess. I have emphasized those theories and discoveries which have most influenced the development of human psychology. 'Animal' is not used in the technical sense, but, following normal colloquial usage, refers to mammals other than human beings, birds, reptiles and amphibians. Except for a brief period at the beginning of this century the study of invertebrate behaviour developed independently and effectively had no influence on psychological thinking.

The detailed account begins around 1870 because this is when the behaviour of animals was first viewed within the context of a well-developed and widely accepted theory of evolution. From this point one can trace a continuous tradition which was often maintained by personal contact between contributors from different generations in a way that had not happened previously. My original plan had been to cover a full century. However the increasingly narrow concentration on conditioning that occurred after 1930 meant that it would not be possible to discuss the work of the next four decades without providing a full introduction to its theories and technical vocabulary. Up to this point it is feasible to attempt, as I have tried throughout the book, to describe what various animal psychologists did, using everyday English in a way that I hope will make the book comprehensible to a reader for whom experimental psychology is completely unknown.

There was another reason why 1930 seemed a good point at which to stop. A dramatic increase in the scale of animal psychology occurred around this time. It was therefore impossible to give subsequent developments the same comprehensive and detailed treatment that I found necessary for earlier studies.

My final comments concern the decisions that had to be made as to whether to concentrate on the work itself, on the people who had the ideas and put time and energy into their research, or on the intellectual and social context in which this took place. I found it difficult to maintain any kind of general rule about this. In asking myself such questions as 'Where did theory *A* come from?' 'Why did *X* choose to concentrate on *P* and not *Q*?' 'Why was idea, *R*, that had so much promise, not followed up?' and so on, I found that sometimes the answer was to be found mainly in the work itself, sometimes in what was happening in that person's life and sometimes in terms of very general changes in intellectual climate or in the institutions supporting such studies. Consequently the book switches from one level to another in a way that may seem unpredictable, but not, I hope, confusing.

In the past, histories of psychology have concentrated on general theories and large scale bodies of research, adding items of biographical information, usually of a respectful kind, concerning a few major figures. In general they have refrained from discussing such mundane, yet often crucial, questions as, for example, who paid for the animals, space, equipment and labour. Clearly such accounts provide a misleading view of science, both by ignoring such pragmatic factors and by undervaluing the contributions made by the many productive researchers who failed to acquire the status of a hero in psychology's list of honour. The past few years have seen the entrance of professional historians of science into this area and they have begun to unearth some fascinating archival material which bears on the various practical and political constraints that operated upon some of the studies described in this book. I have been fortunate in being able to benefit from this work. However, to concentrate on such factors and give only a passing glance to the content of the theories and research that constitute a science is also a mistake. I hope that something else which will emerge from the following pages is that, when people have held strong beliefs about the nature of animals' minds and when these beliefs – or challenges by critics – have impelled them to leave the domain of words and find out what an animal does in some specific test, then very often the result has been a surprising one: they have discovered something that drastically changes their own beliefs and on some occasions such a discovery has had a profound effect on the way that a whole generation has viewed the human mind.

1
Mental evolution

The high standard of our intellectual powers and moral disposition is the greatest difficulty which presents itself, after we have been driven to this conclusion on the origin of man.
Charles Darwin: *The Descent of Man* (1871)

For millions of years man has regarded his fellow animals with great interest as sources of food, danger, power, amusement or companionship. At various times and places he has also regarded them with more detached curiosity. Most often this curiosity has been directed towards the physical characteristics of animals. Over the centuries the increased understanding of the anatomy and physiology of non-human animals produced by such curiosity has made an incomparable contribution to our knowledge of the human body. Less frequently this curiosity has been concerned with the way that other animals move within and act upon their worlds. Do animals have minds, as men have minds? Do they possess anything resembling human intelligence: can individuals from species other than man learn from experience, think or communicate? Can they feel pain, or pleasure? Can any being, but a human being, act in ways that can be judged right or wrong? Are some species more man-like, in ways other than physical resemblance, than are other species?

Attempts to obtain answers to questions like these have been made here and there over the centuries. Such attempts took on new importance in the middle of the nineteenth century, when theories concerned with the evolution of life on this planet changed man's perception of the way he was related to other living things. Until then, interest in such questions had been sporadic and disconnected. But from this time began a continuous tradition in the study of behaviour and the animal mind.

The beginnings of this tradition in Victorian England of the 1870s coincided with the separate birth of attempts to change the study of the human mind from its traditional position as a speculative sub-branch of philosophy into something that more closely resembled a natural science, a subject which would deserve the title of scientific psychology. These two developments soon became entwined, and one consequence, nearly fifty years later, was the rise within North American psychology of a movement known as behaviourism. This movement came to dominate American psychology for many decades. This dominance persisted during a period in which psychology was regarded as a relatively minor subject – or better, a subject of dubious intellectual content – elsewhere in the world. In our time behaviourism continues to affect our world by directly influencing the way we teach our children and treat those we consider mentally disordered.

Beyond the specific effects of the behaviourist movement the century-old tradition of what may more generally be termed animal psychology has had a deeply pervasive effect on how we see our fellow men. The general acceptance in our day of one theory of evolution, that of Charles Darwin (1809–1882), has been accompanied by less widespread acceptance of his view that the differences between the psychology of man and those of other animals are differences of degree, and not of kind. Whether this particular belief has been accepted or not, the general view of man as a part of nature implied by Darwin's theory has meant that following his work the study of animals has had a new kind of relevance to the understanding of man.

The chapters that follow cover a period of sixty years. They describe the changes that occurred in the way animal behaviour was studied and in the kinds of question that were asked about the mental functions of non-human species. Some of these changes were the results of new discoveries and techniques or of events in the lives of the people that pursued such questions. Other changes reflected major shifts in attitudes towards the study of mind, in subjects only indirectly related to psychology or in the social and institutional conditions in which psychological research was carried out. Consequently we shall sometimes be looking in detail at the lives and outlooks of particular individuals and at other times at such events as the emergence of a new kind of university in North America or the way that the First World War affected psychology, according to whatever seems most illuminating in trying to understand how and why various studies of animals were undertaken.

During the period covered by this book there were recurring clashes between two distinct scientific traditions, the evolutionary and the physiological. For almost twenty years before Darwin first published his theory of evolution, experimental physiologists, mainly working in German universities, had been making a series of important discoveries about the nervous system. A general theoretical concept for much of this work was the idea of the reflex. Eventually this concept was extended in a way that many hoped would provide a generally adequate explanation of why animals, and possibly people too, act in the way that they do. However, this did not have a major effect on the study of animal behaviour until the beginning of the twentieth century and so discussion of the physiological tradition is postponed until the fourth chapter.

The present chapter looks at some of the British scientists and philosophers whose work contributed to the beginnings of animal psychology. Darwin himself was clearly the most important figure and the publication of his book *The Descent of Man* in 1871 the most important point of departure. However, there were other important developments in the early 1870s which did not arise at all directly from concern with the problems of evolution. Interest in the nature of learning processes came mainly from philosophy, as seen in the work of Alexander Bain (1818–1903). It was also given major prominence in the second edition of Herbert Spencer's (1820–1903) influential *Principles of Psychology* published in 1870. Animal psychology later came to place overwhelming emphasis on experimental methods; the first use of an experimental paradigm for studying animal behaviour that had any continuity was by Douglas Spalding (*c.* 1840–1877), who was much more concerned with the issues raised by the philosopher-cum-psychologist Bain, than with the evolutionary questions to which Darwin's work was directed. Although Thomas Huxley (1825–1895), the final contributor discussed in this chapter, was a close friend and colleague of Darwin and for a time at the centre of the debates on evolution, his subsequent influence on animal psychology did not stem directly from his contributions as a prominent Darwinian.

Nevertheless, Darwin's views on evolution provided the most important starting point and the debates that followed were major elements of the intellectual environment in which animal psychology developed. These are described in the following section, where Darwin's views would receive fuller discussion if these were not already so adequately described elsewhere. In contrast, the contributions of the other men described in this chapter are less widely known.

Charles Darwin and *The Descent of Man*

In 1831, when HMS *Beagle* left Devonport in England to sail for South America and the Pacific Ocean, the physical extent of the earth and its present geography had in general terms been well charted. The main purpose of the voyage was to provide more detail about islands and coastlines on the other side of the globe.

The age of the earth, and the variety and origins of terrestrial life, were not understood. Since the beginning of the century the discovery of fossils and the study of geological strata had begun to extend man's temporal horizon. At least within scientific circles there was rapidly weakening belief in the estimate, derived from Biblical texts, that the earth had existed in its present form for about four thousand years. Attempts to reconcile the geological evidence with the account of the earth's origin found in the book of Genesis were of major concern to scientists of that era and of considerable public interest.

Speculation on the past history of life on earth posed even more of a threat to orthodox Christian belief. But the few theories of evolution that appeared early in the nineteenth century received little public debate.

Until the earth's physical history was better understood there was no secure framework for such theories. For the moment, increased understanding of the way that the structures and behaviour of living organisms were so well adapted to their particular environments provided further evidence for a divine creator, the argument for God's existence based on the perfect design of nature. The most systematic attempt to describe and explain changes in species, that of Jean Lamarck, was largely rejected in his native France and in England, where it became widely known just before the *Beagle* set sail. As well as being seen as a dangerously atheistic doctrine, it was considered scientifically unacceptable. Lamarck had died in poverty and scientific disrepute. At his funeral his daughter is said to have cried out: 'My father, time will avenge your memory!'[1]

In one way her prophecy proved to be correct in that his name became attached to an idea that he regarded as one of the less central aspects of his theory. The 'principle of the inheritance of acquired characteristics', alternatively known as the 'law of use and disuse', was to retain considerable currency for well over a century. According to this principle, the effects of an individual organism's interaction with its environment on its structure can be inherited to some degree by its descendants. In the present context, the important aspect for psychology of this Lamarckian principle is the idea that specific actions that an animal has acquired during its lifetime, and which have become habitual, may become at least partially instinctive in its offspring.

Among the few civilians on board the *Beagle* was a young man of twenty-two, whose student years at Cambridge had allowed plenty of time for his enthusiasms as a naturalist and had provided good training in geological investigation. The voyage provided ample opportunity for Darwin to make himself completely familiar with the current state of geological theory from the books he had brought with him. The scientific expeditions he undertook in Argentina and Chile removed any of his remaining doubt about the antiquity of the earth's crust. It became quite clear to him that conditions on earth had remained essentially the same for millions of years. Just as fascinating as the rocks and strata of the southern part of the continent were the varied and exotic species he encountered, and their occasionally striking resemblance to creatures whose fossilized remains he now collected. He began to grapple with the problem of understanding the origin of species.[2]

Darwin's interest in the relationship between man and other animals was apparent long before he found a satisfactory answer to the species question. Among the other civilians on board the *Beagle* were three natives of Tierra del Fuego, taken as hostages on an earlier expedition and now to be returned. In their few years of exile the three Fuegians had learnt some English and Spanish and acquired many European habits and manners. One of them, Jeremy Button, with his good humour and sympathy towards anyone in distress, had become a universal favourite on board the ship. Darwin's contact with these three left him ill-prepared for his first sight of Fuegians who had never left the island: 'It was without exception the most curious and interesting spectacle I ever beheld: I could not have believed how wide was the difference between savage and civilized man: it is greater than between a wild and domesticated animal, inasmuch in man there is a greater power of improvement'.[3]

Fig. 1.1. Jean Lamarck

The contrast between Jeremy Button and the people Darwin encountered during the *Beagle*'s stay in Tierra del Fuego made a deep impression on him. Questions of why the Fuegians' way of life remained so wretched, or why they should have remained in so inhospitable a land, seem to have occupied Darwin for years after his visit. In his early comments upon such matters a very Lamarckian viewpoint was prominent: in his account of the *Beagle*'s voyage he reflected that, since there was no reason to believe that the population was declining, the Fuegians must enjoy a certain kind of happiness to render life worth having, and concluded: 'Nature by making habit omnipotent, and its effects hereditary, has fitted the Fuegian to the climate and production of his miserable country.'[4]

In Darwin's notebooks and scattered references in early published works one can find reflections on the relation between man and other animals, evidence for his early acceptance that man was to be included within a general theory of evolution and indications of the important part that the Lamarckian principle played in his thought about psychological questions.[5] Yet four decades were to pass after the *Beagle* first set sail, before he wrote directly about such issues.

In the years following his return to England,

Darwin was busy publishing accounts of the work carried out on the voyage and with various kinds of biological research. In the orderly seclusion of his country house in Kent he worked only intermittently on the problem of evolution. He was already gaining a reputation as a leading scientist from other work by the time that, late in 1838, he first began to develop the principle of natural selection as a general theory of evolution. A brief essay on the theory was written for publication in the event of his death and the ideas were tried out on a few close associates. Meanwhile he continued to amass evidence and develop the theory in detail.

Darwin was by no means the only person concerned with the question of species. A younger man, Alfred Wallace (1823–1913), began as a collector of beetles as Darwin had done. Lacking Darwin's private fortune, Wallace had made a career as a collector of tropical specimens, largely butterflies, first on the Amazon and then in the East Indies.[6] He had corresponded a little with Darwin on the subject of evolution, but the latter in reply had been careful to refrain from describing his own particular theory. Then in 1858, lying in bed with fever on a small island in the Moluccas, the idea of natural selection as the mechanism of evolution occurred to Wallace, as it had twenty years earlier to Darwin. The paper describing the theory was sent by Wallace, with a note requesting that, if Darwin considered it to have any merit, it should be forwarded to the foremost society for the study of natural history, the Linnaean Society. Darwin, greatly perturbed, consulted the friends who had already read his unpublished essay. The delicate problem of priority of publication was then amicably resolved by arranging that Wallace's paper and one hastily prepared by Darwin should be presented at the same meeting of the society, in the absence of both authors. Neither paper mentioned the evolution of man, and neither paper roused great interest.

This event spurred Darwin to complete at least an abridged version of the book that he had been planning for so many years. This was published in 1859: *On the Origin of Species by Means of Natural Selection, or the Preservation of Favoured Races in the Struggle for Life*. The book had two aims: first, to demonstrate that evolution had taken place and, second, to argue that the primary mechanism of evolution was natural selection. The first purpose was clearly achieved: within scientific circles at least, there has never since been any serious questioning of the view that life on this planet is constantly changing and that species that exist now have their origin in very simple forms of life that existed millions of years ago.

Fig. 1.2. Charles Darwin at about the time of writing *The Origin of Species*

The second purpose was less easily achieved.

Darwin's theory is one that emphasizes diversity, in the form of chance variations from one generation to the next. In 1859 these variations were seen as always being infinitesimally small. But some slight variability was all that was needed, given that particular variations added just a little to the chance that the individuals possessing them would survive. With sufficient time the environment could exert a constant pressure in selecting individuals with these favourable characteristics and a new species would emerge.

This theory raised problems that were seen as increasingly serious as the years passed. Because of the difficulties described below, by the end of the century Darwin's account of evolution in 1859 was widely held to have been disproven and to be of historical interest only. It was not until the 1930s that natural selection re-acquired, and has since maintained, its central role in evolutionary theory. Thus, although animal psychology developed against the evolutionary background provided by Darwin, it did so at a time when his major theoretical contribution was thought to be of decreasing importance. Disturbing criticism had already begun when Darwin at last came to write on the evolution of the human mind and it affected what he had to say.

For natural selection operating on small and occasional variations to produce the intricate forms of life that exist now – the vertebrates, for example, with their complex brains and sense organs – demands a vast amount of time. In 1859 Darwin was satisfied, because of the geological evidence, that enough time had been available. The evidence from nineteenth-century physics disputed this. Calculations based on principles of thermodynamics whittled down the duration of the period when the earth could have supported life from the thousand million years demanded by Darwin's 1859 theory to perhaps only twenty or thirty million years. By 1870 the physicists' estimates had become widely known and accepted. The temporal horizon that had receded for the first half of the century now advanced, pressing biologists to find more rapid processes of evolution.

Darwin needed no critics to point out another problem, the absence of any accompanying theory of heredity. The occurrence of variation, slight differences between parents and offspring, was assumed, but not explained; no account was attempted of the much more salient aspect of heredity, the similarities between parents and offspring and the stability of species. For a while it seemed, as Darwin hoped, that one could understand evolution without making many assumptions about the process of inheritance. The more hostile reviews by fellow biologists immediately following the publication of *The Origin of Species* had in general increased Darwin's confidence that his theory was correct. But in 1869 there appeared a challenge to his ideas that he found particularly perturbing. This was a review, by an engineer, not a biologist, that included both a clear presentation of the arguments from physics regarding the limited age of the earth and an essentially mathematical demonstration which, based on the current 'blending' view of heredity, showed that new species could not arise by the continual selection of small variations.[7] One of the examples from this paper described the chance arrival of a European on some desert island where his particular characteristics equipped him to survive much better than its original inhabitants. Given that he exploited his superiority to the full, it appeared to follow from Darwinian theory that eventually the island would be peopled by a new race with European characteristics. This was absurd; the obvious outcome was that within a few generations there would be little trace of his arrival, a process of 'blending' would have taken place whereby the favourable characteristics he possessed were steadily diluted generation by generation. In the absence of a theory of heredity it was not at all clear how to refute such an argument.

This example leads to a third major issue that was added to those of time and of heredity in the debates on evolution preceding *The Descent of Man*: does the principle of natural selection explain the evolution of man? Publication of *The Origin of Species* had been long delayed by Darwin's belief that the favourable reception of a revolutionary idea requires cautious preparation, and also, maybe, by a fear of possible persecution and distaste for the public controversy that he knew would follow, given the religious beliefs of English society at that time. To prepare a theory then that could be seen as encouraging the spread of atheism was perhaps equivalent to advocating a point of view today that would encourage paedophilia. The arrival of Wallace's paper in 1858 had ended this delay, yet still there was no discussion of man. The *Origin* contained a single sentence referring to the question, which promised that the theory would shed light on human origins. Although veiled, the implication that man was as much the product of biological evolution as any other species was immediately perceived and provided the basis for most of the hostility and the intense public interest generated in the early 1860s. This was fanned when the implication was stated openly in 1862 by Thomas Huxley, by then known as Darwin's foremost champion, in his book, *Evidence as to Man's Place in Nature*. In this book Huxley concentrated on the anatomical similarities between the human brain and those of the great apes.

Two years later Wallace published a paper that applied the theory of natural selection to human evolution.[8] Two of the problems he discussed suggested a solution that emphasized the distinctiveness of man. One was the intellectual chasm that, despite the physical similarities, appeared to divide man from the great apes. The other was the indication from the limited amount of human fossil evidence then available that man had existed in his present form for a remarkably long time: these prehistoric ancestors who had lived in the same world as the mammoth and other mammals now long extinct did not appear to have been very different, in body or in size of brain, from the Victorians who discovered their remains. Wallace's explanation in 1864 was that natural selection had first produced a series of physical changes leading to the achievement of an upright posture by man's ancestors, and at that point a slow development of his brain had taken place. Once this had reached a sufficient level man's consequent control over his environment made him a species uniquely free from the pressures of natural selection.

The ability to make fire, clothes, tools, shelters and devise varied forms of social organization meant that the human race could triumphantly survive enormous changes in climate and habitat without the aid of any further physical change. Wallace saw these abilities as depending upon both the possession of a well-developed brain and, as importantly, upon the continuity of cultural traditions. The human brain is not so complex an organ that each individual can be expected to discover unaided the advantages of rubbing two dry sticks together in an appropriate way or how to construct a bow. Human evolution meant, for Wallace, the gradual accumulation of knowledge and skills. In studying the human mind the important question was to understand the mechanisms, not of heredity, but of cultural transmission.

Wallace did not regard the handing on of skills by one generation to the next as a particularly human trait. In the 1860s he combined his concern with mental evolution with an interest in the topic of nidification, the construction of nests. He came to believe that the way birds build their nests, and the way they sing, is much more dependent on learning than was commonly held. To a large extent these are skills that each individual has to acquire, mainly on the basis of imitation. Such learning by imitation provides for their continued persistence.

Fig. 1.3. Alfred Wallace prior to his departure to the East Indies

Wallace saw a great deal in common between the construction of shelters by birds and by man. Nonetheless it was clear to him that the human brain was entirely different from that of any other creature. What selection process could have led to its development?

According to Darwinian theory, a bodily organ is improved only to the extent that it provides an advantage over direct rivals in the struggle for survival. Darwin had not been impressed by the mental abilities of the Fuegians he had encountered; no large discrepancy between their minds and their level of existence was obvious to him. For Wallace it was otherwise. His travels had made him much better acquainted with non-European peoples than his fellow evolutionists. In 1869 he published a further paper on the subject, concluding that natural selection could not have produced the human brain: 'Natural selection could only have endowed the savage with a brain little superior to that of an ape, whereas he actually possesses one but very little inferior to that of the average member of our learned societies'. In his struggles with other species, millions of years earlier, what benefit to man followed from the capacity for acquiring complex language, inventing abstract logic and mathematics, creating music and art or developing a high moral sense? It seemed to Wallace that 'an instrument has been developed in advance of the needs of its possessor', and the only way he could see that this might have happened was by the intervention of some higher intelligence in the development of the human race.[9]

For Darwin to admit such a 'miraculous addition' was to undermine all that he and Wallace had achieved. A paper by Wallace once again added to the pressure on Darwin to make his own views widely known and in 1871 these appeared in *The Descent of Man*. The first two chapters were concerned with the physical resemblances between man and other mammals, discussing issues that were now less controversial and evidence that was often already familiar. The following two chapters are the ones of most interest here. They were entirely devoted to a comparison between the mental processes of man and those of other animals. Beginning with recognition of the great differences between even the inhabitants of Tierra del Fuego and the 'most highly organized ape', he set out to show that nonetheless there was no fundamental difference between man and the higher animals in their mental faculties.

He first considered the objection that there is a qualitative difference between man and other animals that can be expressed by the statement that the behaviour of other animals is entirely guided by instinct, while that of man by reason. He argued first that in almost any species the behaviour of an individual is in part instinctive and in part dependent on the individual's past experience. Furthermore, it appeared that instinct and learning are not inversely related across species: the beaver displays some of the most complex forms of instinctive behaviour of any mammal and is also quick to learn. It was clear to Darwin that there are gradations of intelligence between different species and that in some, notably the apes, behaviour occasionally demonstrates intelligence of an almost human level. As examples, he cited various reports of chimpanzees and orang-utans using tools such as stones to crack nuts and sticks used as levers. The human mind could be understood as a further step, even if a large one, in the evolutionary development of intellectual functions that could be observed in animals.

Another prominent objection was that based on language. To some of Darwin's critics it was self-evident that human language was so different from any form of animal communication that it could not be the result of evolution. Darwin's reply was to point out that several basic elements of language exist in the non-human world – the development of song in birds resulting from both learning and from an instinctive tendency, vocal mimicry in parrots and other birds, repertoires of calls in monkeys indicating various affective states – and that such elements, combined with a high development of mental powers, could well have led to the development of human language. The parallels that seemed to exist between biological evolution and what was known then about the historical development of languages added to his argument.

Finally, Darwin conceded that the most important distinguishing feature of the human mind was its moral sense, or conscience. He argued that the development of such a feature was the inevitable result, given the existence of certain basic instincts such as parental and filial affection, that followed from the evolution of intellectual powers and of language. 'Conscience' was by no means some fixed mental attribute, which was either possessed or not possessed; he pointed to the changes that had occurred recently in 'even the most civilized nations' in attitudes towards slavery, the status of women and indecency.

The central importance attached to intelligence, in the sense of skills in solving practical problems, should be noted. As we have seen, other distinguishing characteristics such as language and conscience were seen as inevitable by-products of the evolution of intelligence. But, to return to the problem that had caused Wallace's appeal to a higher intelligence, how had this occurred? Darwin's answer was to stress two mechanisms that had been given little prominence twelve years earlier. Both had the added attraction of facing the other two major objections to his theories, those of time and of blending; in principle they made possible a greater rate of evolutionary change, and also could prevent the rapid swamping of a newly emerged and favourable variation by cross-breeding within the species.

One mechanism was the Lamarckian principle. Since his days on the *Beagle* Darwin seems never to have seriously questioned his belief that the skills, habits and ways of thinking which an individual develops in his own lifetime are to some small degree passed on to his children, as part of their biological inheritance. The impression made by Jeremy Button's rapid acquisition of European manners appears to have left the conviction that mental evolution could be a rapid process. With circumstances favouring the practice of intellectual skills, frequent usage of the brain would produce cumulative changes over a relatively small number of generations. Although Fuegians might now be mentally inferior to any other members of the human race, it was conceivable that appropriate environmental changes could bring them to the level of Europeans in the not too distant future.

The second mechanism was that provided by sexual selection. The last, and major, part of *The Descent of Man* was devoted to this topic. It seemed to Darwin that many of the peculiarly human characteristics that perturbed Wallace might well have developed in this manner. As well as the possible sexual advantages conveyed by intelligence, the strange human lack of much bodily hair could, for example, have arisen in the same way as the exotic colouring or complex song of various birds, characteristics which do not obviously increase the chances of survival for their individual possessor.

The importance assigned to acquired characteristics and to sexual selection in the *Descent* marked a change of emphasis from the *Origin*, but change through survival of the fittest remained the central idea of Darwinian theory. In 1859 the aim had been to change man's view of nature from that of a harmonious world containing related, but distinct, forms of life, forms that had been reached under the guidance of some predetermined purpose or goal, to one of continuous flux governed by laws of chance. This had involved breaking down the boundaries that appeared to separate species from species. In 1871 the aim was to remove a further barrier, that between the human and animal mind. For Lamarck the inheritance of acquired characteristics was guided by purpose; it aided ascent on the ladder of life, the *scala naturae*, stretching up from the simplest organisms to man and beyond. For Darwin there was no such scale; he cautioned himself against using 'higher' and 'lower', against the idea that one species can be compared in 'degree of evolution' to some quite different species. Instead there was the symbol of the irregularly branching tree of life. In his use of the Lamarckian principle the teleological element was absent. Any development of the human mind that stemmed from what previous generations had learned was just as much the outcome of an entirely mechanistic interaction between an organism and its environment as the development of any physical organ.

The arguments for mental continuity between man and other animals put forward in the *Descent* were far from overwhelming. Reviewers were quick to point out the dubiously anecdotal nature of the evidence. For Darwin, it was sufficient to conclude that 'the difference in mind between man and the higher animals, great as it is, certainly is one of degree and not of kind'. Yet his tone is not that of pronouncing the final word on an old subject, but rather of pointing out the way that a new field of enquiry might develop.

The Spencer–Bain principle

One kind of criticism that greeted the publication of Darwin's theory in 1859 was that it did not conform to the standards of scientific explanation as conceived in current philosophies of science.[1] Such matters were one of the concerns of an intellectual tradition distinct from that of the group of geologists and naturalists to which Darwin belonged. This philosophical tradition was vigorously represented by a man only three years older than Darwin, John Stuart Mill (1806–1873). His *System of Logic* of 1843, with its analysis of induction and the logic of experimental research, was later to have a great, but indirect, influence on the study of animals.

A more immediate influence on animal psychology was the philosophy of mind developed by the succession of empirical philosophers that had preceded Mill. The two most pertinent aspects of this philosophy were its belief that the human mind develops from an initially unformed state – the *tabula rasa* – as a result of an individual's experience, and the argument that the process underlying this development is the formation of associations between ideas, through the perception of related occurrences of events in the world. To these basic principles Mill added the suggestion that a compound idea, formed by the association of two simple ideas, might have different properties from those of its constituents, just as common salt has properties entirely different from those of its elements, sodium and chlorine. Mill's arguments on what came to be called 'mental chemistry' encouraged animal psychologists later in the century to retain a belief in the fundamental role of associative processes and yet also, in a sometimes ill-defined way, a belief that the functions and development of the human and animal mind have different degrees of complexity.

Mill considered these issues from the familiar viewpoint of philosophy. But one of his closest associates was a man who has been described as 'in a certain sense the first psychologist'.[2] If Alexander Bain deserves the title, it is because he was the first person to devote almost his whole life to the study of the mind. The two books he wrote in the 1850s, *The Senses and the Intellect* in 1855 and *The Emotions and the Will* in 1859, were the first systematic attempt to detach the study of psychological problems from its position as one of the traditional concerns of philosophy and establish it as a natural science.

Physiological research over the previous decades had begun to make the nervous system far less of a mystery than it had been before the nineteenth century. The principal theme of Bain's psychology was a detailed examination of the relationship between neural processes and psychological phenomena. This examination began, as was still considered necessary at that time, with discussion of proofs that the brain is the principal organ of mind. Mental and bodily events were held to occur in a completely parallel fashion, with no causal connection between them. In his analysis of psychological processes Bain relied mainly on the kind of introspective evidence and associationist framework used by Mill. But, also like Mill, Bain advocated the use of experimental methods in the study of mind. However he never became an experimentalist himself and the real beginning of experimental psychology came later in Germany.

The original contribution made by Bain which is of most interest here stemmed from his attempt to understand the 'instinctive germ of volition', or – in more familiar terms – the origins of voluntary action. Since his solution for what he termed this 'grand difficulty' was to have a long and influential history,

Fig. 1.4. Alexander Bain

and is still current, for example, in the form of behaviourism advocated by B. F. Skinner, it deserves close attention.

Bain was very much concerned to draw a clear distinction between reflexive actions produced by the nervous system in response to some external event, the kind of activity emphasized in the physiology of that era, and what he termed spontaneous activity; 'the exercise of active energy originating in purely internal impulses, independent of the stimulus produced by outward impressions, is a primary fact of our constitution'.[3] The existence of this spontaneous activity was seen as an 'essential prelude to voluntary power', where 'volition is a compound, made up of this and something else'.[4] The 'something else' must be an element that transforms the random nature of spontaneous activity, giving it the directed property of voluntary behaviour and connecting it with perception of the emotions of pleasure and of pain.

Pleasure and pain were discussed at length in Bain's psychology and provided a central theme in Mill's political and ethical philosophy. Modifying the criterion of utility used by his father, James Mill, and by Jeremy Bentham, Mill applied his 'greatest happiness principle' to the problems of politics and of personal morals discussed in his books, *On Liberty* of 1859 and *Utilitarianism* of 1863. An action that is good is one that increases the sum total of human happiness, or one that decreases the sum total of human misery. The principle was applied to questions concerning the forms that political institutions should take, the kinds of social or legal constraints that could be justified and the decisions an individual should make in his personal life. Mill did not discuss the psychological processes that make such decisions possible.

A theory as to how emotion and action came to be linked was first outlined by Bain in 1855 and then developed in his discussion of the will in 1859. It seems to have been prompted when, accompanying a shepherd, he watched the first few hours of life of two lambs. Bain noted that vigorous initial movements appeared to be completely random, but that when chance contact occurred, first with the mother's skin and then, after two or three hours, with her teat, the lamb's actions became progressively more directed in character. 'Six or seven hours after birth the animal had made notable progress . . . The sensations of sight began to have a meaning. In less than twenty-four hours, the animal could at the sight of the mother ahead, move in the forward direction at once to come up to her, showing that a particular image had now been associated with a definite movement; the absence of any such association being most manifest in the early movements of life. It could proceed at once to the teat and suck, guided only by its desire and the sight of the object.'[5]

Bain provided other examples – of a human baby learning to remove a needle pricking its skin or to gain warmth when chilled – to suggest that this process of 'trial and error', as he called it, was the universal means by which voluntary control over spontaneous activity is first achieved. He was unable to suggest the nature of the physiological mechanism that enabled this process to occur. 'I cannot descend deeper into the obscurities of the cerebral organization than to state as a fact, that when pain co-exists with an accidental alleviating movement, or when pleasure co-exists with a pleasure-sustaining movement, such movements become subject to the control of the respective feelings which they occur in company with. Throughout all the grades of sentient existence, wherever any vestiges of action for a purpose are to be discerned, this link must be presumed to exist. Turn it over as we may on every side, some such ultimate connexion between the two great primary manifestations of our nature – pleasure and pain, with active instrumentality – must be assumed as the basis of our ability to work out ends.'[6]

These first two books were written by Bain during the years when he worked as a free-lance journalist in London and repeatedly failed to obtain any kind of university appointment. The rest of his life, before and after this period, was spent in Aberdeen. As a child he seems to have been as precocious as Mill, whose early educational attainments had amazed his father's associates. But Bain's situation was a very different one from the bookish atmosphere that Mill breathed from birth. Bain was the son of a weaver and from an early age had to work at the loom to earn money for an irregular education. As a Scot he was fortunate to live in possibly the only country 150 years ago where a boy from this kind of background could obtain a university education without very much difficulty, providing that he showed intellectual promise. After graduating he moved to London where he joined the circle around Mill. In 1860, at the age of forty-two, he finally obtained an academic post and returned as Professor of Logic to Aberdeen, where he remained for the rest of a long life.

Bain shared with Mill a surprising lack of interest in the debates of the 1860s over evolution. The observations on the behaviour of newborn lambs were exceptional in that there are few other references to animals in Bain's two books. This is in complete contrast to another author of a book on psychology that was also published in 1855. In the first edition of his *Principles of Psychology*, Herbert Spencer proclaimed that 'mind can be understood only by showing how mind is evolved'. Well before *The Origin of Species* appeared he had coined the phrase 'survival of the fittest' which was later adopted by Darwin.

Spencer's background was very different from that of either Bain or Darwin. As a boy he received an intermittent education from an uncle and from his father, a school master in the industrial town of Derby. From this education Spencer retained a life-long interest in a variety of subjects, especially scientific and political ones, a tendency to form strong opinions guided more by independent judgement than by custom or by careful study, and a critical attitude towards orthodox beliefs. By 1837, when Spencer was seventeen, the boom in railway building had begun in England. Spencer's childhood instruction in intellectual self-help was then followed by more formal training as a civil engineer over the next ten years when he worked for various small railway companies.

In 1848, on the strength of a few published articles, he went to London to work as a journalist, first as sub-editor on *The Economist* and then, like Bain, as a free-lance. His interest and ideas on psychology and on evolution were stimulated almost entirely by what he read. Mill's *System of Logic* was one important book. As Spencer acknowledged, he was never prone to study human nature in the concrete as well as in the abstract.

A few intellectual friendships were also highly important. He became close friends with Huxley, who served as a willing, but highly critical, sounding board for Spencer's ideas. Huxley commented that 'Spencer's idea of a tragedy is a deduction killed by a fact'[7] and continually advised the suspension of judgement in the absence of adequate evidence – advice rarely heeded by Spencer, whose attitude towards empirical evidence is indicated by his comment that he studied some topic so that 'I should be able to furnish myself with such detailed facts as were requisite for the setting forth of general conclusions'.[8]

Other friends of Spencer's included George Henry Lewes and Marian Evans. Lewes was a fellow journalist with a remarkably wide range of activities that encompassed literary criticism, writing biographies, acting and science. He wrote a lively book that summarized in a popular, but very informed, way the discoveries in physiology from continental universities. As will be seen later, this book's influence extended as far as Russia. Evans was also a journalist at the time. But after Spencer had made it clear that his friendship for her could never be any more than an intellectual one, he introduced her to Lewes and, partly as a result of the relationship that subsequently developed between them, she became George Eliot. Despite his own considerable achievements Lewes has remained known primarily for the support that enabled Eliot to write her novels.[9]

From his voracious reading and the exchanges with his group of friends during the early 1850s Spencer acquired a general vision of the world that was to have a more pervasive effect on nineteenth-century thinking than that of any other philosopher of his era. Much of what has since been called 'Social Darwinism' and the beliefs about the human mind which, later in this book, are termed 'Psychological Darwinism' could more appropriately be given the title 'Spencerism'. The key idea in Spencer's vision was that every aspect of the world is continuously changing and that the direction of this change is from simple to complex. An individual animal starts as an

Fig. 1.5. Herbert Spencer at about the time of writing *The Principles of Psychology*

embryo and develops into a highly differentiated organism with countless specialized and interdependent parts; societies begin as small groups in which any member can take over most of the duties and functions of any other member and historically in Western Europe and North America have developed the intricate co-ordination demanded in an industrialized nation; similarly, the living world began with a few simple forms of life from which increasingly complex creatures have steadily evolved.

Like Bain, Spencer's major theme in his *Principles of Psychology* was to examine the relationship between psychological phenomena and physiological processes. Bain was mainly interested in the way that detailed knowledge of the nervous system could provide an understanding of human perception. Spencer's interest was very different. The *evolution* of the nervous system and the general principle of a steady progression from its simple, undifferentiated state in primitive organisms to the complex and specialized structure of the human brain provided the framework for Spencer's psychology. The evolution of mind was intimately related to this progression: with increasing 'heterogeneity' came increasing capacity for movement and for sensing distant events, and for Spencer mind and motion were identical – although it is not very clear what he meant by propounding this identity.

Within this framework the behaviour of animals assumed a key importance. He proposed that the evolution of the nervous system resulted from increasing complexity in the way that reactions occur to external events. His scale for these went from reflexes to instincts, then to memory, and finally to behaviour based on reasoning. The process responsible for this progression was the law of association, now viewed as a fundamental principle of nature: 'Hence the growth of intelligence at large depends on the law, that when any two psychical states occur in immediate succession, an effect is produced such that if the first subsequently recurs, there is a certain tendency for the second to follow it.'[10]

One of Spencer's examples will serve to illustrate how this law was supposed to operate. Take a simple organism that contracts when touched, and also has some form of primitive sensitivity to light. If a shadow consistently falls across the creature before it is touched, the visual stimulus will become associated with the tactile one and hence will itself begin to elicit an anticipatory contraction response. Given that this pattern of events is a constant element in the environment of the creature, and given the Lamarckian principle which Spencer never doubted, its descendants would acquire a more complex nervous system which ensured that shadows elicited contraction, independently of an individual's experience.[11] By such means – the inherited effect of the kind of learning that Pavlov was later to study under the title of 'conditioned reflexes' – instincts arose, defined by Spencer as 'compound reflexes', in which complex configurations of stimuli can evoke a finely co-ordinated series of movements, as in a newly hatched bird which immediately pecks at and captures an insect.[12]

The same law, producing ever more complicated relationships between external events and resultant actions and hence a more highly developed nervous system, led to the emergence of memory, whose origins lay in incipient actions, and then reason. And with this progression comes increasing consciousness. For the individual, an action might be made on the basis of highly conscious thought, but if the action is frequently performed, then it becomes increasingly automatic and independent of consciousness and, for the individual's descendants, might eventually attain the status of a mere reflex. When first learning some piece of music on the piano, a beginner might concentrate on every note. When more expert, the pianist might be just as able to carry on a conversation when playing the piece as when walking. And if the pianist and his offspring persist in such habits, then musical ability might begin to run in the family.[13]

The above ideas make an interesting comparison to those held by Darwin and Wallace. In *The Origin of Species* Darwin was very concerned to show why a Lamarckian principle of inheritance was wholly inadequate as a basic mechanism of evolution and assigned it a very minor role. As we have seen, twelve years later in 1871, it was given much more importance in *The Descent of Man*, where Darwin's discussion of behaviour and of psychological issues became close to that of Spencer.

In another respect too, Spencer was a Lamarckian – characteristically Spencer's initial interest in the whole topic began when an authoritative dismissal of Larmarck's theory persuaded him of its truth – in a way that Darwin always rejected. For Spencer evolution was unidimensional; although he did not use the term, the *scala naturae*, now the ladder of progress, lay beneath his whole approach. 'Evolution' meant 'progress' and 'more evolved' meant 'more complicated' and 'better.' He dealt with *the* evolution of *the* nervous system, not the various ways in which different forms of neural organization have evolved; and, in parallel with this, *the* evolution of *the* mind. Thus, for Spencer, 'mental continuity' meant a single continuum from the conditioning of a simple reflex to human intelligence. For Darwin it had no such meaning: 'mental continuity' implied simply that there was no especially abrupt jump from the animal to the human mind and that at least the rudiments of various human psychological processes could be seen here and there on the surviving branches of the tree of life.

Unlike Wallace, Spencer did not see much similarity in mental capacity between Europeans on the one hand and primitive or prehistoric man on the other, nor later did he appreciate Wallace's point about cultural evolution. With his single scale Spencer showed no hesitation in attributing higher mental qualities to the 'large brained European',[14] while claiming, for example, that 'amongst most of the lower races, acts of generosity or mercy are incomprehensible';[15] he stated that differences between men and women were based on the degree of organization of their brains;[16] and that the history of mathematics and astronomy directly reflects the development of the capacity for rationality in the human brain.[17]

In the 1850s Spencer's ideas evoked only limited interest: partly because he failed to see that the 'survival of the fittest' could serve as a primary process for evolutionary change and instead emphasized a Lamarckian viewpoint which was already familiar; and partly because, as someone lacking in scientific training and without the scientific reputation that Darwin first steadily acquired, his ideas on evolution tended to be seen as idle speculation. His reaction to Darwin's detailed arguments for organic evolution in *The Origin of Species* was pleasure mixed with chagrin that he himself had failed to see the potential importance of natural selection. Its publication created an intellectual climate that was now much more receptive to his own more general ideas. People were now eager to read about evolution.

By this time Spencer had decided on his life's work, to produce a comprehensive philosophy of science and of the human situation for this new age; successive volumes would examine the problems of biology, psychology, sociology, education and ethics within an overall framework constructed on the basic principles of what he termed his 'synthetic philosophy'. These volumes were to illustrate the global nature of evolution whereby heterogeneity emerges out of homogeneity.

The ambitious plan for a synthetic philosophy became feasible in financial terms as a growing readership for Spencer's work developed in the 1860s in the United States as well as in Britain. His personal habits also made the scheme possible. The impressive output of many of his fellow Victorians suggests unquenchable energy and concentration, supported by wives and servants to minimize domestic distractions. But for sheer quantity of published words Spencer had few rivals. As a young man he had avoided encumbering himself with such 'hostages to fortune', as he put it,[18] as a family and he remained throughout a long life a bachelor whose only passion was his philosophy. In addition, his productivity was increased by the then very unusual method of dictating most of his books. Thus he was able to make rapid progress with his grand scheme, reaching psychology in time to produce an extensively revised version of his earlier book, now appearing as the fourth and fifth large volumes of the synthetic philosophy, for publication in 1870 and 1872. These were largely dictated on a boat, in between vigorous bouts of rowing on a lake in a London park.

The general theme of that part of the first edition concerned with mental evolution was at least simple and coherent: a single process of learning, based on the principle of association, which ensured development along a single route. In the second edition Spencer introduced a second kind of process to explain how the behaviour of an organism adapts to its environment, without explaining how this process fitted into the scheme already described. The process was the one already described by Bain and, although unattributed, it seems very clear that Bain was Spencer's source.

In the second edition Spencer followed Bain in discussing the topics of pleasure and pain, but from an evolutionary point of view that now reflected the impact of Darwin's theory. Pleasure was defined as a state of consciousness that an animal seeks to prolong, and pain a feeling that an animal seeks to get out of consciousness and to keep out.[19] Pleasures are in general correlated with healthy activities and pains with biological injury. This was seen as an obvious consequence of natural selection, since species which persisted in pleasures that were uncorrelated with events of positive biological significance would tend not to survive. However, although these correlations are likely to be very high in relatively undeveloped species, in the higher animals and particularly in the case of man the natural connection between emotions and the biological utility of events that accompany them can become deranged. Spencer proposed that such derangements occur because of a long time lag from a change in a species' environment to an appropriate corresponding adjustment in what is inherently pleasant or painful for that species. Since man's habitat has changed very rapidly and since human society has – in Spencer's view, very foolishly – acted to resist the process of natural selection, the evolutionary connection between emotion and biological utility has become very distorted in man.[20] What we enjoy is no longer necessarily good for the human race.

The final part of the 1870 volume began with the statement that the earlier account of mental evolution, based on the law of repetition of experiences, was not enough. Spencer decided that it still left open the question: 'By what process is the organization of experiences achieved?'[21] The answer that he provided in terms of presumed processes in the nervous system is hard to understand. But the key idea is that of Bain, now expressed as follows. 'On the recurrence of the circumstances, these muscular movements that were followed by success are likely to be repeated; what was at first an accidental combination of motions will now be a combination having considerable probability.'[22] Bain felt that current knowledge about neural processes in the brain did not permit any useful suggestion as to what physiological mechanisms made his principle work. Spencer explained Bain's 'autonomous activity' as the result of 'a certain diffused discharge to the muscle at large' from 'the ganglionic plexuses', but his account of how a successful autonomous action is learned was ambiguous. In one paragraph he stated that such learning depends on 'a consciousness sufficiently developed to perceive the connexion between a muscular act and its immediate effect'. And yet in the next he provided a very different analysis, essentially the one set out clearly, thirty years later, by Thorndike, whereby the pleasurable sensations that closely follow the successful action are accompanied by a 'large draught of nervous energy'; this in turn increases the permeability of the 'line of nervous communication', through which the diffused discharge had passed just previously to produce the action.[23]

One hundred years later Spencer's books on psychology appear confused as well as barely readable. Bain's are clear and sensible. One suspects that their contemporaries shared this view. Darwin and the biologists had little respect for Spencer's judgement on subjects in which they were expert, but were prepared to respect his opinions on topics that were unfamiliar. In one burst of enthusiasm Darwin wrote that he considered Spencer to rank with philosophers such as Leibnitz and Descartes: but then spoiled the compliment by admitting that he had read nothing by either of these philosophers.[24] Mill's circle seem to have held a similar opinion, in their case having least respect for his philosophical ideas. Nevertheless Spencer gained considerable support, including indirect financial help, from a wide circle of intellectual acquaintances, which was based, one suspects, on widespread respect for his originality and the view that, even though most of his ideas might be foolish or turn out to be wrong, he produced so many that some had to be right.

Very few copies of Spencer's first edition or of Bain's books on psychology were sold. But by 1870 Spencer had acquired a considerable reputation and the second edition became one of the most influential and widely read books on psychology in English-speaking countries for at least twenty years. Consequently Bain's idea concerning the basis of voluntary action, the increase in frequency of some movement that occurs when it has been followed by a pleasurable event, became widely known until 1911 as the 'Spencer–Bain principle'.

Douglas Spalding's experiments on instincts

In Spencer's discussion of animal behaviour the emphasis given to instincts was at variance with the strongly environmentalist viewpoint dominant in British philosophy of the 1860s. Neither Mill nor Bain accepted the common belief that animals possessed the ability to perceive distance or spatial relationships at birth. Bain's own observations of the lambs had convinced him that such abilities might be acquired very rapidly during the first hours of life.

In 1862 Bain's lectures at Aberdeen were attended by a young man of about twenty-two, named Douglas Spalding, who had no previous formal education and who was earning his living by mending slate roofs. Spalding was greatly interested by the lectures, but at the same time puzzled that issues such as instinct had been debated for so long with arguments based on speculation and plausibility and not on empirical evidence. At the end of the year he left for London where he trained as a lawyer and contracted the tuberculosis which was to cause his early death. In 1868, while travelling in France and Italy, he met Mill, now living in retirement at Avignon, and was greatly impressed. Some time during this period he began to carry out a remarkable series of experiments on the topics Bain had discussed. The place and circumstances of his early research are not known.[1]

This work was first described at a scientific meeting in Brighton in August, 1872. The ability of a very young chick to move about its world without bumping into obstacles, to peck accurately at small objects or to locate the source of sounds were, according to the kind of view suggested by Bain, partly based on sensory experience gained during the period after hatching when the bird was capable of little movement. After removing part of the eggshell, Spalding slipped hoods over the eyes or inserted wax into the ears of a number of chicks, just before they emerged from their eggs, to deprive them of this initial experience of visual or auditory sensations. When, two or three days later, their neuro-muscular system had developed sufficiently and they were able to make co-ordinated movements, the hoods or the wax were removed. Spalding reported that these chicks were just as capable of directing accurate pecking movements towards and consuming insects, of avoiding obstacles and, at the sound of a call from an unseen mother hen, of moving in an appropriate direction as chicks that had not suffered any sensory deprivation. The unplanned arrival of a wild hawk, and subsequent tests with a tame one, suggested that fear reactions to certain kinds of specific visual stimuli were also innate.

Whereas these examples seemed to Spalding to be completely instinctive, the development of other reactions involved some element of experience. He referred to the latter as 'imperfect instincts'. One interesting example was what Spalding described as the 'following reaction'; with the rediscovery of this phenomenon sixty years later, it has been known since as 'imprinting'. If chicks or ducklings saw Spalding when their hoods were removed within a day or two of hatching, they would subsequently follow him around thereafter. However, if three or four days elapsed before the hoods were removed, no tendency to follow him developed and instead the chicks would show marked signs of fear on seeing him. There appeared to be a limited critical period in which this kind of attachment to some other, arbitrary creature could develop.

Spalding also noted other cases in which behaviour was modified by brief early experience. Initially chicks pecked at their own excrement, but quickly stopped doing so. Neither chicks, nor even ducklings, showed any sign of innate visual recognition of water: this came only after direct contact with water had occurred. However, unlike Bain or Spencer, he does not seem to have been particularly interested in the way that experience modifies behaviour. His concern was much more with finding out the extent to which behaviour depends on the maturation of inherited organization of the nervous system. In later studies he showed that co-ordinated flight and good spatial perception were immediately displayed by swallows and other birds, in whom wing movements had been prevented by cylindrical collars for a period immediately after hatching. He repeated his visual and auditory deprivation studies with piglets, with similar results to those obtained earlier, and commenced some experiments on cross-hatching in fowls.

Spalding's concept of instinctive behaviour was very close to Spencer's; he accepted the view that the origin of instincts was to be explained by the 'doctrine of inherited acquisition'. Although the results of his studies were seen as lending support to Spencer, Spalding continued to feel a greater general intellectual affinity with Mill and Bain. The differences between Spencer's and Mill's circles were not limited to the question of innate behaviour in animals. They also held opposed political views. Some comment on such issues may be appropriate at this point.

Spencer's views on society were as radical as his views on science, and equally influenced by the implications he drew from his evolutionary theory. For him progress in society was to be achieved by a 'genuine liberalism' which maximized individual liberty and minimized interference from the State; vaccination, and care for the infirm or insane, only served to promote the regression of the human race; economic and social differences between races, sexes or classes were part of the natural order, a necessary part of evolution. As for political institutions, enthusiasm for a democratic system based on universal suffrage was not included within Spencer's type of liberalism: 'representative government is the best possible for the administration of justice and the worst possible for everything else'.[2]

In contrast, extension of representative government in England by electoral reform was one of the causes in which Mill and his circle were very actively involved. Darwin and the biologists were wary about any social or ethical conclusions that might be drawn from theories of evolution: Karl Marx proposed to dedicate the English translation of *Das Kapital* to Darwin, but this was politely declined. Although most of the scientists felt this caution over matters in which they judged themselves to have little expertise, an exception was Huxley, for whom there was no limit to the range of subjects on which he was prepared to become expert. But even Huxley wrote little on such matters until much later when, in reaction to the growing influence of Spencer's social philosophy, he came to argue with increasing vigour for dissociating scientific knowledge from political and moral judgements.

Publication of Darwin's *Descent of Man* evoked an editorial from *The Times*, predicting a rapid collapse of morality should this view of man's origins gain wide acceptance. Yet the public and private conduct of men like Darwin, Wallace or Huxley measured up to the strictest canons of Victorian propriety. Trained as scientists to doubt intuitive explanations of natural phenomena, they were nonetheless able to prize the ethical intuitions of a Christian gentleman as possessing a universal validity which needed no underpinning of religious belief. Another set of ideas perceived as a source of potential danger to conventional morality was the philosophy of utilitarianism. In his writings Mill always stressed that the happiness principle was concerned with the higher pleasures and in his life gave no sign of enjoying lower ones. Neither Darwin nor Mill ever expressed publicly their lack of belief in religion. Both judged that to do so would markedly hinder the wide acceptance of their ideas on evolution or philosophy.

Mill's younger associates appeared to have felt less constrained by contemporary standards of respectability. One, Lord Amberley, campaigned vigorously for birth control, thus ensuring that he lost the only

Fig. 1.6. John Stuart Mill in old age

election for which he stood. His wife shared his views and herself became notorious for working for the emancipation of women. Whether Spalding was also involved in such political activity is not known, but in his book reviews he did not hesitate to express uncompromisingly materialistic views of the human mind and a lack of belief in Christianity.

In 1873 the Amberleys employed Spalding as a tutor to their elder son. All his later research was carried out at their house, Ravenscroft, in the countryside of the Wye Valley, with Lady Amberley acting as his research assistant. The circumstances attending the abrupt ending to this research are known because sixty years later the Amberley's second son, Bertrand Russell, became sufficiently curious about his parents to locate and edit their papers. At Ravenscroft Spalding's young animals mixed with the Amberley children in wandering through house and garden. While Spalding designed experiments and instructed the Amberleys on carrying them out, the Amberleys in turn discussed the problem of Spalding's celibacy with the result that, as

Fig. 1.7. Animal psychology's first research assistant, Lady Kate Amberley, with Bertrand Russell on her lap

a matter of principle, Lady Amberley took him to bed at intervals. Research was punctuated by winter expeditions to the Mediterranean for reasons of health. An infection contracted on one such journey soon ended this era with the death of Lady Amberley. Two years later Lord Amberley died, designating Spalding and another tutor as legal guardians to his two sons. The will was challenged by the grandparents, who had disapproved of Spalding from the start and who wielded considerable power; Lord Amberley's father had twice been Prime Minister. Spalding left England and a year or so later died in France at the age of 37.[3]

If Spalding had been able to continue his work, he might well have become the founder of the twentieth-century science of ethology. The research he published was recognized as outstanding in its time and later it received lengthy discussion in William James' highly influential psychology textbook of 1890. But, with psychology's increasing preoccupation with the nature of learning, in the twentieth century his findings were forgotten and had to be rediscovered; and his legacy, that of providing a starting point for an experimental approach to animal psychology benefited the study of learned, and not innate, behaviour.

Thomas Huxley and animals as conscious automata

There was one man of that era who shared very similar views, but held a very different social status, to Spalding and who similarly felt no hesitation in expressing them. If some forthright statement of what he saw to be true was likely to cause bitter controversy, then Thomas Huxley positively relished the prospect. And when such controversies died away Huxley remained in a more unassailable position than ever before.

Almost paradoxically Huxley had become a pillar of English society by the 1870s. His public reputation stemmed mainly from his vigorous attacks on some of the beliefs most cherished by his generation. At the same time he possessed, and was well-known to possess, the virtues most highly valued in the mid-Victorian age: honesty, courage, intellectual brilliance, mastery of written English and of public speaking, energetic involvement in public affairs and a capacity for unremitting hard work. It was these virtues – together with additional ones of humour, personal charm and political shrewdness – rather than his beliefs or his considerable scientific achievements that had gained for him a unique position.

Huxley wrote very little about the evolution of intelligence. He did not suggest any theory to explain the adaptiveness of behaviour. He did not perform any notable experiments on the behaviour of animals. Nonetheless the voyage of HMS *Rattlesnake*, a crucial event in Huxley's life, was almost as important to the subsequent history of animal psychology as the earlier voyage of HMS *Beagle.* Of Huxley's many and varied contributions, two in particular are notable here. He was the main source, partly through his student Lloyd Morgan, for the very sceptical attitude which first came to permeate animal psychology and then, via the behaviourist movement, a great part of American psychology in the middle of the twentieth century. Referring to his first name he once wrote: 'Why I was christened Thomas I do not know; but it is a curious chance that my parents should have fixed for my usual denomination upon the name of that particular Apostle with whom I have always felt most sympathy.'[1] In addition, Huxley was the first to point out quite clearly the implication that seemed to follow from believing, on the one hand, that non-human animals are highly complex biological machines and, on the other, that there is complete continuity between man and other animals. The logical consequence was, he argued, that 'we are conscious automata', where the relation between consciousness and the physical basis of mental processes is much the same as the relation between the sound of a clock's bell and the workings of its parts.

The conditions of Huxley's childhood were much like those of Spencer and of Wallace. He too was the son of a schoolmaster. But as a young man, he saw more of the poverty and squalor of urban life in Victorian England than either of these two, or many of his future intellectual peers. The scenes he witnessed as apprentice to a doctor in the East End of London remained vivid throughout his life.

Long hours of study in his spare time enabled him to win a scholarship to a London hospital, where in 1845, now aged twenty, he passed the first part of the medical examination. Needing to earn a living, Huxley managed to obtain an appointment as an assistant-surgeon in the Navy and was posted to HMS *Rattlesnake*, which set sail in 1846 to chart the coasts of North-East Australia and of New Guinea. There he used his time to study the anatomy of marine organisms and packeted home to England the results of his research. As he commented later, his approach was not really that of a naturalist like Darwin, but was more akin to that of an engineer, asking of living organisms what is the structure and how does it work.

When the *Rattlesnake* returned four years later, Huxley was surprised to discover that he was already well-known in scientific circles. His reports had made a good impression and, furthermore, the ideas they contained were of direct relevance to the question of species. For the next four years Huxley wrote further papers on marine anatomy that confirmed his reputation. Dismissal from the Navy was followed within a few months by a teaching appointment at the Government School of Mines, where he remained for thirty-five years while this institution changed, partly as a result of Huxley's efforts, from being a small technical college to become the Imperial College of Science and Technology. Membership of the Royal Society had followed soon after his early publications. Despite lack of wealth and of family influence, his personal qualities gained him election to exclusive London clubs, while his increasing effectiveness as a frequent public speaker at meetings of working men's societies made him one of the best known popularizers of contemporary scientific thought. By 1858, when still a young man, he had become one of the most eminent scientists in England and one of the best comparative anatomists in the world.

It was natural for Darwin to include Huxley among the small group of people whose reaction to the theory of natural selection was to be tested prior to any publication. There was some trepidation on Darwin's part. Darwin later claimed that in comparison with Huxley he felt quite infantile in intellect, while Wallace commented that Huxley gave him a feeling of awe and inferiority which neither Darwin nor Lyell produced. In addition to his intellectual prowess, his thorough knowledge of anatomy and his lack of sympathy for previous ideas about evolution, there was Huxley's pugnacity. Darwin knew that Huxley could be formidable in the role of critic and perhaps guessed that he might be even more devastating in the role of supporter. To Darwin's relief Huxley showed a positive interest in natural selection and regarded it as an interesting hypothesis that deserved encouragement. It was only when Huxley came to read *The Origin of Species* on its publication in 1859 that the full force of the theory made its impact on him. 'How stupid not to have thought of that!'[2] he exclaimed, and joined Darwin's circle as an ally: 'And as to the curs which will bark and yelp, you must recollect that some of your friends, at any rate, are endowed with an amount of combativeness which (though you have often and justly rebuked it) may stand you in good stead. I am sharpening up my claws and beak in readiness.'[3]

Despite the prominent intellectual and social position he had now earned, Huxley maintained his

Fig. 1.8. Thomas Huxley at about the time of first sharpening his claws on Darwin's behalf

distrust of authority and his hatred of humbug. It was the desire to expose intellectual dishonesty, and not any contempt for religion, that provoked Huxley's encounter with Bishop Wilberforce in 1860. Huxley had been reluctant to attend the session at the British Association meeting in Oxford which was to include a paper discussing Darwin's ideas. Once present, Huxley's anger was stirred by an attack on evolutionary theory delivered by the bishop. The latter made the fatal mistake of being facetious: he enquired whether Huxley was descended from an ape on his grandfather's or on his grandmother's side. 'The Lord hath delivered him into my hands!' muttered Huxley to a startled neighbour. In replying, Huxley carefully answered the scientific points the bishop had made and then turned to the matter of ancestors: ' . . . it would not have occurred to me to bring forward such a topic as that for discussion myself, but I was quite ready to meet the Right Reverend prelate even on that ground . . . If then, said I, the question is put to me would I rather have a miserable ape for a grandfather or a man highly endowed by nature and possessed of great means of influence and yet who employs these faculties and that influence for the mere purpose of introducing ridicule into a grave scientific discussion – I unhesitatingly affirm my preference for the ape.'[4]

Darwin could remain contentedly in the countryside knowing that he had an able champion to take his part in any future disputes: 'How durst you attack a live bishop in that fashion? I am quite ashamed of you! Have you no respect for fine lawn sleeves? By Jove, you seem to have done it well.'[5]

In the decade that followed, Huxley kept fully to his promise. As well as claws and beak, Huxley's papers and books, his work on the comparative anatomy of the human and primate brain, and his lucid public lectures played an enormous part in the acceptance of Darwin's ideas. During the same decade he became ever more deeply involved in a variety of other commitments. Though his name was anathema in most British households, the government appointed Huxley to a number of Royal Commissions of Enquiry. Administrative duties to do with institutional and educational aspects of science increased. The concern for working class education which inspired many of his lectures led him to seek election to the London School Board: once elected, he pushed through a programme which set the pattern of education in English elementary schools until the end of the Second World War.

Among Darwin's circle, delight over Huxley's scientific contributions was mixed with increasing dismay that his talents and industry seemed too widely spread. As Spencer put it, Huxley was becoming a man who was continually taking two irons out of the fire and putting three in.[6] At the end of 1871 overwork and financial worries brought on the Victorian malady of dyspepsia and for a while Huxley could neither work nor think. Long holidays were necessary the following year. For once he missed the summer meeting of the British Association, an annual event he had attended for twenty years, and while Spalding was presenting his important first paper in Brighton, Huxley was with his family in Devon.

One of Huxley's new irons three years earlier had been to join the 'Metaphysical Society', a small grouping of leading politicians, eminent churchmen and scientists which met to discuss philosophical issues, in the hope that a spirit of mutual tolerance would ease any friction caused by their strongly divergent viewpoints. It was in this context that Huxley felt constrained to invent a label for his outlook on scientific and religious beliefs. He explained that he had discovered fairly early in life that it was a sin for a man to presume to go about unlabelled: 'The world regards such a person as the police do an unmuzzled

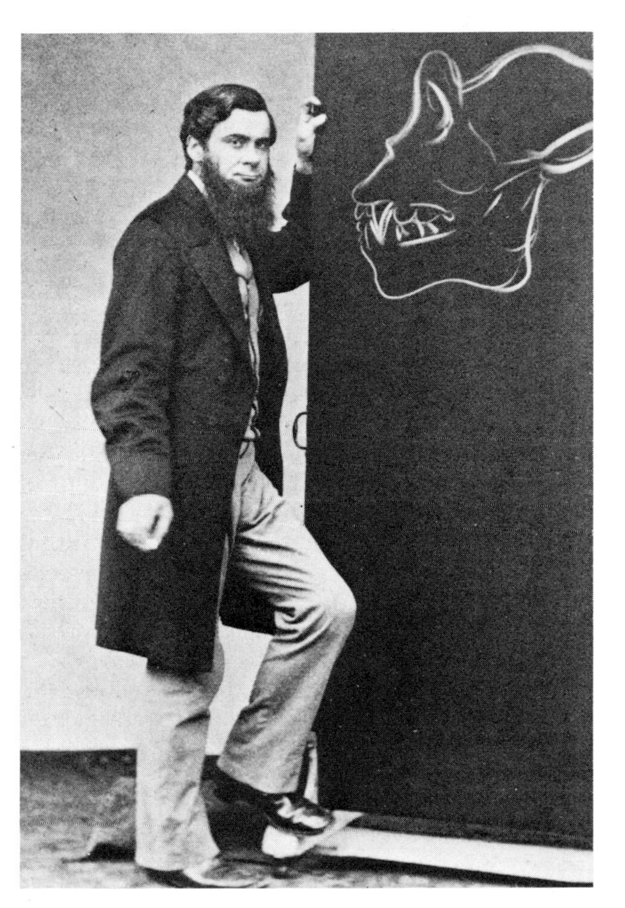

Fig. 1.9. Thomas Huxley lecturing on primate anatomy

dog, not under proper control. I could find no label to suit me, so, in my desire to range myself and be respectable, I invented one: and, as the chief thing I was sure of was that I did not know a great many things that the -ists and the -ites about me professed to be familiar with, I called myself an Agnostic.'[7]

The Metaphysical Society also provided the occasion for Huxley to develop his ideas about the minds of animals. These were presented publicly in an invited address at the 1874 meeting of the British Association in Belfast under the title, 'On the hypothesis that animals are automata, and its history'.[8] Early in the paper Huxley discussed the change in man's attitude to life that had occurred in the early seventeenth century. Harvey's discovery of the circulation of blood had added great weight to the idea that the physical process of life can be explained in the same way as other physical phenomena, and from Descartes' writing came the idea that the behaviour of animals can similarly be explained by principles derived from the study of inanimate matter. Huxley argued that increasing knowledge about the nervous system gain in more recent years had only served to strengthen Descartes' claim that animals are machines, or 'automata'. Huxley was *not* disposed to accept the full Cartesian doctrine: that brute animals are mere machines, devoid not only of reason, but of any kind of consciousness. Though he could see no way in which it could be positively refuted, or confirmed, he felt prejudiced against it for other reasons. In the absence of any prospect of settling the issue of whether animals possess consciousness by appeal to empirical evidence, he implied ethical considerations dictate that one should treat animals like 'weaker brethren'. Descartes was correct in regarding them as automata, but they may well be more or less conscious, sensitive automata. As for what it means to attribute consciousness or free-will to an animal, he continued: 'The consciousness of brutes would appear to be as related to the mechanism of their body simply as a collateral product of its working, and to be as completely without any power of modifying that working as the steam-whistle which accompanies the working of a locomotive engine is without influence upon its machinery. Their volition, if they have any, is an emotion indicative of physical changes, not a cause of such changes.'

Such a view of the psychology of animals, expounded by the famous advocate for the case for continuity between man and animals, left only one inference that the audience could draw. At first feigning some reluctance to go further, Huxley characteristically went on to make that inference fully explicit. The final part of his paper provides a fine example of his whole style and so is quoted here at length.

'Thus far I have strictly confined myself to the problem with which I proposed to deal at starting – the automatism of brutes. The question is, I believe, a perfectly open one, and I feel happy in running no risk of either Papal or Presbyterian condemnation for the views which I have ventured to put forth. And there are so very few interesting questions which one is, at present, allowed to think out scientifically – to go as far as reason leads, and stop when evidence comes to an end – without being speedily deafened by the 'drum ecclesiastic' – that I have luxuriated in my rare freedom, and would now willingly bring this disquisition to an end if I could hope that other people would go no further. Unfortunately, past experience debars me from entertaining such hope . . .

'It will be said that I mean the conclusions deduced from the study of brutes are applicable to man, and that the logical consequences of such applications are fatalism, materialism, and atheism – whereupon the drums will beat the *pas de charge*.

'One does not do battle with drummers; but I venture to offer a few remarks for the calm consideration of thoughtful persons. It is quite true that, to the best of my judgement, the argumentation which applies to brutes holds equally good of men; and, therefore, that all states of consciousness in us, as in them, are immediately caused by molecular changes of the brain substance . . . We are conscious automata, endowed with free will in the only intelligible sense of that much abused term – inasmuch as in many respects we are able to do as we like – but nonetheless parts of the great series of causes and effects which, in unbroken continuity composes that which is, and has been, and shall be – the sum of existence.

'As to the logical consequences of this conviction of mine, I may be permitted to remark that logical consequences are the scarecrows of fools and the beacons of wise men. The only question which any wise man can ask himself, and which any honest man will ask himself, is whether a doctrine is true or false. Consequences will take care of themselves; at most their importance can only justify us in testing with extra care the reasoning process from which they result.'

The paper drew almost as sharp a public reaction of indignation as the publication of the *Origin* fifteen years earlier. But this time no full campaign followed the initial shock. Instead, two years later, Huxley included a longer discussion of his views on mental continuity and on free will in a restrained and scholarly book that he had been invited to write as an introduction to the philosophy of David Hume.[9]

At the end of the book Huxley examined the question of what is meant by our belief that most of our actions are voluntary, if our actions and our mental processes are completely determined by the past and present state of our nervous system, and the question of moral responsibility for such actions. Basing his argument on Hume, he suggested that 'volition is the impression which arises when the idea of a bodily or mental action is accompanied by the desire that the action should be accomplished'.[10] Thus, it is desire that distinguishes a voluntary action. But, since our desires are also determined by past events, we have no more real choice over voluntary than over involuntary actions. 'Half the controversies about the freedom of the will . . . rest upon the absurd presumption that the proposition, "I can do as I like", is contradictory to the doctrine of necessity. The answer is: nobody doubts that, at any rate within certain limits, you can do as you like. But what determines your likings and dislikings? Did you make your own constitution? Is it your contrivance that one thing is pleasant and another is painful?'[11]

If our actions are set by present circumstances and desires shaped by past events, how can they be judged as good or evil? Huxley's reply was that a man's moral responsibility for his acts has nothing to do with their causation, but depends on the frame of mind which accompanies them. On what then are moral criteria based, whereby some actions, accompanied by certain frames of mind, are approved, whereas others are condemned? The answer given by Huxley must have surprised a public that saw him as the arch materialist and hardest-headed scientist of the time. It must have disappointed those, like Spencer, who were confident in the deduction of ethical principles from evolutionary theory or those, like the utilitarian philosophers, who attempted to provide a rational basis for ethics from consideration of the consequences of actions. 'Morality', answered Huxley, 'is based on feeling, not on reason; though reason alone is competent to trace out the effects of our actions and thereby dictate conduct. Justice is founded on the love of one's neighbour; and goodness is a kind of beauty. The moral law, like the laws of physical nature, rests in the long run upon instinctive intuition, and is neither more or less "innate" and "necessary" than they are.'[12] He went on to make clear that this did not mean that one man's moral intuitions are necessarily as good as those of another. The failure by some to understand geometry does not diminish its value. And he suggested that, just as a few gifted individuals like Newton have had profound intuitions about physical laws, or in art and music, so equally there have been men of moral genius. He ended at this point, and did not discuss the problem – one perhaps more obvious now than one hundred years ago – that, whereas in science there are fairly widely agreed criteria for distinguishing the intuitions of a Newton from those of a crank, in art and in ethics there can be considerable disagreement.

An earlier chapter of Huxley's *Hume* is entitled 'Mental phenomena in animals'. Although this contains no explicit discussion of whether one can distinguish voluntary from involuntary behaviour in animals, the possibility of making this distinction follows from its arguments, echoing those of the earlier Belfast address and quoting Hume in support, for continuity between the mind of animals and of man. The implication is that, for animals too, a voluntary action is one that is accompanied by desire for that action. The troublesome problem of determining an animal's desire, in contrast with the easy task of recording its action, is not addressed. In its place Huxley repeats his arguments on the impossibility of settling whether another creature is conscious or not, and on the irrelevance of the contents of consciousness to understanding the machinery of the mind.

As part of his case for mental continuity in the book on Hume, Huxley also found further support in evidence from the 'new science of comparative psychology'.[13] However, he did not go into the nature of this evidence and this was a science to which he made no direct contribution. Some scientific work in zoology was continued, along with lecturing, administrative work, educational projects, textbooks and, now and then, a sharp public dispute.

His visits to Darwin had long become occasional. During Darwin's final illness in 1882 Huxley wrote to offer medical advice. To the note of thanks in reply, Darwin added: 'I wish to God there were more automata in the world like you.'[14]

Summary

The idea of studying the behaviour of animals in order to understand the human mind and its evolution was independently developed and propounded by Darwin and by Spencer. Of these two Darwin was, at least until well into the 1870s, by far the more influential. In developing his theory that the primary cause of evolution is the wholly mechanistic process of natural selection, in *The Origin of Species* he merely hinted at his belief that the origin of the human race was no exception. By 1872, when the reason for this belief were spelled out in *The Descent of Man*, it was important for Darwin to make a strong case for mental continuity between man and other animals.

A central factor was the increasing theoretical divergence between himself and Wallace. For Wallace natural selection remained the only process of evolution and so, since he failed to see how various human characteristics – particularly the human mind – could have evolved in such a fashion, he came to believe that some non-natural influence had intervened in the course of human evolution. This issue, together with other problems that included those of heredity and of shorter estimates for the age of the earth, led Darwin to place more emphasis on secondary mechanisms. One was sexual selection, which was treated at length in *The Descent of Man*. The other was the Lamarckian principle that the use or disuse by an individual of its organs, including its brain, affects their inheritance by its offspring.

The Lamarckian principle had been a central concept in Spencer's early ideas as expressed in his first book on psychology published in 1855. In 1870, when he published the first volume of a revised version of the book, he had also accepted from Darwin and Wallace the idea of natural selection as a mechanism of evolution. Furthermore, the serious attention to ideas on evolution and great public interest that had been won by Darwin ended Spencer's relative obscurity and led to his growing influence. With Spencer's acceptance of natural selection and Darwin's increasing reliance on Lamarckian inheritance there were strong similarities in their views on mental evolution, but one important difference remained. In Spencer's psychological theories the notion of evolution as a linear progression, with existing species and races providing living evidence of continuity, is never far away. Even in his treatment of mental evolution Darwin usually managed to maintain his view of evolution as an irregularly branching process producing marked discontinuities in forms of life that have survived to the present day.

The views of Darwin and Spencer on the evolution of intelligence provided the main basis for the subsequent development of comparative psychology, which in turn served as the main route by which evolutionary theory made its impact on psychology as a whole. At the same time other issues were debated which were to become prominent points of controversy during the growth of animal psychology and the emergence of behaviourism.

A major factor in the Lamarckian emphasis given by Spencer and Darwin to their treatment of behaviour was the resemblance they perceived between instincts and learned actions that have become habitual. This resemblance is reflected in colloquial English even today: we talk of someone's well-practised skill having become 'second nature' to him, or describe the driver of a car, reacting quickly in some emergency, as 'instinctively' changing gear. The task of distinguishing habit from instinct remained an important one until the end of the century.

The early 1870s was a time when a strong environmentalist view on such matters was prevalent, mainly because of the influence of the philosophical tradition represented by Mill and Bain. In succeeding decades, the general outlook swung towards the other, hereditarian, extreme. One element that was to contribute towards this swing was the research carried out by Spalding. Its importance lay in the example he set of using experimental methods to attack questions concerning animal behaviour. The aspect of his findings of most interest to Spalding was the demonstration that the extreme environmentalist claims of Bain were wrong: no prior sensory experience was necessary for a chick to interact appropriately with some features of its environment. Other aspects of his results, those to do with 'imperfect instincts', could have been cited in partial support of Bain's argument for the important and rapid effects of early experience. However, no experimental studies directly concerned with the nature of learning in animals were performed at this time.

The lack of empirical work on learning was not due to an absence of theoretical ideas. In the first edition of Spencer's *Principles of Psychology* the principle underlying the transition from reflex to instinct was held to be a process of learning based on contiguity and frequency: if one stimulus regularly and immediately precedes another, then the first will come to elicit the reaction formerly evoked only by the second. A second process of learning was also described, first in 1855 by Bain and then in 1870 by Spencer. According to the Spencer–Bain principle, any spontaneous action which is accidentally, but immediately, followed by those changes in the state of the brain correlated with subjective feelings of pleasure, or of a decrease in pain, thereafter becomes more likely to recur. However, for the time, the statement of these principles stimulated neither close attention to the differences between them nor empirical investigation.

The Spencer–Bain principle of learning arose in the context of Bain's analysis of voluntary action. In Bain's philosophy physical events and mental events occur in parallel; the task of psychology is to study subjective experience and discover its relation to the workings of the brain. He firmly rejected the view of his former student, Spalding, that a scientific study of mind should not, and cannot, concern itself with the contents of consciousness, a view that was also given the weight of Huxley's authority. Whether animals 'possess consciousness', and if so, what means exist for detecting it, became another central issue in comparative psychology.

Fifty years after the events described in this chapter a view of psychology similar to those held by Spalding and Huxley became the starting point of the behaviourist movement. The view was that psychology can make progress only if it restricts itself to the study of behaviour; there is no way in which a true science can be based on the analysis of subjective experience. The objection was raised that, if animals including man are to be treated as machines whose behaviour is completely determined by past and present events, then there is no sense in which a man's actions can be free, and no basis for morality. The same objection was raised in the 1870s, in reaction first to *The Descent of Man* and then to Huxley's Belfast address. Darwin was circumspect on such matters and Spencer obscure, but Huxley made his position clear: voluntary actions are as much determined by prior causes as any other event in nature; and, as for reasons for judging an action to be good or evil, the methods of science can no more establish these than they can answer questions concerning consciousness.

2
Intelligence and instinct

Man's body may have been developed from that of a lower animal form under the law of natural selection; but . . . we possess intellectual and moral faculties that could not have been so developed, but must have another origin.

Alfred Wallace: *Darwinism* (1889)

By 1880 there was widespread acceptance of the general idea that life had evolved. To this extent the arguments that Darwin had marshalled twenty years earlier had had their intended effect. There was, however, considerable disagreement among those now known as the 'Darwinians' over the process or processes by which evolution had taken place. The purist position was that evolution was solely the result of natural selection. This was Wallace's view, except that he appealed to supernatural intervention to account for the human mind. Darwin himself believed that all aspects of human evolution could be explained in terms of the same processes that had led to the evolution of any other species, but attributed much more importance to Lamarckian inheritance as a means of accelerating evolution than when he had first written *The Origin of Species*.

From around the time of Darwin's death in 1882 such differences were much more openly debated, possibly because Darwin's immense prestige had discouraged dissent, but more probably because there was no longer any need for the evolutionists to present a united front to the outside world. Over the next twenty years the study of animal behaviour was largely inspired by disagreements over fundamental aspects of evolutionary theory and its development was particularly influenced by the split between those agreeing with Darwin's later views and those agreeing with Wallace. Broadly speaking, in the 1880s animals were studied because of interest in the evolution of the human mind, while in the 1890s it became of key theoretical importance to settle the question of whether any behavioural tendency acquired by an individual animal could be passed on genetically, or by some other means, to its offspring.

This chapter describes the work of three men, George Romanes (1848–1894), Conwy Lloyd Morgan (1852–1936) and Francis Galton (1822–1911). As with other evolutionists, their range of interests was broad, including geology, physiology and what would now be called genetics. What distinguished Romanes and Morgan was that their main interest was in the minds of animals. For both of them this interest developed because they wished to understand evolution. But a number of problems stood in the way. One was methodological; for explanations of animal behaviour the familiar ways of collecting, assessing and interpreting evidence that had served for geology and anatomy were applied, but began to seem inadequate. There was a further kind of problem that also did not arise in the case of the other natural sciences related to evolutionary theory: before any progress could be made on the animal mind, the philosophical issues surrounding the concept of mind had to be faced. A third preliminary task was that of clarifying the concepts of intelligence and of instinct. Both words had rather different meanings to the ones they have acquired over the past century. *Intelligence* was relatively straightforward. To refer to an action as intelligent was in general understood as indicating that its performance showed some beneficial effect of past experience. A generally accepted definition of *instinct* was more difficult to find.

A central part of this third problem was the continuing uncertainty over mechanisms of heredity. Where this bore on human evolution, the heredity issue was closely related to the question of the age of the earth. Wallace had been quick to accept the physicists' arguments for limited geological time. Since he was prepared to appeal to supernatural intervention in the origin of man, he did not need vast amounts of time. To the extent that orthodox Darwinians, resistant to Wallace's views on man, reluctantly accepted shorter estimates for the age of the earth, the

Lamarckian principle appeared more important as the only natural process of evolution capable of producing rapid changes. Whereas the ideas discussed in the last chapter were developed in the context of expanding geological time, during the period covered by the present chapter time was contracting.

The mid-Victorian era in Britain of the 1860s and 1870s was a period of vigorous economic growth and of relatively egalitarian attitudes towards class and racial differences. During the last two decades of the nineteenth century, hereditarian views became steadily stronger. The huge contrast between the power and technical achievements of societies in North America and Western Europe and those of the rest of the world came more and more to be seen as directly reflecting inherent differences in the psychology of different races. It was the age of massive and deliberate European colonization of many other parts of the world; in 1885 the Congress of Berlin met to decide how Africa should be apportioned among the various European powers.

The connection between such events and the study of animals was by no means diffuse and indirect. The leaders of Victorian society were very concerned to examine the moral and intellectual basis for their actions. Many of those who held political power maintained a close and informed interest in a wide range of current scientific and philosophical developments. Thus, Spalding's modest reports on his research on young birds had an influence that no comparable work could enjoy today. His results first surprised and then inclined many readers to consider that the individual future of a human mind might also be largely determined at its birth. The answers to questions about consciousness, intelligence, reasoning and instinct in non-human animals were of considerable interest beyond the still small group of people who spent a good deal of time on such matters.

The third man whose work and career is described in this chapter, Francis Galton, did not study animals, but nonetheless his work provided an important part of the context in which animal psychology developed. He was the one member of the Darwinian circle whose main interests were in mechanisms of heredity and human psychology. In contrast to many of his fellow Darwinians Galton's strongly hereditarian views on human differences were not based on Lamarckian inheritance. Thus, when this was seriously challenged in the 1890s, Galton's arguments on this topic remained intact and gained in prestige. His outlook was in stark contrast to that of Morgan who emphasized the importance of early learning and followed Wallace in stressing the importance of cultural transmission in man. This was the beginning of a marked division within psychology between the tradition founded by Galton which concentrated on the measurement of human differences and that of animal psychology which attempted to understand general processes of learning.

Fig. 2.1 Charles Darwin at about the time of meeting Romanes

The systematic classification of anecdotal evidence by George Romanes

Three years after he had published *The Descent of Man* Darwin, now sixty-five years of age, was impressed by a letter he read in a scientific journal and invited the author to visit him at his home in Kent. The author, George Romanes, was greeted with outstretched arms, a bright smile and the exclamation: 'How glad I am you are so young!' A deep friendship developed from this meeting, made up of fatherly kindness on one side and uncritical reverence on the other. To contemporaries it was seen as Darwin passing on his mantle to Romanes.[1]

At the time of this meeting Romanes was a student of physiology. Following a rather nomadic childhood and desultory education, Romanes had studied mathematics and science at Gonville and Caius College, Cambridge. He was a religious young man. He did not read any of Darwin's books until after he had obtained his degree in 1870. His wife later wrote that reading Darwin had had an extraordinary effect on him and led to increasing religious scepticism over the next eight years. Nevertheless, three years after reading Darwin, his first claim to public notice came from writing a prize-winning essay on 'Christian Prayer'. Following his degree Romanes remained at Cambridge to work in the newly-founded physiological laboratory. A considerable private income allowed him to devote himself to scientific research.

The promise that Darwin had seen in the young man appeared justified. During summers spent in Scotland Romanes carried out careful work to determine whether jelly-fish possess a nervous system and continued with some highly original experiments on the nature of reflexes. This work was impressive and he rapidly ascended the ranks of the British scientific establishment, acquiring its ultimate recognition by being elected to the Royal Society at the relatively early age of thirty-one.

Fig. 2.2. George Romanes

Meanwhile, his interest in the evolution of mind was developing and he began a collection of reports on animal behaviour. The animal mind had become an extraordinarily popular topic in the 1860s and 1870s. Countless letters flowed in to scientific and popular journals, reporting striking observations of animals that suggested unsuspected mental capabilities. In addition, Romanes corresponded directly with contacts in all quarters of the globe and received from Darwin the latter's own collection of notes on behaviour; almost to Darwin's relief, it seems, that someone was taking on the job whose importance he had stressed. 'It is so much more interesting to observe than to write', Darwin noted in a letter to Romanes,[2] and carried on with his observations of plants and earthworms in his garden. Apart from this collection and a few suggestions – for example, that Romanes keep a monkey at home and study its behaviour – Darwin did not make a direct contribution to Romanes' work.

Romanes saw his task as one of making sense out of the confused flood of evidence on the behaviour of animals, first by systematically classifying observations, and then deducing general principles for a theory of mental evolution. Since the organized presentation of the evidence alone provided sufficient material for a fair-sized book, this was published first. The book, *Animal Intelligence*, appeared in 1882 a few weeks after Darwin's death. Romanes was nervous that, if this work was judged in isolation from the theoretical interpretation to be supplied in a later volume, it would be considered 'but a small improvement upon the works of the anecdote mongers'.[3] This fear later proved to be justified. *Animal Intelligence* was widely read and is still frequently cited today; his later books attracted far less attention. Romanes had a great influence on the subsequent development of animal psychology, but his successors saw him only as the archetypal purveyor of anecdotes about animals.

In fact, Romanes was a good deal more critical than many of his contemporaries and more given to doubt highly implausible stories than Darwin. The dismay he felt at popular books on the subject, consisting of a loose stringing together of animal stories, and the disrepute they were likely to bring to the new science of comparative psychology, provided reasons for writing *Animal Intelligence*. From the vast amount of material available he first selected contributions from observers 'well-known as competent'. For Romanes there was no one of greater competence than Darwin and, if Darwin reported something, then this made it firm evidence; although even Romanes felt uncomfortable in accepting that snails possess the ability to communicate complex information to each other, he did so since an observation supporting this claim was included by Darwin in *The Descent of Man*.[4] Since the criterion of the writer's known competence allowed through only a small amount of evidence, Romanes also accepted reports that in his judgement were clearly based on careful observation and those describing an ability in some particular species which had been noted by a number of independent observers.[5]

The text suggests that other factors also influenced his selection. There are some odd cases which were included, one can only judge, because the informant would have been offended by their omission; for example, an anonymous young lady known to Romanes reported 'that her two younger sisters (children) are in the habit of feeding every morning with sugar an earwig, which they call "Tom", and which crawls up a certain curtain regularly every day at the same hour, with the apparent expectation of getting its breakfast'.[6] A factor much more important than this appears to be the social status of the observer. An account of a tribunal of rooks, sitting in judgement of a jackdaw accused of some misdemeanour, is reproduced when the informant is a bishop and a similar story is told by a major-general.[7]

Romanes was no doubting Thomas. He showed little of Huxley's willingness to suspend belief, or the attitude of Spalding who had wryly noted that 'the many extraordinary and exceptional feats of dogs and other animals seemed to be constantly falling under the observation of everybody except the few that are interested in these matters'.[8] The difference between the believers and the sceptics was not confined to matters of animal behaviour. The supernatural also enjoyed a vogue at this time. It attracted the interested curiosity of Wallace, Spencer and, to some extent, Darwin. When told of some message relayed from beyond the grave, Huxley merely commented that, if this were typical of conversation in the spiritual world, it provided another good argument against suicide;[9] a similar attitude, expressed in a different idiom, to that of Groucho Marx's reply, when asked if he had any questions for the Great Spirit beyond: 'Yeah, what's the capital of North Dakota?'

Darwin shared this lack of belief in the supernatural, but he did not feel that a highly sceptical attitude was beneficial for a man of science. 'I am not very sceptical', he once wrote, 'a frame of mind which I believe to be injurious to the progress of science. A good deal of scepticism in a scientific man is advisable to avoid much loss of time, but I have met with not a few men who, I feel sure, have often thus been deterred from experiment or observations which would have proved directly or indirectly serviceable.'[10] This is another respect in which Romanes followed Darwin.

Romanes' contribution lay in something more than providing his successors with an approach that they could vigorously reject. In many cases the observed behaviour was interesting and believable; it was the observer's rich interpretation in terms of current ideas from human psychology that was wide open to objection. Later animal psychologists rejected the interpretations, but mined Romanes' collection for ideas to use in devising situations where the behaviour of animals could be studied more carefully. His comments on the mechanical abilities of cats provided an important example, and so are worth quoting at length: 'In the understanding of mechanical appliances, cats attain to a higher level of intelligence than any other animals, except monkeys, and perhaps elephants.' Having discussed the fact that he knew of only one report of a dog operating a latch, whereas he had received some half-dozen reports of cats doing so, he continued by relating a typical instance. 'I may add that my own coachman once had a cat which, certainly without tuition, learnt thus to open a door that led into the stables from a yard into which looked some of the windows of the house. Standing at these windows when the cat did not see me, I have many times witnessed her *modus operandi*. Walking up to the door with a most matter-of-course kind of air, she used to spring at the half-loop handle just below the thumb latch. Holding on to the bottom of this half-loop with one fore-paw, she then raised the other to the thumb piece, and while depressing the latter, finally with her hindlegs scratched and pushed the doorposts so as to open the door. Precisely similar movements are described by my correspondents as having been witnessed by them. Of course, in all such cases the cats must have previously observed that the doors are opened by persons placing their hands upon the handles and, having observed this, the animals forthwith act by what may be strictly termed rational imitation.' He goes on to argue that the process of 'rational imitation' must involve understanding of the mechanical properties of the door and considerable reasoning power.[11]

Romanes believed that a science contains two parts, which can be considered in isolation from each other: one part consisting of a body of factual information; the other of principles deduced from these facts. His model for the science of comparative psychology was comparative anatomy. The number of teeth possessed by a given animal, the shape of a skull or the length of a particular bone are indisputable facts. From consideration of the anatomical resemblances between different species a theory of physical evolution can be derived, and this provides the means for obtaining a true classification of anatomical data.

Animal Intelligence was intended as a compendium of facts about the animal mind. Romanes suggested that a theory of mental evolution would provide a classification scheme very different from one based on physical similarity. He was particularly impressed by the wide variation in psychological abilities shown by different species of rodents: 'In no other group do we meet with nearly so striking an exemplification of the truth that zoological or structural affinity is only related in a most loose or general way to psychological or mental similarity'.[12] Nevertheless the organization of *Animal Intelligence* followed traditional zoological categories, starting with simple organisms and working through familiar orders, genus by genus and species by species. Treatment of the mental life of insects, particularly that of ants and bees, was detailed and fairly restrained. A considerable amount of careful and intelligent experimental work on these animals had been recently performed, notably by Sir John Lubbock who was a close friend of Darwin and Huxley and managed in a relaxed manner to combine research with a very active political career. Lubbock's fame rested on an unlikely combination of contributions: research on insects and the introduction of the English Bank Holiday. In at least one respect Lubbock's work made Romanes err on the side of caution. Lubbock had attempted to test a prevalent belief that a bee can indicate the location of a new source of food to other members of the hive. He failed to obtain any evidence in support of this belief. Subsequent research has shown that Romanes' consequent decision – despite his generosity in the case of Darwin's snails – to reserve judgement on the possibility of communication between bees was unnecessary.[13]

Although ants and bees were given careful attention by Romanes, most non-mammalian species were treated very briefly in *Animal Intelligence*. On reaching mammals there was little restraint. Like Darwin, Romanes was intrigued by accounts of the building of dams, canals and bridges by beavers. In the case of dogs, and then of monkeys, Romanes had no hesitation in attributing to them a wide range of complex mental processes, ranging from hypocrisy and deceit to the geometrical knowledge that the chord of a circle is shorter than the arc and an understanding of the mechanical principle of the screw.

The justification for inferring that animals possess such human-like emotions and intellectual abilities was discussed briefly in his introduction to *Animal Intelligence*, where Romanes explained what he meant by mind and what criterion he used for deciding that a given type of animal possessed a mind. These issues were treated at greater length in his second book, *Mental Evolution in Animals* of 1884, whose principal aim was to develop a theory of mental evolution from the factual basis laid down in *Animal Intelligence*. This theory and Romanes' more general ideas on animal psychology are discussed in the following section.

Romanes on mind, instinct and intelligence

How can we know whether a monkey experiences colours in the way that we do, whether a dog dreams, whether a worm anticipates the consequences of its actions, or whether a jelly-fish feels pain? The answer given by Romanes was that we have to make an inference from the creature's behaviour, in just the same way as we infer from other people's behaviour that they have subjective experiences like our own.

In the introductions to both books he pointed out that the word 'mind' is ambiguous. 'By Mind we may mean two different things, according as we contemplate it in our own individual selves, or as manifested by other beings. For if I contemplate my own mind, I have an immediate cognizance of a certain flow of thoughts and feelings, which are the most ultimate things – and, indeed, the only things – of which I am cognizant. But if I contemplate Mind in other persons or organisms, I can have no such immediate cognizance of their thoughts and feelings; I can only infer the existence of such thoughts and feelings from the activities of the persons or organisms which appear to manifest them. Thus it is that by Mind we may mean either that which is subjective or that which is objective.'[1] Our subjective experience, 'consciousness', provides the only direct way of understanding the workings of our own minds and the basis of our actions. When we perceive that the activities of other people resemble what we do ourselves, then, on the basis of analogy, we attribute to them minds like our own. And the same holds with regard to animals: to the extent that their behaviour is analogous to ours, then they possess minds. But what aspect of the activity of other beings leads us to believe that they are conscious? Romanes suggested that the only firm criterion to be found was the presence of an ability to make choices, and this was demonstrated if an individual's action was influenced by previous events in its life. Thus, the question of whether a particular animal has a mind can be answered by determining whether modifications of its behaviour occur as a result of its own past experience.[2] Romanes felt that this was probably too strict a criterion, and that many forms of life incapable of benefiting from experience might nonetheless possess a glimmer of consciousness. But he was unable to suggest any objective method for testing this possibility.

In his choice of learning as the test for the presence of mind, Romanes rejected the Huxleyan notion of animals, and man, as automata, which, in any case, he believed 'can never be accepted by common sense'.[3] Although not explicitly discussed, it is clear that for Romanes machines cannot benefit from experience. It was some decades before technology produced counterexamples.

Despite the clear opening statements on the ambiguity of the word 'mind' and its basic reference to subjective experience, the point of view pervading Romanes' books is that known, since the time of Descartes, as dualism. 'Consciousness' has replaced 'soul' or 'spirit'. There are two kinds of events in the world, physical and mental. Whereas in the kind of parallelism adopted by Bain the two kinds of event do not interact, in Romanes' system the physical actions of our bodies can be directly governed by mental processes.

The evolution of mind could then, in principle, have taken a very different course from that of the evolution of bodily structures. And, indeed, this is what Romanes decided. His theory of mental evolution, promised in *Animal Intelligence*, was summarized in the diagram that is reproduced in Figure 2.3. A large part of *Mental Evolution in Animals* was devoted to explaining the meaning of this diagram which, he claimed, 'is not so much a product of any individual imagination, as it is a summary of all the facts which science has been able so far to furnish upon the subject'.[4]

Although this figure uses the Darwinian symbol of the irregularly branching tree, the usage is very different from that of Darwin. What appear on the branches are not various orders and genera, but psychological terms. These also appear in the column immediately to the right, while in the next column appear the names of different kinds of animals. This linear ordering seems a most inappropriate way of representing the complexities discussed in *Animal Intelligence*. The wide range of mental abilities displayed by different species of rodents appears to have been forgotten. The impressive evidence of learning abilities in bees, which in *Animal Intelligence* were seen to be superior to those of many fish and reptiles, fails to lift the category of insects and spiders above level 20.

These ascending scales are less reminiscent of Darwin than of Spencer, who is cited frequently by Romanes. His whole approach to the animal mind is not merely anthropomorphic, but also anthropocentric, in the sense that mental evolution is seen as an orderly progression leading towards its culmination in the human mind. As indicated in the third column, the development of the human mind, from ovum and spermatozoa to the baby of fifteen months, was viewed as retracing the history of mental evolution. At three weeks the baby's mental abilities have reached the highest level attainable by insect larvae, at four months that attainable by reptiles and at fifteen months begin to excel those of the apes and the dog. Unlike Darwin, Romanes had no hesitation in using the terms 'higher' and 'lower'.

In another important aspect Romanes' steadfast respect for Darwin failed to reflect the spirit of Darwinian theory. The key ideas of natural selection are the role of chance and the rejection of apparent purpose. Yet, in the whole collection of observations presented by Romanes, there is no suggestion that an animal might, by chance, have happened to behave in a manner appropriate to the situation, and every opportunity to assert that an action was accompanied by conscious intent is taken. Similarly there is no reflection of Darwin's inherently historical approach; Romanes showed little inclination to view the present behaviour of an animal as the outcome of developments within its own past. Morgan was later to see behaviour in this way and in doing so was more of a Darwinian than Romanes.

In some important respects, then, Romanes owed less to Darwin, and probably more to Spencer, than at first appears. Some of the findings from Romanes' earlier physiological research had appeared to support Spencer's speculations on the origin of nervous action. And Spencer's *Principles of Psychology* was for Romanes the most important book on the subject. Nevertheless Romanes differed sharply from Spencer on two important issues. These were the distinction between reflex and instinct, and the origin of reasoning.

The notion of instinct common in the middle of the nineteenth century was nicely expressed in a didactic, but very popular, children's book some thirteen years before *The Origin of Species* was published. 'Instinct in animals, William, is a feeling which compels them to perform certain acts without thought or reflection; this instinct is in full force at the moment

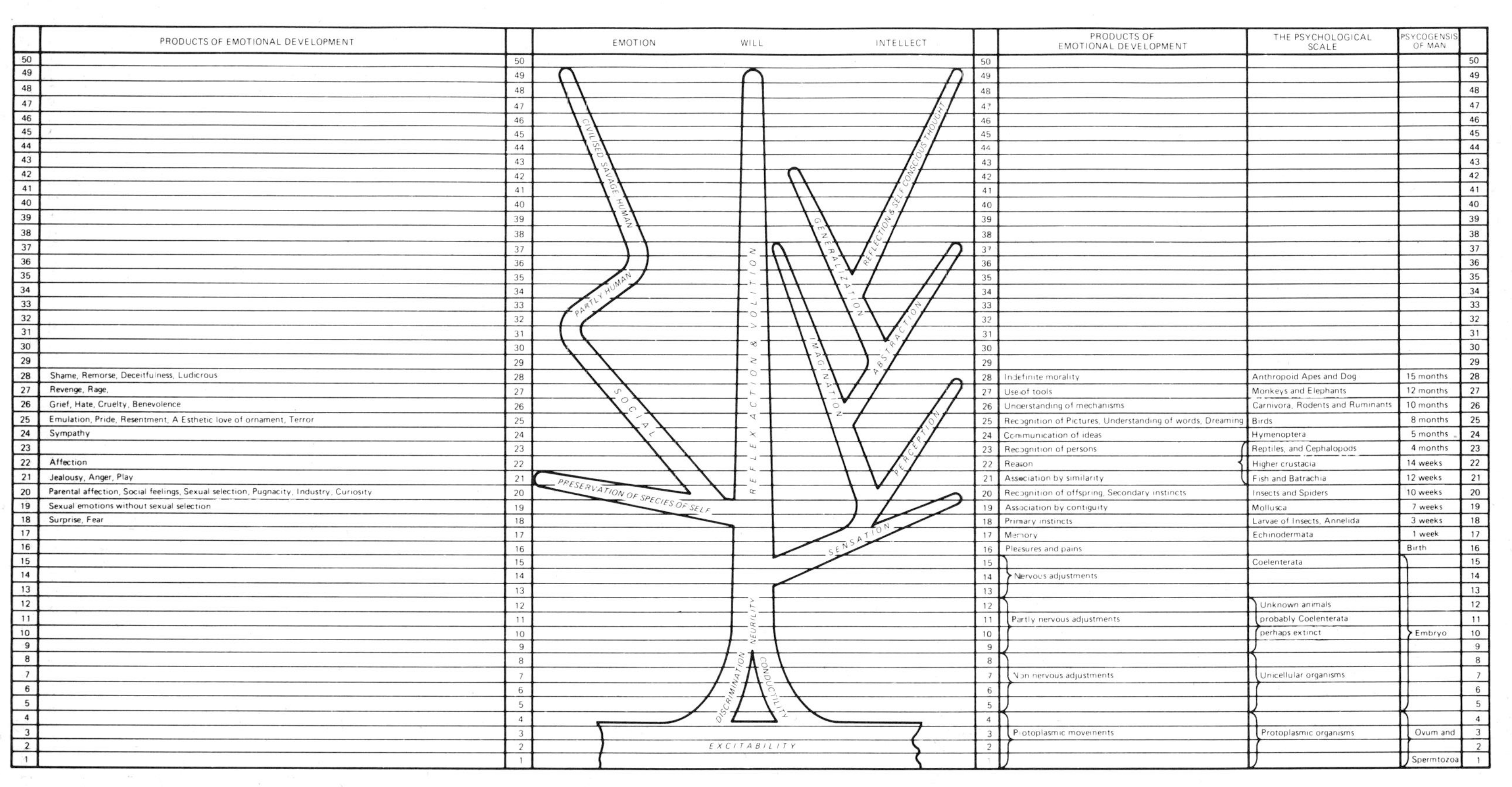

Fig. 2.3. Romanes' tree of mental evolution with accompanying scales to illustrate the relative intelligence of different species and stages of mental development in human beings

of their birth; it is the guidance of the Almighty's hand unseen; it was therefore perfect at the beginning, and has never varied. The swallow built her nest, the spider its web, the bee formed its comb, precisely in the same way four thousand years ago, as they do now.'[5] This passage in Marryat's *Masterman Ready* is followed a few pages later by some conclusions concerning the relative intelligence of different species which anticipate those reached by Romanes, but which are explained by appeal to religious rather than evolutionary principles. 'Although the Almighty has thought it proper to vary the intellectual and the reasoning powers of animals in the same way that he has varied the species and the forms, yet even in this arrangement he has not been unmindful of the interest and welfare of man. For you will observe that the reasoning powers are chiefly, if not wholly, given to those animals which man subjects to his service and for his use – the elephant, the horse and the dog; thereby making these animals of more value, as the powers given to them are at the service and under the control of man.'[6]

The idea of instinct as a mysterious feeling and source of creative energy had been dismissed by Spencer. Instead Spencer had defined an instinct as a 'compounded reflex'; that is, an invariable sequence, or chain, of simple reflexes. On this view instinctive behaviour is always automatically triggered by appropriate events or objects in an animal's surroundings. Spencer's definition ignores the possibility that whether some particular instinctive action occurs might depend on the internal state of the animal.

Romanes decided that the idea of an instinct as a 'compounded reflex' failed to take account of the complexity of instinctive behaviour; the examples of instincts provided by Spencer were for Romanes too simple to deserve the label. The alternative definition offered by Romanes was that 'instinct is reflex action into which there is imparted the element of consciousness'.[7] Its significance rested on the distinction he drew between a sensation and a perception: sensation being the direct result of a simple change in physical stimulation, whereas perception is produced by some process of inference acting upon sensation. Thus, if most members of a species react in certain specific ways to a change in the level of illumination, to a sudden noise or to a tap on the knee, these are reflexes. On the other hand, if a reaction is affected by the direction and distance of a stimulus source or determined by specific objects, then it is an instinct.

Romanes' view of instinctive behaviour was very much influenced by Spalding's research.[8] The fact that young chicks, without prior sensory experience, can move in an appropriate direction towards the sound of an unseen mother hen, and can use their visual sense to avoid obstacles, indicated to him that perceptual processes relating to the size and distance of objects were involved. Hence Spalding was correct in describing the reactions as instinctive.

A century later, 'instinctive' has come to be more or less synonymous with 'innate'. It is important to realize that this was not the usage suggested by Romanes. In defining instinct as 'reflex plus consciousness' he was not employing the criterion of learning as an index of consciousness. Like Spalding, Romanes also included as instincts reactions whose emergence required the support of certain kinds of events in the young animal's environment, and similarly termed these 'imperfect instincts'. The results from studies that, following Spalding, he had carried out on cross-species fostering in birds – a hen foster-mother given the eggs of a pea-fowl to hatch, for example – were important for his views on the plasticity of instincts.[9] Another topic discussed in this context was that of birdsong. Romanes regarded this as certainly instinctive. The extensive evidence on birds learning the songs and calls of their species, or more unnatural sounds, was seen as illustrating the modifiability of instincts. Romanes noted that, in the case of birds, the effects of learning by imitation are not so deeply engrained as those which are stamped in by heredity. In contrast he suggested that in mammals, including the early years of human childhood, the faculty of imitation played a much more important part in the perfecting of instincts.[10] In the same vein Romanes found it quite appropriate to discuss the 'instincts of a gentleman' as a set of imperfect human instincts requiring an appropriate environment for their proper development.[11]

Romanes was also very much concerned, like Darwin and Spencer, with the origins of instinctive behaviour. As we have seen, Spencer viewed all instincts as habits which over the course of generations had become progressively more hereditary. In his early writings on the subject Darwin had emphasized the role of natural selection acting on behavioural patterns that initially occurred on the basis of chance; for him a key example telling against a purely Lamarckian theory of the origins of instincts was that completely innate, but highly organized, behaviour patterns are found in asexual members of certain kinds of insect species, such as ants.

Romanes adopted an intermediate position by which natural selection and acquired characteristics were given equal weight. He illustrated his view by means of the curious diagram shown in Figure 2.4.

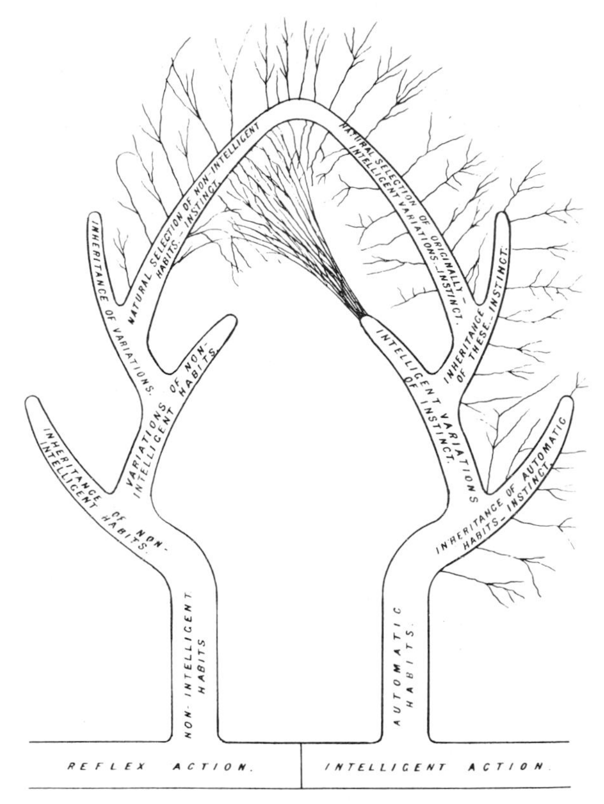

Fig. 2.4. Romanes' tree diagram illustrating his theory on the dual origin of instincts

The left-hand half of this diagram represents the evolutionary development of 'primary instincts', whereby natural selection operates on basic reflexes to produce increasingly complex 'non-intelligent habits', which are stereotyped patterns of behaviour elicited by specific perceptions and not modifiable by an individual's experience. The evolution of the second type, 'secondary instincts', is shown in the right-hand half of the figure and represents the Lamarckian element. Behaviour initially learned by an individual, 'intelligent action', becomes through repetition 'automatic habit' and eventually is heritable. Such secondary instincts do not have the same fixed quality of primary instincts and, being subject to modification by experience, allow the possibility of 'intelligent variations'. Although never explicitly discussed, Romanes clearly believed that the time span for the operation of the evolutionary processes seen in this diagram was relatively brief. He cites in full a report, communicated to Darwin by a Dr Huggins, FRS, concerning the inherited fear of butchers acquired over a small number of generations by a family of English mastiffs.[12]

As we saw earlier, in *Masterman Ready* a few species are regarded as capable of reasoning. Probably a more common view in the middle of the nineteenth century was that no animal is able to reason; reasoning was seen as a uniquely human characteristic, and human behaviour alone seen to be based both on instinct and reason. In his attempt to develop Darwin's arguments on mental continuity Romanes found it necessary to redefine reason as well as instinct.

The distinction between sensation and perception was as important to his discussion of reason as it was to his concept of instinct. He was familiar with contemporary studies of visual perception by German psychologists which had indicated the important role of unconscious inference. The size and distance of an object that we see is inferred from a variety of cues of which we are usually totally unaware. For Romanes 'reason' was based on the perception of relations, and human reasoning resulted from the evolution of ever more complex perceptual processes. In this view he again differed from Spencer. Thus, Romanes' tree of mental evolution, Figure 2.3, shows the limb of 'abstraction' stemming from the branch of 'perception', and not from the trunk representing evolution from 'reflex' to 'action and volition'.

By the end of the nineteenth century Romanes was remembered only for the observations catalogued in *Animal Intelligence*. The seriousness with which he treated such evidence was scorned. Moreover, the Lamarckian principle, which played so large a part in his treatment of instinct, was discredited. In the twentieth century the behaviourist movement in the United States rejected, not only his methods, his dualistic approach and his ideas on instinct, but his ideas on perception as well. In doing so the behaviourists returned to Spencer's linear progression leading from reflex, via instinct, towards reasoning. In Europe comparative psychologists continued for a few decades to take very seriously the idea, which Romanes was the first to discuss in the context of animal psychology, that the basis of intelligent action is closely related to the processes of perception.

One important topic is hardly discussed by Romanes. Possibly reflecting the swing from an environmentalist to a hereditarian outlook that had occurred over the past decade, nothing was said about the nature of learning processes. This is curious in that, as discussed earlier, the one clear objective criterion for the presence of consciousness offered by Romanes was the effect on an individual's choice of actions of its own past experience. Although the double-tree symbol of instinct, shown in Figure 2.4, was soon forgotten, it can be seen as foreshadowing a later division. The primary instincts of the left-hand side became the main preoccupation of what was later to become the European-based science of ethology. The right-hand trunk, containing habits and the effects of experience, suggests the future emphasis given to learning by animal psychologists in North America.

At first it seems puzzling that an able scientist, who had performed excellent experimental work in physiology, did not make a more notable contribution to the study of the animal mind. It was a subject which he considered to be of central importance and to which he devoted a major part of his adult life. Yet those who had been unimpressed by the evidence Darwin had cited in support of the argument for mental continuity were unlikely to be convinced by Romanes' work. And, independent of theories of evolution, it remained unclear in what direction animal psychology might advance from the base described by Romanes, beyond the further collection and more detailed classification of anecdotal evidence.

One answer to the puzzle lies in Romanes' doubt about the relevance of the scientific methods familiar to him as a physiologist to problems of psychology. On occasion he would undertake some semi-experimental work himself. The studies of young birds, inspired by Spalding, have already been mentioned. Another involved collecting a number of cats from their homes on the periphery of Wimbledon Common and driving them in his coach to the centre of the common. The cats were then released, while Romanes, to the amazement of passers-by, stood on the roof of his coach to watch whether they headed off in an appropriate direction. The cats did not display much understanding of what direction to take.[13] No mention of this occurs in his books where he discusses the cat's sense of direction. Though he acquired a monkey, as Darwin had suggested, the literally painful and time-consuming work of caring for it and observing its behaviour was entrusted to his sister. But he felt that in general the role for experimental verification of his analysis of the mind of animals and its evolution was very limited. 'In cases where such verification is not attainable', he asked, 'what are we to do?' His answer was that there was no alternative, other than to abandon such study altogether, but to adopt the methods of human psychology: 'In the science of psychology nearly all the considerable advances which have been made, have been made, not by experiment, but by observing mental phenomena and reasoning from these phenomena deductively.'[14] Thirty years later his successors were less impressed by the advances achieved by deductions from data based on introspection.

Another factor was his near devotional loyalty to Darwin, reflected in both the lack of scepticism with regard to informal evidence and his unquestioning acceptance of the Lamarckian principle. By the 1880s Wallace's opposing view, that the principle of natural selection alone could explain all but human evolution, was also very vigorously expressed by the German biologist, August Weismann. The latter's 'germ' theory of heredity, which ultimately led to the development of modern genetics, became known among British biologists at this time. Weismann was highly critical of any Lamarckian mechanism of evolution and his arguments were beginning to be accepted by younger biologists, but not by Spencer nor by Romanes. Presumably because the Lamarckian principle, as part of what had become orthodox Darwinism, had become so integral a part of his own system, Romanes spent most of the energy that ill-health allowed him in his last few years in attempting to find evidence confirming the principle.

Lloyd Morgan and the cinnabar caterpillars

Romanes chose a successor, just as Romanes himself had been chosen by Darwin. And, as had happened just a decade earlier with Darwin and Romanes, the relationship between Romanes and Conwy Lloyd Morgan began with a short note to the journal *Nature*.

In *Animal Intelligence* Romanes had cited some reports on the behaviour of scorpions that suggested a tendency for this animal to commit suicide under conditions of severe stress.[1] The note by Morgan, sent from South Africa, reported some experiments that threw doubt – characteristically, it turned out later – on this particular suggestion by Romanes. By placing various noxious substances on a number of scorpions and observing their subsequent behaviour, Morgan came to the conclusion that the movement of their tails was a reflex action, normally serving to remove sources of irritation, and not an attempt by the scorpions to sting themselves mortally. The latter interpretation, one offered by his servant as well as by Romanes' correspondents, was likely to occur only to an inexperienced and casual observer, Morgan suggested.[2]

In his last years Romanes had come to consider Morgan to be 'the shrewdest, as well as the most logical, critic that we have in the field of Darwinian speculation'.[3] This opinion was largely based on a book that Morgan had published in 1890, *Animal Life and Intelligence*. As its title suggests, this was concerned with the same set of topics as Romanes' major works and in it Morgan paid tribute to Romanes' pioneering contributions to comparative psychology.

The book did not mark any great departure from the lines laid out in *Animal Intelligence* and *Mental Evolution in Animals*. Yet some points may have raised misgivings in Romanes' mind about the direction his successor might take. Already in 1890 Morgan displayed a very much more critical attitude towards anecdotal evidence than Romanes. He suggested that the human-like aspects of some of the behaviour reported of pet animals threw a misleading glamour over what were not more than special tricks.[4] What was needed 'is always to look narrowly at every anecdote of animal intelligence and emotion, and endeavour *to distinguish observed fact from observer's inference*'.[5] Turning, for an example, to one of the tests carried out on the monkey studied by Romanes' sister, Morgan maintained that what the monkey discovered was not, as Romanes claimed, 'the principle of the screw', but that the action of screwing produced the results he desired – 'a very different matter!'[6]

Morgan was also critical of Romanes' analysis of instinctive behaviour. To employ as a defining criterion the presence of an element of consciousness is unhelpful, Morgan suggested. What is required is some *objective* criterion for deciding whether some reaction is to be classified as instinctive.[7] And as for Romanes' analysis of emotions in animals: 'Throughout the sections of Mr Romanes' work which deal with the emotions, I feel myself forced at almost every turn to question the validity of his inferences'.[8]

What perhaps outweighed any possible misgivings, was the fact that Morgan, at this stage, endorsed a Lamarckian view of the inheritance of acquired habits and dispositions. Yet, even on this point, Morgan's attitude was closer to a Huxleyan suspension of judgement than Romanes' own fervent belief in the principle. Morgan discussed the familiar example, provided by Spencer, of how, through constant extension over generations, the long neck of the giraffe might have evolved. But he then pointed out that inordinately developed parts of the body can be similarly found in certain soldier ants, where inherited use can play no part in evolution since such soldiers are sterile.[9] However certain kinds of behavioural evidence did appear to support the Lamarckian principle and Morgan decided that 'the balance of probability is here on the side of some inheritance of experience';[10] further observation and discussion, and a receptive rather than dogmatic attitude, were needed.

And so, despite the fact that the heir was in this case only four years younger, Morgan was summoned during Romanes' final illness and given instructions on various unpublished papers.[11] These instructions were loyally carried out. However, for Morgan, warm regard for a predecessor and friend did not have to imply reverence for his work. The shrewdness and logical criticism led to an almost complete dismantling of Romanes' system.

A major influence in Morgan's intellectual development was Huxley. Morgan's contact as a young man with Huxley came about as an indirect result of his father's financial incompetence. When Morgan was a boy he lived with his family in a country village – now suburban Weybridge – on the outskirts of London. His father worked as a solicitor in the City. For generations Morgans were educated at Winchester College and Oxford University before joining the family legal firm. The father differed from his forebears in his enthusiasm for the Welsh origins of the family and his lack of caution. Morgan was baptized as 'Conwy Lloyd' and continued to use the signature 'Lloyd Morgan' throughout his life, even though the Welsh connection was remote.

The father's interest in investment in railway expansion took him on a tour of England that ended with little remaining of the family's financial assets. Since there was no money left to pay for an education at Winchester and Oxford, Lloyd Morgan was sent to a local grammar school and from there, following the aptitude he had shown for scientific subjects, to the London School of Mines for training as a mining engineer.[12] He arrived there in 1869, during the period when Huxley was transforming this institution from one which offered purely technical training to one providing a general education in contemporary science that included highly innovative teaching in biology.

As an adolescent Morgan had been persuaded to read Bishop Berkeley by a local clergyman. A developing interest in philosophy was added to his hobby of bird-watching. He was given an article by Herbert Spencer to read and was thrilled by its vision of biological evolution as a key to the scientific understanding of man. These interests were not displaced by engineering and chemistry during his two-year course, but were further amplified by Huxley's lectures. When the course ended, instead of beginning a professional career, he took on temporary employment as travelling companion to a family embarking on a few months' tour of America. This journey was an important event in his life and allowed ample opportunity for reading Darwin for the first time.

On returning to England he took up an offer

Fig. 2.5. Conwy Lloyd Morgan at about the time of his return to England

Huxley had made to him earlier of a position for one year as a research associate in biology at the School of Mines. Unfortunately this appointment appears to have coincided with the period of Huxley's breakdown and there was no direct contact between them. A long period of searching for some permanent appointment, while taking on a series of temporary jobs, ended when he successfully applied for a teaching post in South Africa. The subsequent five years were spent in a small college near Cape Town. In addition to lecturing in physical sciences, English literature and constitutional history, he continued to read philosophy, Spencer, Darwin and, now, Romanes. He also began to observe carefully various examples of animal behaviour and carried out the study of self-destructive tendencies in scorpions already referred to. The founding of a new college in Bristol provided an opportunity to return to England in 1884 and there he became Professor of Geology and Zoology. Over the next nine years this position allowed him to publish the work which established his growing reputation as an authority on animal intelligence and as an able contributor of ideas on evolution.

By 1893 Morgan was middle-aged – forty-one years old by then – and had obtained a respectable academic position and reputation, but felt that he had made no distinctive contribution of his own. In that year he received a letter which prompted him at last to 'work upon some definite inquiry with a definite end in view'.[13] By the end of the next four years he had made a distinctive contribution.

In his book of 1890, *Animal Life and Intelligence*, Morgan had cited extensively, as Romanes had also done, the work of Spalding and the conclusions that the latter had drawn from his experiments. An American friend, Mann Jones, wrote to say that his own research threw some doubt on these conclusions and led him to regard with some suspicion much that was written about the 'philosopher's chick'. If Morgan were to repeat these studies, he might also come to share the suspicion.[14] Morgan responded to the suggestion and began to repeat the kind of research that Spalding had undertaken, hatching chicks, ducklings and pheasants in incubators and observing their early encounters with the world.

Spalding had been impressed by the degree to which behaviour was independent of the individual's experience. In contrast, Morgan's interest was aroused by the way that such early experience had far-reaching consequences for the individual's subsequent development; he became concerned with the processes by which Spalding's 'imperfect instincts' became perfected. This concern prompted three major conceptual developments. It led Morgan to return to the problem, faced earlier by Bain, of analysing the nature of learning processes. It changed his attitude towards the kind of interpretation given by Romanes to examples of complex behaviour: from scepticism about the direct inference of mental capacities and emotions from an animal's behaviour, Morgan came to see that a relatively simple psychological process operating in an appropriate environment could generate the performance of a variety of complex actions. Finally, it led him to reject eventually any Lamarckian inheritance of behaviour and develop a concept of instinct very different from that held by Romanes and very similar to that held today.

These developments are discussed later in this chapter. To understand Morgan's ideas on processes of learning, on complex behaviour and on instinct, it is helpful to look at the kind of investigation carried out by Morgan after receiving the letter from Mann Jones.[15]

He took eggs two or three days before they were due to hatch and placed them in an incubator. After hatching, the ducklings or chicks remained in the drawer of the incubator for a further twelve or eighteen hours and then Morgan began to test them. Once again it was observed that a chick would initially direct pecking responses at a wide range of small objects and that this range became rapidly modified. If the chick pecked its own excrement, then it would subsequently attempt to wipe its bill and show other signs of distaste; after one or two such experiences it no longer pecked at its droppings. Morgan found that the accuracy of pecking was initially not nearly as fine as he expected from his reading of Spalding. Also, he was surprised to find that there appeared to be no innate recognition of water; incidental insertion of the beak into water appeared to be necessary before a thirsty, but inexperienced, chick or duckling would drink.

Following Spalding's report of the instinctive terror shown by young birds at the cry of a hawk, Morgan tested the effects of various stimuli to find out what kind would elicit signs of fear. He discovered that any sudden, loud noise would do so, including a sharp chord on the violin. And he doubted whether his chicks had any inherited acquaintance with violins.[16]

One reason for some of the discrepancies between these and Spalding's results appears to be that Morgan was far less careful about completely restricting the sensory experience of his young birds before they were tested. In addition, he seems to have missed Spalding's point that the reaction of an inexperienced bird to a given stimulus can depend very critically on the length of time since it was hatched. We have seen that Spalding obtained the 'following reaction', the phenomenon of imprinting, in two- to three-day old birds, but not in birds that were a few days older than this. Thus, although Morgan found to his surprise that two ten-day old chicks were completely indifferent to the clucking of a hen,[17] he might not have obtained the same result if he had tested them at an earlier age and had ensured that they had had no prior exposure to any sounds. More careful reading of Spalding would have revealed to Morgan that there was no reason to be surprised. Twenty years earlier Spalding had also failed to obtain any reaction to the sounds made by a hen from ten-day old chicks.

These criticisms do not, however, detract from the importance of his observations on the way early, and often very arbitrary, events in the young birds' lives affected their behaviour. In one study the only access to water given to two ducklings was in a shallow tin placed on a black tray, where they would drink and throw water over their backs. On the sixth day Morgan gave them the black tray and the tin, but without water. They sat in the tin making exactly the same sort of movements as before, though less vigorously as time passed.[18] The way that such arbitrary associations might work in a natural environment to produce adaptive behaviour, in the face of considerable changes in the ecological niche of a species, was illustrated by a study involving caterpillars. Having given some chicks the opportunity of pecking at and consuming caterpillars of an edible species he called 'loopers', Morgan introduced some caterpillars of the cinnabar moth, which have striking black and gold bands. These appeared to be distasteful to the chicks, evoking much the same reaction as excrement; following just a few pecks at them, they were afterwards consistently avoided.[19]

Such observations were seen by Morgan as examples of the Spencer–Bain principle, whereby behaviour is modified by its immediate consequences. His version of this principle was: 'What we term the control over our activities is gained in and through the conscious reinforcement of those modes of response which are successful, and the inhibition of those modes of response which are unsuccessful. The successful response is repeated because of the satisfaction it gives; the unsuccessful response fails to give satisfaction, and is not repeated.' It is interesting that observations on the effects of taste aversion learning in chicks were important in the origin of reinforcement theories; seventy years later a revival of interest in this, by then neglected, topic again contributed to the development of learning theory.

Another observation provided Morgan with a more complex instance of what he now termed, following Bain, 'trial-and-error learning with accidental success'. He constructed in his study a small pen with newspaper walls, insecurely propped against various objects, and placed a week-old duckling in it. The duckling pecked at the walls, and then seized and pulled at a corner of one newspaper, thus making a breach through which he promptly stepped out of the pen. When replaced in the pen, he made for the same corner and quickly escaped again.[20] Morgan's reason for viewing this as a more complex example of trial-and-error learning than that of the cinnabar caterpillar was because of the initial lack of any relationship between whatever it was that released the action of pecking and the consequence of that action (the possibility of escape from the pen). This is an important distinction and is discussed more fully below.

All the more systematic studies undertaken by Morgan employed young birds. But just as valuable to him were the less formal observations he made of the behaviour of his pet fox terrier. This dog proved to be a useful inspiration for Morgan's revaluation of the evidence collected by Romanes.

In one test Morgan wished to evaluate its perception of spatial relationships. The procedure he used was to throw a stick, about nine inches long, to the other side of a fence whose vertical rails were only six inches apart, and send the dog to retrieve it. This was repeated time after time with only slow improvement in the dog's ability to return through the fence without getting the stick caught. The test was conducted again, but with a shorter stick which had a crook in it. When this caught on a rail, the dog persistently tugged at the stick, showing a similar lack of perception of the spatial relationship which might have allowed him to easily disengage the crook. His performance continued to be unimpressive until with a wrench he broke the crook off and brought the stick through the fence. This was observed by a passer-by, who had stopped for a couple of minutes to watch the proceedings, and who commented to Morgan: 'Clever dog that, sir; he knows where the hitch do lie'. A characteristic interpretation of two minutes' chance observation, noted Morgan.[21]

As for the superior understanding of mechanical appliances, which had been attributed by Romanes to cats and, to a lesser extent, dogs on the basis of their skills in opening doors, Morgan was convinced by his observations of the *acquisition* of such skills that these were also no more than the product of trial-and-error learning. The example he described most fully was that of the same fox terrier, Tony, learning to operate the latch of a gate, the one shown in Figure 2.6. When in the garden, Tony would place his head through the rails, gazing into the road. On one occasion he pushed his head under the latch and lifted it. The gate swung open and, after a short delay, Tony moved back from the fence, looked around and bolted out of the gate. Each time he found himself in the same situation the number of times he poked his head through an inappropriate part of the fence decreased. After three weeks he would go with precision to the right place and without any ineffectual fumbling put his head beneath the latch.[22] A performance of the kind so impressive to Romanes' correspondents was achieved, but this still contained a component indicating its origin, since, as Lloyd Morgan noted, 'even now he always lifts it with the back of his head and not with his muzzle which would be easier for him'.[23]

Fig. 2.6. A product of trial-and-error learning: Morgan's dog, Tony, operating the latch of a gate

In terms of the development of methods for animal psychology Morgan's work with young birds seems cruder than that of Spalding. In the way that he studied his dog, the approach was not very different from that, say, of Romanes' sister, except in two respects. One was Morgan's persistence. He would patiently observe how an animal behaved when it was repeatedly exposed to some situation and this allowed him to study the course of development of some action. The other was the degree to which he was able, in a detached and analytic manner, to follow his maxim of distinguishing between an animal's actual behaviour and the interpretation of that behaviour that a human observer is prone, often unwittingly, to supply.

Morgan on comparative psychology and theories of learning

The studies described above were reported in *An Introduction to Comparative Psychology*, which was published in 1894, just after Romanes' death. The book turned out to be the most widely cited of the several that Morgan wrote, but is by no means the most readable. It was the only one in which a major theme was the mind–body issue and which attempted, in the light of his views on this issue, to define the nature of comparative psychology.

In *Animal Life and Intelligence* of 1890 he was careful to distinguish between a dualistic view of mind and matter, and his own position, which he termed a *monistic* one. Just as a curved surface can be convex from one point of view and concave from another, phenomena can have two aspects, the subjective and the objective. The idea that mind and matter refer to separate kinds of entities – implicit in much of Romanes' early writing and explicit just before his death – was firmly rejected.[1] Morgan quoted with approval two new terms suggested by Huxley, which are now used in a very different sense: 'neuroses', referring to molecular events in the brain, and 'psychoses', referring to concomitant states of consciousness.[2] According to monism a neurosis is an event described from an outer, objective view and the corresponding psychosis is the same event viewed from inner, subjective experience.

Four years later the idea of monism was again discussed, but now Morgan was more concerned to distinguish it from a purely materialistic view of mind. He was still emphatic that 'man, as an organism is one and indivisible, no matter how many aspects he may present subjectively or objectively' and that 'mind is not extra-natural or supra-natural, but one of the aspects of natural existence'.[3] However, the Huxleyan terminology was abandoned and the message of Huxley's Belfast lecture that, in Morgan's words, 'the body is the real substance, the mind being one of its properties' was firmly rejected.[4]

In *An Introduction to Comparative Psychology* the problem of whether animals are conscious automata is discussed at length. In deciding that they are not, Morgan came to put much more weight on the criterion of learning than Romanes had done. He pointed out that the word 'automaton' is ambiguous and argued that, in its interesting sense, an animal that demonstrated the beneficial effects of its past experience was therefore not an automaton. 'It is well to bear in mind that the words "automaton" and "automatic" are used in two different senses. An "automaton" is defined as "a self-moving machine" and "automatic" as "having the power of moving itself". If now we lay stress on the *self-moving*, we have one sense in which the word is used. In this sense I am quite prepared to regard myself and animals as conscious automata. But if stress be laid on the *machine* then the word automatic acquires the connotation of mechanical uniformity of action. In this sense I do not regard myself and animals as automatic – the automatic act may be *accompanied* by consciousness, but the controlled act is *guided* by consciousness.'[5] He went on to make clear that a 'controlled act' is one that reflects the influence of experience and to suggest that the objective processes accompanying such control are uniquely located in the cerebral hemispheres. Thus, the only vertebrate animals that may deserve the description 'automata' are those that have had their cerebral hemispheres removed.[6]

Morgan distinguished two kinds of psychology. One is based on introspection and the careful analysis of our subjective experience. The other, comparative psychology, starts from objective observations, whether of other animals or of other human beings. A comparative psychologist is required to make two kinds of induction, where induction is seen as the development of hypotheses on the basis of facts and the testing of such hypotheses in the light of further facts. One kind, and the most basic kind, for a psychologist, is to induce the laws of the mind from introspection; the other is to derive theories for objective data that are compatible with those based on introspection.[7] Morgan noted that 'there are, I am well aware, many people who fancy that by the objective study of animal life they can pass by direct induction to conclusions concerning the psychical faculties of animals'; he wished it to be quite clear that this was an erroneous fancy and that a thorough understanding of subjective psychology was necessary for the comparative psychologist.[8]

The distinction Morgan made between introspective and comparative psychology may be illustrated by an example he used to show that there is a delay between an objective event and the corresponding subjective experience. He described a sequence of stills from the pioneering work on high-speed photography by Muybridge which showed one girl pouring water over another girl sitting in a bath. In one frame the second girl is seen covered with iced water, but still motionless. In a later frame the girl is seen leaping from the bath with an expression he described as appearing 'to be the index of somewhat forcible states of consciousness, the iced water being somewhat unexpected'.[9] From Morgan's point of view the girl's behaviour is to be explained, not in terms of her

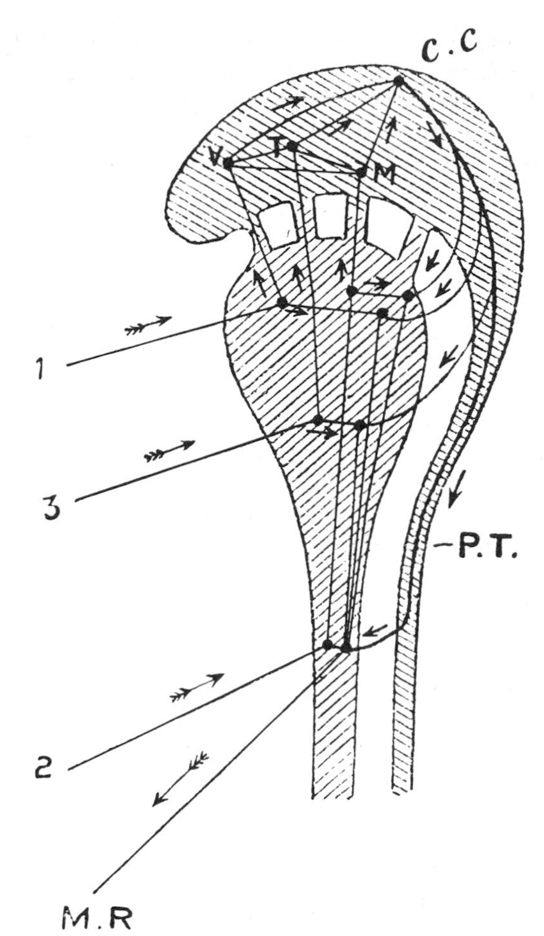

1. *Retinal stimulus.* 2. *Pecking stimulus.* 3. *Taste stimulus.* M. R. *Motor response in pecking.* V. *Visual centre.* T. *Taste centre.* M. *Centre of motor consciousness.* C. C. *Control centre.* P. T. *Pyramidal tract. The impulses from the control centre pass down the pyramidal tract, and go either to the co-ordinating centres, or to the spinal centre for motor response, or to both.*

Fig. 2.7. Morgan's sketch of the neural connections formed in a chick's brain when it learns not to peck at a distasteful caterpillar

subjective feelings, but ultimately in terms of events occurring in her nervous system. The kind of explanation to be reached by such an application of the comparative method is, however, to be guided and to be compatible with an account in subjective terms supplied by a competent observer trained in the analysis of his own experience when like events have happened to him.

As applied to the more prosaic problem of explaining why a chick no longer pecks at a cinnabar caterpillar, this point of view led, via a discussion of the laws of association that could be derived from introspective evidence, to the formulation of a theory of learning. This is illustrated in Figure 2.7 by the diagram Morgan used to explain his theory. The essential idea is as follows. First, there occur in close succession activation of neural centres corresponding to the sequence of two psychological events: perception of the distinctive visual markings of the caterpillar and experience of the unpleasant taste. As a result, a neural connection is established between the representations in the chick's cerebrum of the two events. The primary, and innate, effect of seeing the caterpillar is, like that of any small object within close range, to elicit a pecking movement in its direction. Once the neural connection is established, the visual stimulus now also has the further effect of activating the taste centre, which in turn – because the taste in this case is inherently unpleasant – acts in an *inhibitory* manner on the control centre and prevents the occurrence of pecking.[10] Although not expressed in the diagram, Morgan suggested that, with repeated exposure to caterpillars with such markings and continued inhibition of the pecking response, the primary effect would progressively weaken so that avoidance would become automatic, without the need for cerebral control.

Morgan admitted that this was probably far too simple an account of what was happening in the chick's brain. Nevertheless, it marked a considerable advance over the only comparable previous attempt, that by Spencer. As described in the previous chapter, Bain had simply maintained that there was a general principle of learning whereby the subsequent probability of occurrence of some action was determined by whether its immediate consequences were pleasant or unpleasant. Where Bain had refrained from speculating about the basis for this descriptive principle, Spencer had sketched a theory that intermixed concepts from human psychology and from neurophysiology.

Morgan's theoretical approach was different in two respects. First, his monism led him to be careful in separating 'introspective' and 'comparative' accounts of how the Spencer–Bain principle operates. And second, he saw that any given example of the principle, or of trial-and-error learning with accidental success, might be based on any of a number of alternative accounts. The chick, which first pecks at a cinnabar caterpillar, receives an unpleasant taste and then refrains from pecking at objects of the same or similar appearance, provides one kind of example of trial-and-error learning. The consequences of the action have decreased the probability that the action will occur in the future. But in this particular case Morgan's explanation of why this occurs does not make any assumption that the animal had learned a connection between its behaviour and the consequence. Instead he assumed that the change in the chick's behaviour occurred as a result of the formation of a connection between neural centres representing two perceptual events, the visual appearance of the caterpillar and its taste. On the other hand, the kind of trial-and-error learning exemplified by the duckling that escaped from the newspaper pen, or by the dog that operated the latch of the garden gate, could not, it seemed, be explained as based simply on an association between two perceptual events. Morgan believed that in these cases it was necessary to assume the formation of a connection between the neural representation of an action and that of the consequent event.[11]

The idea that there are at least two kinds of trial-and-error learning appears to derive from the distinction that Bain had first drawn, between innate responses for which some external eliciting event can be found and those which appear to be autonomous. Morgan included a similar discussion in *Animal Life and Intelligence*,[12] although not in the *Introduction to Comparative Psychology*. In the case of the cinnabar caterpillar, the object that elicits the pecking response in the first place is the one in which the noxious substance is located. In the case of escape from the pen, pecking at the walls presumably first occurred for reasons that had nothing to do with escaping. An explanation of why the effective response is acquired in the second situation must contain, Morgan argued, some form of knowledge that this particular response is followed by its particular outcome. The previous theory does not contain this element, but he failed to sketch a theory that might make the idea more specific.

In this context the use of the word 'successful', instead of 'pleasant', in Morgan's version of the Spencer–Bain principle quoted above, is confusing. In normal usage 'successful' denotes the realization of an outcome that is to some extent anticipated. If someone takes up a floor board in order to get at some faulty wiring and happens to find a ten pound note, the result may be pleasure, but it is inappropriate to describe the action as successful. On the other hand, it would be appropriate to use the word 'success' to describe a subsequent tearing up of other boards which produced further notes. As related in the following section, Morgan doubted whether any animal was capable of making the kind of inference needed in order to anticipate the result of an action, when the animal had never previously performed that action in a similar situation. Thus, it would have been clearer to retain the terminology of pleasure and pain in describing the Spencer–Bain principle.

The distinction between two kinds of trial-and-error learning was related to his evolutionary views on learning. For Morgan instinct provided the outline sketch of an animal's behaviour, while experience filled in the detail; high in the scale of mental evolution the original outline might become indistinguishable. The model of Figure 2.7 indicates the first step by which individual experience carries on from the results of evolution. As suggested by Spalding's term 'imperfect instincts', natural selection is assumed to operate so as to produce increasing co-ordination of a response system; but only as far as a stage where the environment of the young animal will normally bring it to perfection. This perfection is achieved by an associative mechanism of the type previously illustrated. At this stage the only flexibility, or intelligent capacity, present is the ability to form an association between two external events occurring close together in time in the animal's perceptual world. If the second event has innate properties of pleasure or pain, as a result of the evolutionary process suggested by Spencer, and if, in addition, there are innate connections between the neural structures underlying the perception of an emotional event and those involved in the organization of the instinctive response, then a great deal of adaptive behaviour can be explained.

The more complex cases of trial-and-error learning indicated to Morgan a further step of increasing freedom from what later came to be called biological constraints on learning. This step occurred when an animal was no longer confined to acquiring associations between two external events, but in addition could acquire associations between its own actions and the external events that followed them.

For Morgan, as for Romanes, it was quite appropriate to talk of mental evolution as well as of organic evolution, but for him the two processes were completely interdependent. His monistic point of view led logically, he believed, to the view that, just as in objective terms there had been evolution from simple organization of matter to complex organic structures, there had been a parallel evolution of subjectivity. Morgan described mental evolution as progressing from 'metakinetic' aspects of inorganic matter, via 'infra-consciousness', to full human consciousness.[13] In very Spencerian terms he wrote that he conceived man 'to be the self-conscious outcome of an activity, selective and synthetic, which is neither energy or consciousness, which had not been evolved, but through the action of which evolution has been rendered possible; which is neither object nor subject, but underlies and is common to both'.[14] He expressed his strong antagonism to the idea that mind was a product of organic evolution. And, although laying such stress on the part played by learning in psychological development, Morgan wished it to be clear that he did not endorse the view of the 'Empiricists' who 'are apt to regard psychological genesis as wholly the results of the conditioning effects of the environment and to make the individual mind a mere puppet in the hands of circumstances'.[15]

In reaching these conclusions on the evolution of mind Morgan was guided by the logic of his monistic point of view, and appears to have been little influenced by the problem of religious belief that was so important for Romanes. Morgan's attitude to religion seems to have been similar to that of Spencer, who felt that there was no fundamental conflict between science and religion and that they were ultimately compatible in the realm of 'the Unknowable'; an attitude that has been compared to that of the husband proposing that, as a basis for future marital harmony, he should take the inside of the house, while his wife took the outside.[16]

Morgan's canon, psychological complexity and instinct

The only feature of Morgan's *Introduction to Comparative Psychology* to be cited with regularity over the eighty years or so since it appeared has been the statement of a guideline for comparative psychology, which he termed his canon. Over the preceding years his sceptical attitude towards the kind of interpretation offered by Romanes' correspondents had developed further and, aided by the kind of observational study described earlier in this chapter, had changed into a belief that the complexity of an animal's behaviour rarely, if ever, indicates psychological complexity. To his earlier reluctance to accept that Romanes' monkey understood the mechanical principle of the screw was added rejection of claims that a horse, which was observed to take a zig-zag course up a steep hillside, thus demonstrated acquaintance with the principle of the inclined plane;[1] that a dog, heading to cut off a rabbit scuttling in a circular path, understood the chord of a circle to be shorter than the arc;[2] that a hen bird's choice of the colourful male suitor shows an aesthetic sense;[3] or that even dogs demonstrate some understanding of moral laws.[4]

The canon was stated at various places in the book, the most common formulation being: 'In no case may we interpret an action as the outcome of the exercise of a higher psychical faculty, if it can be interpreted as the outcome of the exercise of one which stands lower in the psychological scale.'[5] But he felt that *process* was probably a better term to use than *faculty*.[6] The canon can be seen as simply the application of the general law of parsimony to explanations of behaviour. Nevertheless, Morgan did not justify it on these terms, but on the grounds of evolutionary theory. If a particular process is sufficient to allow the development in a given species of appropriately adaptive behaviour, then there is no selective pressure for the evolution of a more complex process. In cases where there was firm evidence from one situation that a species possessed some complex process, Morgan was prepared to be generous in some other situation where the behaviour of this species could be interpreted either in terms of the same process or in terms of a simpler one.[7] Many later applications of Morgan's canon were to be far less generous. A common interpretation has been to view it as a general rule of psychology that, even when an appropriate complex process may be within the capacity of some animal, adaptive behaviour is likely to be based on a simpler process, if this is sufficient for the occasion.

The canon's admonition is without value unless it is accompanied by a description of 'the psychological scale' and thus of what constitutes a relatively complex, and what a relatively simple, psychological process. The major theme of the *Introduction to Comparative Psychology* is to discuss in turn processes of increasing psychological complexity. The ordering more closely resembles the linear progression, developed almost forty years earlier by Spencer, than Romanes' branching tree. The importance given by Morgan to the perception of relations was directly derived from Spencer. Adhering rather rigidly to his conception of comparative psychology, Morgan first described a process on the basis of introspective evidence in one chapter, and in the next from the objective point of view gained from observing the behaviour of animals. Running throughout is discussion of a more philosophical nature, mainly related to the mind–body issue, in what Morgan himself recognized was probably a wearisome manner. In addition, there are side excursions to consider processes that cannot be fitted into the main progression. Consequently, although the psychological scale demanded by the canon is provided by the book, it is not readily apparent. Morgan did not include a concise description for the convenience of the reader, nor a summary diagram of the kind produced by Romanes.

The book deals first with simple associations, distinguishing – as already described – between those involving two sense impressions and the more complex case of associations between actions and their outcome. The development of associations, based on correlations between events in the physical world, makes possible a synthetic process whereby, for example, the perception of objects with shape and definite locations comes to replace that of sensations. So far these are all processes that animals also share.

The next more complex process, involving the perception of relations, is similar to what Romanes called primitive reasoning. For Morgan, this set the dividing line between the animal and human mind. Had his dog passed the test of getting the nine-inch long stick through the six-inch wide gap, in a way that could not have arisen from trial-and-error learning, this performance would have ranked the dog's mind at the 'rational' level. As it was, Morgan could find no firm evidence showing that any animal was capable of perceiving relationships.

Romanes, late in his life, had carried out a study of a chimpanzee, Sally, at the London Zoo. With the aid of a keeper, Romanes trained the animal to count out a correct number of straws on demand, or at least to perform at a level that he judged satisfactory. This would have indicated what Morgan meant by perception of relationships, but Morgan was distinctly unimpressed on visits to watch this performance.[8] The idea that systematic testing and objective recording of performance might help to settle such a difference of opinion was still not very common.

Despite the lack of convincing evidence favouring the perception of relationships by animals, Morgan left open the possibility that it might be found by future research, most likely if apes or monkeys served as subjects. On the whole, though, he regarded this as unlikely, since the only function he saw for this kind of perception was in communication. He could find nothing comparable to human language in animals 'other than indicative communication, which is primarily suggestive of emotional states, and secondarily (and probably only incipiently) suggestive of particular objects'. Since even primates did not appear to possess the 'power of descriptive intercommunication', they were probably incapable of perceiving relationships.[9]

The remaining stages were based on his understanding of human psychology and were uninfluenced by observations of animals. The next more complex process involved abstract relationships, where the perception of relations had become independent of particular instances. If Sally had shown an ability to count, not just straws, but any set of items, her mind would have been classified at this level. Or, to take examples from situations used more than forty years later, one might test whether an animal can perceive abstract relationships by attempting to train it to select the middle-sized object from a set of three objects of different size, or to select that member of a set of objects which matches some sample, under conditions where the set of objects used on each trial is continually varied.

Morgan was very confident that no animal was capable of perceiving abstract relationships. It was this peculiarly human ability that made possible conceptual understanding and the phenomenon of insight: 'when conceptual thought has been reached, when systematic generalizations have taken form in the mind, the significance of a situation, or given presentation in relation to the system, may arise as rapidly and directly as the meaning of a situation does for sense-experience. Hence comes those flashes of insight which are so difficult to explain on any psychology that is based merely on associationism'.[10]

The final level of complexity was reached with the concept of self. Morgan was not only clear that no animal possessed such a concept, but was doubtful whether self-consciousness could be found in more than a vague form even in schoolboys and peasants.

Morgan also distinguished levels of complexity in other kinds of processes. In general two levels were discussed: a simple form which could be found in animals and did not involve the perception of relations, and a complex form which was restricted to man and did involve the perception of relations. The topic of communication, mentioned above, provides one example. Can animals communicate with each other? Morgan's answer was: yes, in the 'indicative' sense of conveying their emotional state or suggesting the presence of some object, but no, in the 'descriptive' sense of providing information about relationships between events.

Another important example was memory. Can animals remember? In one sense he accepted the claim supported by the weight of evidence gathered by Romanes and others, and also by his own observations, that they do possess memories. Scores of reports indicated that in a wide range of species animals showed signs of recognition of an object or person that they could not have had any contact with for many years; or performed an appropriate response in a situation that had not been encountered for some similarly long time. Reports about elephants and

Darwin's welcome on his return from the five-year voyage in the *Beagle* by his old dog were particularly popular examples.[11]

This kind of memory was somewhat disparagingly labelled 'desultory' by Lloyd Morgan. It would now be termed 'associative memory'. For him such reports simply indicated the relative permanence of associations. He compared this kind of memory in animals to that of a student, who can trot out the answer '1596' to the question: 'When was Jonson's play *Every Man in his Humor* written?', but is completely confused when asked whether Shakespeare was still living or Cromwell dead. He contrasted this with memory that involves the recall or recognition of facts which have meaning with reference to a system of knowledge. He termed this 'systematic memory'. Nowadays it would be termed 'semantic memory'. A second student, who might be unable to produce the answer '1596', but who could, on the basis of other knowledge about Jonson and the late Elizabethan age, place the date to within a few years, would be demonstrating the possession of systematic memory.[12] Morgan could find no evidence that animals possessed memory in this second sense of the word; no indication in a non-human species of any knowledge of the temporal relationship between past events.

He treated the questions of whether animals experience emotions, or possess anything that might be termed an aesthetic or moral sense in a similar manner. In each case two levels were distinguished, and in each case possession of the more advanced form, involving the perception of relations, was found to be confined to man.

The *Introduction to Comparative Psychology* was a pivotal book, marking the end of one era and the beginning of another in the study of animal behaviour. One important aspect was the emphasis on learning. For several decades to come many of the topics that Morgan discussed – for example, the analysis of trial-and-error learning and the ability of animals to perform appropriately on the basis of perceptual or abstract relationships – became central issues in the theoretical disputes accompanying the development of animal psychology and of behaviourism. Equally interesting are examples of the modification of behaviour by experience that receive no discussion in this book.

There is what seems today a glaring omission at the simple end of the main scale. The simplest, and most widespread case of a change in behaviour is habituation, a decrease in the intensity with which an animal reacts to some event, when this occurs repeatedly and is unrelated to any other event. Morgan never discussed this phenomenon. One can only guess that, from his point of view, it represented a decrease in consciousness; consequently it would be of little interest to the psychologist and its analysis better left to the physiologist. Whatever the reason, animal psychologists neglected the phenomenon almost as completely as Morgan until very recently.

Another kind of learning, given some importance by Romanes and also discussed quite extensively by other writers of the time, was imitation.[13] Rather curiously no discussion of this topic was included in *An Introduction to Comparative Psychology*. Two years later Morgan made up for the omission. The interest in imitation was related to the development of his views on instinct and to his eventual rejection of any Lamarckian inheritance of acquired behaviour. Early in 1896 Morgan received an invitation to visit North America and give a series of lectures in Boston, New York and Chicago. This provided an opportunity to present his current ideas on instinctive behaviour. The lectures were revised for publication later in the same year in a book, *Habit and Instinct*, in which imitation was now given an important place.

Six years earlier in *Animal Life and Intelligence* Morgan had been inclined to reject the principle of use and disuse in the inheritance of bodily organs, but to accept – at least as a working hypothesis – a Lamarckian element in the origins of instinctive behaviour. As described earlier, although he thus endorsed Romanes' view on the dual origin of instincts, he disagreed with Romanes' definition. Instead, Morgan defined an instinct in a way that was closer to Spencer's 'compounded reflex', as an 'organized train or sequence of co-ordinated activities by the individual in common with all members of the same more or less restricted group'.[14] The definition was intended to cover three classes of instinctive behaviour: perfect instincts, which appear at, or shortly after birth; imperfect, or incomplete, instincts, which require some 'self-suggested trial and practice' – as in the newborn mammals which require some practice in the use of their limbs before they can walk or run; and deferred instincts, that emerge later in the life of an animal, as in the case of those involved in sexual behaviour.

With regard to specific ideas on heredity Morgan had been very critical in 1890 of a theory of inheritance by Darwin, known as 'pangenesis'. At the same time he considered Weissmann's theory, with its strong rejection of Lamarckian inheritance, to be a distinctly retrograde step.[15] Subsequently Morgan became steadily more impressed by Weissmann's theory and arguments. Also, his doubts about the inheritance of acquired habits perhaps grew with his realization that the quality of evidence in its favour was no better than that for higher mental processes in animals. What finally decided Morgan's opinion on this issue were the results he obtained in testing the reaction of young birds to water. Since birds had been drinking water for millions of years and the appearance of water had presumably not changed during this time, an instinctive tendency to drink at the sight of water should have become established – given that the Lamarckian principle is a factor in heredity, if only a very weak one. Why then should a duckling not show any innate recognition, but have to depend on accidental contact between its beak and water? Morgan's answer was that there was no pressure to select such an instinctive reaction; in the vast majority of cases the environment of young ducklings was such that accidental contact with water was assured.[16]

With the decision that natural selection could serve as the sole basis of instinct, Morgan formulated a view of instinctive behaviour, and its relationship to acquired behaviour, that is essentially identical to one held very generally today. An instinct is behaviour which is entirely determined by the congenital organization of the nervous system, being either 'connate' (performed at, or shortly after, birth) or deferred. Whereas a reflex involves a restricted group of muscles and is initiated by a simple external stimulus, instinctive reactions involve the whole organism, are elicited by more complex stimulus events and are sensitive to the internal state of the organism.[17] All that an individual animal can inherit in addition to bodily organs are reflexes, instincts, the capacity for forming associations – some more easily than others – and susceptibility to specific pleasures and pains.

With the rejection of the Lamarckian relationship between habit and instinct, Morgan was faced with the problem of explaining the kind of evidence that naturalists for decades – including Darwin, Romanes and Morgan himself in earlier days – had seen as supporting their belief in this relationship. The crucial cases were those in which the same, highly specific and adaptive behaviour could be observed in most members of a group of animals and in successive generations, and yet it was very difficult to see how the behaviour could have arisen as a random variation on which natural selection had operated.

Morgan proposed that there were two ways in which such uniformity of behaviour could arise, even though the behaviour had to be acquired by each individual member of a species. The first arises in situations where only one particular response can be effective, so that the process of trial-and-error engaged in by each animal always produces the same result. An example he discussed was one that Darwin had provided over twenty years earlier. Darwin cited the particularly efficient method used by bullfinches to extract nectar from primroses as a case of a Lamarckian instinct. Morgan argued that the structure of the primrose ensured that each bullfinch would eventually acquire exactly the same pattern of behaviour.[18] In this case, uniformity of behaviour in a species reflects the special properties of its environment.

The second way depended on learning by imitation. To a degree, Morgan's treatment of imitation can be seen as the development of suggestions already made by Romanes. What made it much clearer and more powerful was release from the Lamarckian principle. A friend of Romanes had taught his cat to beg and this cat had borne kittens, which adopted the same habit without being trained. Perhaps forgetting what he had said elsewhere about the importance of imitative learning, Romanes cited this as another piece of evidence supporting belief in the hereditary transmission of habits.[19] By 1896 Morgan interpreted such anecdotes – to the extent that they could be trusted – as indicating the way in which imitation by the infant of the parents', or other conspecifics', actions could ensure that similar behaviour patterns appeared in generation after generation.

An interesting comparison between the views of Romanes and Morgan is provided by the subject of birdsong. The acquisition by imitation of unusual songs or calls was seen by Romanes as something that interfered with, or at least was superimposed upon, the instinct for the natural song. From his review of the evidence on this matter, Morgan arrived at an account based on a more subtle interaction between innate and experiential factors. He concluded that in many species of birds the characteristic song of the adult bird develops only if, during a critical period of infancy, it has heard songs of this kind.[20] Since the mid-1950s a substantial amount of experimental work on the European chaffinch and the American song-sparrow, generated by a renewed interest in birdsong, has verified Morgan's conclusion. The major point of this example is that forms of learning by imitation provide another means by which constancy of behaviour may be maintained. In this case behavioural uniformity in a group of animals reflects the uniformity of the social environment of the young.[21]

A chapter in *Habit and Instinct* was devoted to imitation. As in his earlier discussions of communication and of memory, Morgan distinguished two different forms. One he termed *instinctive imitation*. This he considered to be simply instincts, in which the perceptual event that evoked the reaction happened to resemble the action. Examples from his own studies included under this heading were: if one of a group of chicks starts to drink from a tin, then the others, without prior experience of drinking, run up and do the same; tapping with the point of a pencil initiates pecking in chicks and young pheasants that are close by; when one chick has learned not to peck at a cinnabar caterpillar, then other members of the brood also avoid it; a danger note sounded by one chick tends to elicit a similar note in others. The behaviour in these examples was 'objectively, but not subjectively, imitative'.[22]

The other form of imitative behaviour was seen as more complex and as involving learning, and thus consciousness. This Morgan termed 'intentional, or conscious, imitation'.[23] It applies when, in addition to an animal behaving in a way that resembles some action observed in another, there is evidence of an attempt to reproduce the result of that action. This was regarded as less widespread than instinctive imitation, but Morgan felt it to be an important factor, one 'scarcely open to question', in animal life, especially among gregarious animals. This was one of the least guarded of Morgan's conclusions and one for which he had little direct evidence. The question of whether animals can learn by a process of intentional imitation became a very open one in the decade that followed.

In 1865 Wallace argued that once the human mind had evolved to a certain level, organic evolution by means of natural selection must have effectively ceased to operate. The point of this argument was obscured by its association with Wallace's subsequent appeal to a supernatural origin of mind. The argument appears to have had little impact at first on those who sought to explain all aspects of evolution in terms of natural causes. The kind of direct analogy between organic evolution and the development of societies and cultures drawn by Spencer was far more influential. By the 1890s a number of biologists, including Huxley and many younger evolutionists such as Mark Baldwin in the United States, reacted to what they saw as the distorted use of evolutionary theory to justify social and ethical values to which they were opposed. In the arguments directed against what was now known as 'Social Darwinism', the distinction between organic evolution and cultural tradition became crucial.

Wallace did not develop the idea of cultural transmission within groups of animals or explore its relationship to his earlier suggestion that the human race had escaped from natural selection. Morgan's examples of tradition in animals – the persistence of distinctive songs in birds or of specific food preferences – were simpler than the skills of nest building considered by Wallace. Nonetheless they suggested to Morgan a way of viewing the human mind that neither required Wallace's appeal to past spiritual intervention nor to the Lamarckian principle of Darwin, Spencer and Romanes.

Morgan discussed human evolution in the final chapter of *Habit and Instinct*. He was in agreement with Wallace's point about the special nature of human evolution and was critical of the idea that various human mental faculties were very directly influenced by processes of natural selection. Such a view was most vigorously maintained by Francis Galton, whose main preoccupation for more than twenty-five years had been human heredity and whose influence was rapidly climbing at the time when *Habit and Instinct* was published. The next section describes Galton's life and work, before returning to the very different position held by Morgan on the relationship between human psychology and heredity.

Mathematics, heredity and Francis Galton

From the vantage point of a century later a prominent aspect of the empirical work by Romanes and Morgan is not so much the informality of their procedures, but the near absence of any kind of mathematical argument in their evaluation of evidence. This was not a matter of ignorance; Morgan, at least, had received training in mathematics in his years as an engineering student.

The lack of mathematics is found also in the work of many other evolutionists who were not involved in animal psychology. This was perhaps partly influenced by the conflict between the Darwinians and the physicists, and by the physicists' use of mathematical deduction to dismiss natural selection because of inadequate time and inadequate hereditary mechanisms. In reponse to these criticisms Huxley expressed his opinion on the role of mathematics in a science in his usual vivid fashion. He stated that 'this seems to be one of the many cases in which the admitted accuracy of mathematical processes is allowed to throw a wholly inadmissible appearance of authority over the results obtained by them'; and went on to compare mathematics 'to a mill of exquisite workmanship, which grinds you stuff of any degree of fineness; but, nevertheless, what you get out depends on what you put in; and as the grandest mill in the world will not extract wheat-flour from peas-cod so pages of formulae will not get a definite result out of loose data'.[1]

Many other biologists appear to have regarded mathematics as a tool they might admire, but did not trust.

Huxley's opinion of mathematics was expressed in the course of an exchange with a man widely regarded as the most eminent physicist of mid-Victorian Britain, William Thomson, later Lord Kelvin. The debate between them was one of the few that Huxley lost – or, rather, for the following twenty-five years was seen to have lost. Early in the 1860s Kelvin had questioned the claims of the geologists and evolutionists to essentially unlimited time. His calculations on the age of the earth had been emphasized by his friend and colleague, Jenkin, in the review of *The Origin of Species* that had particularly troubled Darwin.[2] But at first Kelvin's conclusions had not been widely accepted. After the debate with Huxley, Kelvin's calculations were given increasing respect. Two years later at the time *The Descent of Man* was published, Darwin was referring to him as 'that odious spectre'.[3] Kelvin's original conclusion that the earth was 100 million years old now gained widespread acceptance by geologists and biologists, as over the decades his own revisions produced still shorter estimates, tending towards a mere 20 million years.[4]

Also of increasing influence were Kelvin's views on the role of mathematics in a science. Another friend and collaborator of Kelvin, Peter Tait, whose calculations of the earth's age produced even smaller numbers, condescendingly suggested that the limited use of mathematics in subjects other than physics and chemistry reflected their failure to progress beyond the 'beetle hunting' and 'crab catching' stage.[5] One of Morgan's first publications voiced a protest against the precision Tait claimed for his conclusions and made an appeal to the use of biological evidence in decisions on this question.[6] Geology was the first subject to feel that, to mature, it had to quantify. But other subjects as well began to take very seriously Kelvin's famous dictum: 'When you can measure what you are speaking about, and express it in numbers, you know something about it; but when you cannot measure it, when you cannot express it in numbers, your knowledge is of a meagre and unsatisfactory kind; it may be the beginning of knowledge, but you have scarcely in your thoughts advanced it to the stage of *science*.'[7]

In the 1860s very few people were both ardent Darwinians and great respecters of numbers. One of these few was Francis Galton. In addition, he became very much concerned with establishing psychology as a science. This combination of interests in an original and lively mind led in the 1870s and 1880s to his deep involvement in theoretical problems of heredity and, from this, to another important offspring of evolutionary theory, modern statistical analysis. In discussing the future for psychology he produced in 1879 an almost exact paraphrase of Kelvin: 'Until the phenomena of any branch of knowledge have been submitted to measurement and number, it cannot assume the status and dignity of a science.'[8]

Galton was born in 1822, which made him sixteen years younger than Mill, only twelve years younger than his cousin Darwin, and three years older than Huxley. However, the main effects of Galton's influence were only felt long after the names, first of Mill, then of Darwin and Huxley, had become, if not universally venerated, at least universally known. Galton published one of his most original and important books when he was sixty-seven and Darwin had already been dead for seven years. Galton outlived almost all of his generation.

He showed signs of unusual ability at an early age, but then, while many of his contemporaries were laying the basis of their subsequent renown, decades passed for Galton with little intellectual achievement to show for them. The precocious scholastic achievements Galton displayed as a young child are reminiscent of Mill. Whereas the latter was to attribute these to the extraordinary education he received from his father,[9] Galton was later to see his own early command of the classics as due to good lineage. Early promise was followed by early academic honours. At the age of sixteen Galton, like Huxley, began to study medicine. He went directly to a London teaching hospital, rather than commencing with an apprenticeship, and won a prize in the first part of the medical examination. Medical studies were then interrupted in order to go to Cambridge to study mathematics. These studies were begun partly on the advice of Darwin and were ended prematurely by a breakdown in Galton's health. All formal education ended when a large inheritance removed any financial reason for resuming the study of medicine, or for pondering over alternative professions.[10]

For many young men of that age education was followed by foreign travel. Where Darwin, as naturalist, and Huxley, as assistant surgeon, sailed with the navy to the other side of the world, Galton undertook an aimless expedition to the Middle East inspired by the idea of shooting a hippopotamus on the River Nile.[11] A more serious expedition took place in 1850 when Galton and a fellow explorer spent two years mapping an area in South-West Africa. As happened with more than one pair of Victorian adventurers who set off together to map distant territory, that member who happened to return first reaped the honour and glory. While his companion remained in Africa, Galton presented papers to the Royal Geographical Society in London about their travels and was awarded a gold medal. His first journey gave him respect for Muslim culture, but in general travel did not seem to broaden his mind. His reports did not show a great deal of curiosity about anthropological or biological questions. Apart from Arabs, contact with fellow human beings in exotic places only confirmed early feelings of superiority towards those that were neither of his race nor of his social status.[12] While Huxley turned his cool intellect onto the set of prejudices he had absorbed from the cultural background of his youth – towards non-Europeans, towards Jews, and towards women – and progressively shed them,[13] similar feelings became more deeply entrenched in Galton and his efforts turned increasingly towards justifying them.

His commitment to study the differences between individuals and to demonstrate the importance of their basis in heredity was prompted by *The Origin of Species*. To excitement over its contents was added family pride that it had been written by a cousin. Galton published his own first major work, *Hereditary Genius*, ten years later in 1869. This presented collections of statistics on kinship relations between men of achievement in various spheres – judges, for example – as support for the claim that specific talents are inherited. As he was to point out later, until that time terms like heredity and inheritance had been used largely in a legal context. Only subsequently did they become familiar in biology. Later he regretted the use of *genius* in the title, intending only to refer to people of unusual ability and not to evoke the further connotations of this word.

The book appeared when Mill's influence, and general beliefs inclining towards the potential equality of human beings and the paramount importance of individual experience were still strong. Galton abruptly rejected these beliefs. The truth of the book's general claim that mental ability is almost entirely inherited was quite apparent to him. 'I have no patience with the hypothesis occasionally expressed, and often implied, especially in tales written to teach children to be good, that babies are born pretty much alike, and that the sole agencies in creating differences between boy and boy, and man and man, are steady application and moral effort. It is in the most unqualified manner that I object to pretensions of natural equality.'[14] Galton did not believe that the relative infrequency of outstanding accomplishments among the lower classes could be the result of some aspect of class structure – of, perhaps, not being socially related to the right people, rather than not biologically related to them. After all, Galton argued, social barriers are far less rigid in North America, but it was not apparent to him that there was a correspondingly higher proportion of eminent men on the other side of the Atlantic.[15] When put together with indications that the more talented married later and were more likely to be infertile than the less talented, the implication for the future from his analysis was dire: the intellectual level of society would inevitably decline, unless some positive measure was taken to avert this prospect. In the book Galton first suggested what was later to be known by his term 'eugenics': society should positively encourage breeding among its talented members and discourage it among its 'idiots' and 'imbeciles'.[16] This particular suggestion aroused little interest or attention at the time, but the book as a whole was well received. Darwin, as ever generous with praise for the work of friends or relatives, had to stop reading after fifty pages to pen a letter of congratulations: 'I do not think I ever in all my life read anything more interesting and original . . . You have made a convert of an opponent in one sense, for I have always maintained that, excepting fools, men did not differ much in intellect, only in zeal and hard work; and I still think this is an *eminently* important difference'.[17]

Hereditary Genius displayed Galton's interest in mathematics, but it was of a different kind from that of many mathematicians; he attached unusual value to counting. Three themes ran throughout his life; ingenuity in devising inventions and gadgets, human heredity, and an obsession with numbers. When sitting for a portrait he counted the number of brushstrokes made by the artist and reported the total in a letter published in the journal *Nature*. Later in life he set forth to construct a 'beauty map' of the British Isles by recording in each town he visited the number of women he passed whose appearance was above average quality. In 1872 the results of a similar kind of exercise brought him considerable publicity and classified him in the public mind as another militant free thinker like Huxley. Galton calculated the average life spans of kings, clergymen and missionaries. He found that these were no longer than those of comparable groups such as lawyers, doctors, and other gentry, for whom God's blessing is far less frequently invoked. Galton concluded that there is no scientific basis for belief in the efficacy of prayer.[18] In Brighton that summer a session of the geographical section of the British Association brought the name of Galton to public attention for the second time within the year. On this occasion he displayed his interest in heredity. The highpoint of the proceedings was to be H. M. Stanley's first formal report on his expedition to Ujiji and his meeting with Dr Livingstone. Galton's introduction as chairman expressed his main hope that the report would throw light on the question of whether the explorer had been born in North America, as Stanley himself claimed, or whether, as rumours had it, his origin was Welsh, lower class, and of an illegitimate union. This triggered some bitterness, which persisted until the matter was later resolved by the Queen's announcement that she was willing to receive Mr Stanley.[19]

It was some time before Darwin read more than the first fifty pages of *Hereditary Genius*. At the end of the book Galton discussed current theories of heredity and accepted a modified version of Darwin's theory of pangenesis. He followed up this theoretical discussion by performing some experiments, whose results then led him in 1871 to reject the theory. The paper reporting this conclusion earned a rebuke from Darwin – who pointed out that Galton had misinterpreted the original theory of pangenesis – followed by help and advice on further genetic experiments that involved studying the properties of successive generations of peas.[20]

These experiments were undertaken in complete ignorance of those of Mendel in Austria just a few years earlier. Galton considered, and rejected, the use of ordinary peas – the fortunate choice that Mendel had made – and settled for sweet peas, since in these plants 'the little pea at the end of the pod, so characteristic of ordinary peas, is absent'.[21] The result of this choice was that his work did not provide a foundation for twentieth-century genetics – despite the number of distinguished medals he was later to receive for his contributions in this area – but the starting point for twentieth-century statistics. From the graphs illustrating the quantitative relationship between pea sizes in successive generations came increasing sophistication in quantifying distributions and variation about a mean, the idea of regression and the measurement of correlation.

Galton's earlier enthusiasms, which included meteorology – he devised the first weather map and discovered anticyclones – as well as travel and geography, were replaced by a concentration on heredity and a new interest in psychology. He continued to compile evidence on human heredity, to explore ways of systematically measuring human differences and to advocate a policy of eugenics. He sent questionnaires to leading scientists. He carried out the first survey of twins. In 1884 he founded an

Fig. 2.8. Francis Galton

'Anthropometric Laboratory' to amass data on human physical characteristics, together with psychological measurements such as reaction times and sensory abilities.

For a brief, but fertile, period his interest in psychology became predominant. The studies he carried out were described in *Inquiries into Human Faculty* of 1883. One of his inventions, that has been employed ever since, became known as the 'word association test', in which, taking a set of arbitrary words, he noted down the associations that would occur within four seconds of viewing each word. The analysis of his own associations produced some interesting ideas on the relative frequency and fixity of recent and distant events in memory. A little later he attempted to explore the nature and extent of visual imagery by means of a questionnaire completed by friends, relatives and boys at a public school. One item, for example, required the respondent to recall the breakfast table of that morning and to note, among other things, the degree to which the mental image was coloured. A related project was to investigate the kind of image certain people, especially those capable of rapid mental calculation, reported to be associated with different numbers.[22]

All in all, the sum total of Galton's direct investigations into intellectual processes was not large, given that a dominant interest for much of his life was in differences between the mental abilities of individuals. The reason for this was an unshakeable belief that direct measurement of mental processing was unnecessary, since intellectual ability was closely related to sensory ability. And the latter could be measured very accurately with little trouble. Never one to be accused of keeping too open a mind, this belief was not something that Galton chose to investigate. He noted merely a surgeon's report that two idiot boys had felt no pain when their ingrowing toe-nails were excised.[23] The fact that piano-turners are predominantly male and, further, that only the rare woman can distinguish the merits of various wines, only provided confirmation of the view that inferior discriminative ability indicates inferior intellectual ability.[24]

Galton's statistical work on heredity continued throughout the 1880s. The results were reported in his next book, *Natural Inheritance*, published in 1889. This was notably concise for a book of the period and remarkably clear and readable, considering the technical nature of much of its content. By this time Galton had almost abandoned belief in Lamarckian inheritance and his theoretical views on genetics had moved away from Darwin's pangenesis to a position more closely resembling that of Weismann.

The absence of any satisfactory genetic theory was not seen by Galton as a serious obstacle in the way of a statistical analysis of kinship relationships. Starting with an account of the research on sweet peas, the book then examines the way in which various human characteristics reflect the influences of natural heredity. It examines in turn stature, eye colour, artistic tendency and disease. The human data used for these analyses were considerably less extensive and less precise than those from sweet peas. They were obtained from 'Family Records' filled in by unknown correspondents hopeful of winning the money offered by Galton as prizes for the most satisfactory entries. Galton admitted that this method of collecting data was likely to produce many errors. Optimistically, he thought that they would probably cancel each other out.[25]

Galton's discussion of the characteristic he termed the Artistic Faculty, grouping together in his sample people showing aptitude, or special fondness, for music or for drawing, is particularly revealing about his approach to psychological processes and their inheritance. As in Romanes' treatment of the animal mind, it assumes that some special skill or ability directly reflects the presence of the appropriate 'faculty' in the mind. The absence of such skills or abilities just as directly reflects the absence of the faculty. No justification for the assumptions that such faculties are inherited is offered. Of the various human characteristics discussed in the book, an artistic tendency would seem most affected by environmental influences. Galton did not give his reasons for, say, discounting the effect that parental attitudes or the general situation in a household might have on a child's fondness for, or skill in, music. They were simply dismissed. 'A man must be very crotchety or very ignorant, who nowadays seriously doubts the inheritance either of this or of any other faculty.'[26]

Because of the dubious quality of the data, the empirical conclusions of *Natural Inheritance* are suspect. Yet the mathematics are impressive. To use Huxley's metaphor, it was as if Galton's continuing attempts to obtain wheat-flour from peas-cod had led to the invention of a new kind of mill. Galton was quite clear as to the purpose of developing this mathematical machinery. In discussing the charm of statistics, he proposed – with a slight change of metaphor – that 'they are the only tools by which an opening can be cut through the formidable thicket of difficulties that bars the path of those who pursue the Science of man'.[27]

Natural Inheritance fired the enthusiasm of a number of younger men with much greater mathematical ability than Galton. By the turn of the century many of the standard statistical methods in common use today had been worked out by this group. The outstanding member was Karl Pearson, to whom is owed the term standard deviation, the extensive development of Galton's ideas on correlation and regression, and the chi-squared test.[28]

Pearson's continuation of the work in statistics went with a wholehearted adoption of Galton's beliefs on heredity. In the early 1890s this was still not typical. The reservations expressed by Galton and by Morgan concerning Lamarckian inheritance were by no means generally accepted and the later views of Darwin, together with those of Spencer and Romanes, were still very influential. Until people no longer believed that mastiffs could acquire an instinctive fear of butchers within a few decades or that only two or three generations of good breeding and education could produce a gentleman, the force of Galton's convictions was diluted. To the extent that a turning point can be identified, this occurred in the summer of 1892. Once again disagreement over the age of the earth had wide-ranging consequences.

The occasion was the first meeting at Oxford of the British Association since the clash between Huxley and Bishop Wilberforce of 1860. The presidential address was given by the Marquis of Salisbury, respected as an amateur scientist and former Prime Minister. Invited to reply to the address were Huxley and Kelvin. It was Huxley's last public appearance. Salisbury chose to review fundamental questions that science, with all its achievements over the past century, had still failed to answer. The final question he considered was that of how evolution had occurred. Natural selection was rejected. Kelvin had proved, so Salisbury claimed, that there was insufficient time for evolution to have taken place in this manner.[29]

The problem that had weighed on Darwinians for twenty-five years was thus again given prominence. And presidential judgement was given against them. This time Huxley politely expressed his protest at Salisbury's verdict and the expected fierce rejoinder was not made. The reaction to Salisbury came from elsewhere. The address prompted another former assistant and collaborator of Kelvin to re-examine in a thorough fashion the assumptions that Kelvin had made when first calculating the age of the earth. The foundations for Kelvin's dogmatic conclusions began to look less secure. As controversy on this issue now spread among the physicists, the Darwinians began to see time enlarging once more.[30] In 1896 Lloyd Morgan's friend, the zoologist Edward Poulton, was president of the British Association. His address challenged that of two years before, asserting that biological evidence must take precedence in the evaluation of evolutionary theory: 'Natural selection will never be stifled in the Procrustean bed of insufficient geological time'.[31]

In fact the bed by then was no longer Procrustean. In ignorance of the first discovery of radioactivity in that same year, calculations based on familiar principles now produced estimates exceeding Kelvin's original 100 million years. The pressure that had persuaded Darwin, and hence Romanes, to rely on the Lamarckian principle as an accelerating factor in evolution was eased well before Curie and Laborde reported the release of heat from radium salts in 1903. In 1904 Rutherford applied their discovery to the question of geological time, proposing that radioactivity was the unknown source of energy that Kelvin had been unable to foresee in the 1860s.[32] The public debate that Huxley had apparently lost in 1869 was finally settled in his favour.

Morgan, Galton and British psychology

The year that Poulton re-asserted biology's claim on time was also the year in which Morgan wrote *Habit and Instinct* and finally rejected any Lamarckian influence in the evolution of behaviour. As noted earlier in this chapter, this led Morgan to emphasize the part played by early experience in producing continuity in behaviour from generation to generation. His arguments and evidence were from the behaviour of animals. But it was also clear to him that the rejection of Lamarck undermined the basis for the view of man which had been dominant for the past two decades.

Spencer is not mentioned by name. Nevertheless, it is clearly his theory of human evolution that is attacked in the final chapter of *Habit and Instinct*. The idea that the recent achievements of Victorian science and enterprise reflected the superiority of the Victorian mind, gained by the steady inheritance of increments of improved mental function over the centuries, was no longer tenable. Morgan was inclined to believe that the average level of innate ability during Tudor times was, if anything, slightly higher than at the close of the nineteenth century. What he termed 'mental progress' was due to 'the handing on of the results of human achievement by a vast extension of that which we have seen to be a factor in animal life, namely tradition'.[1] Morgan saw human achievement progressing by leaps and bounds. The steady progress of European science and of sensibility in matters of religion and ethics, which was for Spencer indication of an improvement in man's biological inheritance, became for Morgan an example of an enriched cultural heritage. Each achievement can be stored in 'the social environment to which each new generation adapts itself, with no increased native power of adaptation'.[2] One implication that can be seen in this final chapter – although never explicitly stated by Morgan – is that a primary task for human psychology is to understand how an individual adapts to and learns from his social environment.

After *Habit and Instinct* Morgan's work had very little direct influence on animal psychology. In the same year he became principal of his college. The subsequent administrative tasks involved in its transformation into Bristol University and his position as its first Vice-Chancellor occupied much of his time. In his books he had written of the need to establish a research institute for the study of comparative psychology. Once in a more favourable position he appears to have done nothing to implement such a plan.

Even without the pressure of administrative

duties, Morgan's direct involvement in the study of animal behaviour may well have ended. His principal arguments had been directed against the views of those claiming to find immediate relevance for the understanding of man in the study of animals. Major themes in 1894 were the primacy of introspection over comparative methods in psychology, and the simplicity of psychological processes in animals compared to those in man. In 1896 the final argument in *Habit and Instinct* was that, to a degree unrivalled in any other animal, in man 'evolution *has been transferred from the organism to the environment*'.[3] It followed that to study the behaviour of non-human animals was a very indirect way to understand the human mind or human evolution, despite the authoritative views of Darwin, Spencer and Romanes to the contrary.

Morgan's more general views were out of tune with the times. Within British universities psychology remained at best a sub-discipline of philosophy; and the kind of philosophy dominant in the final two decades of the nineteenth century was hostile to the tradition of Locke, Berkeley, Hume and Mill and to the essentially associationist psychology that Morgan advocated. When, just after the twentieth century began, G. E. Moore and Bertrand Russell began to revive this earlier philosophical tradition, their interests were mainly in ethics and logic, and not in psychology.

Morgan's return to the evolutionary theory of Wallace and early Darwin, with its sole emphasis on natural selection, was also premature. The discovery of Mendel's work, the expansion of geological time, and the study of mutations led to rapid developments in the science of genetics. However, for the next three decades these developments encouraged a view of organic evolution that was opposed to the gradual process described by Darwin and Wallace. Also, during these decades the Lamarckian principle was raised again at times, gaining an extra spurt of life at the end in Stalinist Russia.

Morgan was not the man to influence the intellectual trends of his time. He was a professor at an obscure, and new, provincial college. Despite a fellowship of the Royal Society awarded in 1899, he remained on the outer fringe of the scientific establishment of England. He was not an assertive man. An American friend, Mark Baldwin, on a visit to England described Morgan as contemplatively combing a long beard that was as fine as his logic.[4] The conclusions that Morgan drew from that logic were not expressed in clear, ringing tones.

The position of Francis Galton was very different. For decades he had mixed with the eminent men of British science, and had been seen as one of their number. The impression made by Galton in the 1880s was recorded by a perceptive young lady, Beatrice Webb, whose long friendship with Herbert Spencer allowed her to meet most of the stars of mid-Victorian science. She found Huxley a disturbing person, with a strain of madness and haunted by melancholy: 'Huxley, when not working, dreams strange things: carries on lengthy conversations between unknown persons living within his brain'.[5] But, of them all, the one who stayed in her mind as the ideal man of science was Francis Galton, with his 'perfect physical and mental pose' and 'unique contribution of three separate and distinct processes of intellect': curiosity and rapid apprehension of facts, the faculty for ingenious reasoning, and the 'capacity for correcting and verifying his own hypotheses, by the statistical handling of masses of data'.[6]

For many years Galton had put forward his ideas on human heredity and his suggestions for a vigorous policy to avert the catastrophic genetic future he predicted for mankind. They were presented in a forceful manner, a far cry from Morgan's careful, qualified prose. Yet until the 1890s Galton's ideas on eugenics made little impact. By then statistics had taken great strides; the achievements of two decades of animal psychology were paltry in comparison. With the decline of the Lamarckian principle, the distinction between acquiring and inheriting mental abilities now became stark. Galton's work enjoyed a currency it had not achieved before. It was as if for a long while gas had been seeping into a building so slowly that no one had noticed, until the moment when a light was struck.

As a man of independent means living in London, Galton could keep in close contact with the ideas and intellectual movements of the time. He had no formal connections with any educational or scientific institution. However, Pearson was a professor of mathematics at University College, London. Together Galton and Pearson established an institutional framework, based on University College, for the development of research and education in statistics, human heredity and eugenics. In 1901 a professional journal, *Biometrika*, was started. In 1905 a research fellowship in eugenics was endowed, followed two years later by a research scholarship. Also in 1907 the Eugenics Education Society was founded. In 1909 a laboratory devoted to eugenics research was set up at University College.

By this time Galton was one of the few remaining members of that illustrious band of men who began to possess the legendary character of the knights that attended King Arthur's table.[7] Indeed, in 1909, Galton was granted a knighthood, an honour that neither Darwin, Wallace, Huxley, nor any other of their fellow evolutionists had been offered. Survival was perhaps not quite enough. Spencer had perhaps as much right as Galton to be counted amongst the Darwinians, and Spencer also lived long enough to see the twentieth century. But even before his death in 1903, Spencer's philosophy was out of fashion and his pacifist beliefs, producing a condemnation of Britain's conduct of the Boer War, may have forfeited the chance of receiving an award of such distinction.

By the time Galton died in 1911 the kind of science he had started was enjoying a vigorous life. The tradition begun by Galton and Pearson maintained a dominant influence within British psychology as this slowly developed over the next thirty years. The result was that in the United Kingdom psychology retained for many decades a highly mathematical and hereditarian emphasis.

In 1890 Morgan's *Animal Life and Intelligence* had earned a lengthy review in *Nature* from Wallace. Four years later the same journal devoted a single paragraph to *An Introduction to Comparative Psychology* in an anonymous review that also examined three other books. One of these was Morgan's own *Psychology for Teachers*; another was Wundt's *Lectures on Human and Animal Psychology*, which is discussed in the next chapter. In 1908 only sixteen copies of *Habit and Instinct* were sold in the United Kingdom, although in North America the sales were more than double this figure. From the beginning of the century very few people in the United Kingdom showed much interest in the questions that had been central issues for Romanes and Morgan. Half the century passed before research on animal psychology occurred on any scale in Britain, and this was when the subject returned from the United States.

Summary

The progression from Darwin to Romanes and from Romanes to Morgan represented a continuous, but evolving, tradition in the study of animals. Some themes remained constant throughout this progression. The aim of relating an understanding of the mind to general theories of evolution provides one example. Another, less obvious, was the continued emphasis on the behaviour of the individual and those of its actions which, in human terms, could be called intelligent; alternatively, emphasis might have shifted to the social behaviour of animals or to issues that were later labelled by the neologism *motivation*, but it did not.

Other aspects changed. Romanes followed Darwin in treating very seriously the vast amount of loose, informal data on animal behaviour that became available by 1880. Morgan gave little weight to this kind of evidence. Unless the contributor was skilled in the detached observation of behaviour, the data was regarded by him as almost inevitably, and hopelessly, contaminated by the observer's own preconceptions. Furthermore, persistent observation over a period of time was needed to obtain anything of value. Yet Morgan did not go beyond this position to advocate a fully experimental approach to the subject. His own quasi-experimental work was not at all systematic and, although modelled on that of Spalding, was in some respects less elegant.

Decreasing tolerance for what could be classified as respectable data went with a changing view of the nature of the subject. For Darwin and Romanes the potential science of animal behaviour was a form of natural history, with methods and theories not very different from those of mid-century anatomy or geology. With Morgan it shifted towards becoming a part of psychology. For both Romanes and Morgan, understanding the mind of animals could be achieved only by making inferences based on analogies with the human mind. In the psychology of their day introspection was held almost universally to provide the sole means for analysing mental processes. Thus, Morgan concluded that comparative psychology would inevitably remain a junior partner to a human psychology based on the analysis of subjective experience. Although Romanes' early career was in physiology and Morgan at times suggested ways in which psychological processes might be related to events in the nervous system, neither advocated an alternative kind of partnership, between animal psychology and neurophysiology.

Romanes believed that to ask whether a given animal possessed a mind was the same as asking whether it was conscious. The only kind of objective evidence he could envisage as relevant to this question was whether the animal displayed any capacity for choice. Choice occurred when behaviour was observed to have been influenced by past experience. Yet Romanes paid little attention to the relationship between behaviour and experience.

Morgan adopted the same criterion for consciousness and, in contrast to Romanes, was very much concerned with processes of learning. His own observational work with animals convinced him that all reliable evidence of intelligent behaviour could be interpreted in terms of trial-and-error learning. Furthermore, in applying the Spencer–Bain principle,

Morgan distinguished two kinds of processes, both associative in nature, but one more complex than the other.

His studies of the behaviour of newly hatched birds also led Morgan to relinquish finally any belief in the Lamarckian principle of heredity. On Darwin's death Romanes remained the major proponent of the principle. The ever decreasing estimates of the earth's age obtained during his working life indicated to Romanes that mental evolution had occurred rapidly and that Lamarckian inheritance was a major factor. This belief very much affected the account of instinct given by Romanes. It was probably not just by chance that Morgan's break with Romanes on this issue coincided with a major revaluation of Kelvin's calculations with respect to geological time.

Once Morgan decided that no trace of the habits and associations acquired by an individual animal could be passed on in any biological way to its progeny, he formulated a very much clearer idea of what is meant by instinctive behaviour. He examined the behavioural evidence that once had seemed to support the Lamarckian principle and decided that what much of this evidence really showed was transmission from one generation to another by means of tradition. Consequently he was led to emphasize a further type of learning, imitation.

One further difference between Romanes and Morgan foreshadowed a later division in the study of animals. Both were very much influenced by Spencer's ideas on psychology. But Romanes differed from Spencer on the subject of reasoning; he maintained that this was a faculty that had evolved from the processes of perception. Morgan shared Spencer's view on this topic: as well as doubting whether any animal possessed anything that could be called 'reason', he was inclined to the view that the origins of human reasoning were closely related to the development of language.

To a large extent the development of animal psychology during the 1880s and '90s can be seen as the absorption into a Darwinian mainstream of the other elements discussed in the first chapter. Spencer's psychology was one element. Another was the concern with processes of learning that came from Bain. A third element, only partially absorbed, was the notion from Spalding that questions about the behaviour of animals can be answered only by careful experimentation.

The changes that occurred during this period can also be viewed as an interesting reversal: animal psychology in Victorian England began as a reaction to a certain set of views and in Morgan's hands ended, twenty-five years later, by adopting almost all of them. Darwin's ideas on animal behaviour and mental evolution, as expressed in *The Descent of Man* in 1871, were explicitly opposed to the position Wallace had recently reached on the evolution of mind. By 1896 Morgan's main conclusions were essentially identical to those of Wallace: the rejection of the Lamarckian principle; the importance of early experience and cultural tradition in maintaining uniformity of behaviour; the immense gulf in intellectual capacities between man and other animals; and a view of man as the one species that has escaped from the pressure of natural selection. Morgan, in Wallace's words, placed 'man apart, as not only the head and culminating point of the grand series of organic nature, but as in some degree a new and distinct order of being'.[1]

The major difference between Morgan and Wallace was the element that came from Huxley: a sceptical attitude and a willingness to suspend judgement. Because Morgan could not explain satisfactorily the evolution of the human mind, he did not therefore, as Wallace had done, appeal to supernatural intervention.

Wallace's views on mental evolution – caricatured as presenting man as God's domestic animal[2] – and his sympathetic attitude towards spiritualism had diminished his scientific reputation among British biologists after 1870. Morgan's acceptance of Wallace's other views had little influence on British psychology as it developed after 1900. Much more important were Galton's emphasis on biological, rather than cultural, inheritance, and a methodology that, instead of concentrating on detailed observation of the individual animal or human being, relied on brief tests, large numbers and statistical analysis.

3
Experimental psychology and habits

When we began our consideration of the mental life of animals, we condemned the tendency of animal psychology to translate every manifestation of 'intelligence' into an intellectual operation. The same reproach could be made against certain more or less popular views of our own mentality. The old metaphysical prejudice that man 'always thinks' has not yet entirely disappeared. I myself am inclined to hold that man really thinks very little and very seldom.

Wilhelm Wundt: *Lectures on Human and Animal Psychology* (1892)

The task of obtaining unambiguous empirical evidence on some psychological issue is usually far more difficult than it may seem at first glance. Many psychological phenomena are determined by a range of factors that interact with each other. Thus, whether or not some effect occurs can often depend on the particular context. The study of depth perception provides an example that was of considerable interest in the nineteenth century, in that our judgement of how far away some object lies can be based on a variety of different cues whose relative importance changes from situation to situation. As a consequence of attempting to isolate and understand the variables operating in such cases a methodology has developed which is to a large extent peculiar to the subject of psychology. The ideas, for example, of using tightly specified control conditions or of systematically manipulating one variable while keeping all others constant, as well as the statistical methods that have been developed for the analysis of this kind of experimentation, are familiar ones now. They were not familiar a hundred years ago, when the empirical methods used in other sciences appeared to be of limited use in the attempt to establish psychology as an experimental science.

In the middle of the nineteenth century one guideline was provided by John Stuart Mill's discussion of the logical basis of induction from empirical evidence. Another was the experience gained by physiologists, particularly from experimental work designed to probe the workings of the nervous system. The development of appropriate experimental methods and the beginning of psychology as an academic subject occurred in Germany, where a key role was played by Wilhelm Wundt (1832–1920).

As psychology grew, the range of questions to which experimental methods were applied broadened out from Wundt's initial emphasis on perception and reaction times to encompass such topics as memory and thinking. However, in Germany it was not extended to the study of animal psychology for many years. This was not because of a lack of general interest in Darwinian theory of the kind that had stimulated research on the mind of animals in England, since, as described below, the efforts of Ernst Haeckel (1834–1919) ensured that evolution was debated just as intensely in Germany.

Quantitative experimental methods were first fully applied to the problems discussed by the British evolutionists by Edward Thorndike (1874–1949). Thorndike's experiments were carried out in the United States within a university department of psychology where Thorndike was enrolled as a research student; the primary purpose of the work was to provide him with sufficient material for a doctoral thesis. This represents a huge change from the context in which the ideas and research discussed in previous chapters were developed. Thorndike's work was carried out during a period of extensive reform and expansion of the American university system. Departments as units of organization, the study of psychology within a university, and the awarding of doctoral degrees on the basis of experimental research were all very recent developments.

A large part of the research in animal psychology that followed Thorndike's pioneer experiments was carried out in a similar setting. As with Thorndike himself, the way that his successors worked as scientists was very much influenced both by the institutional context in which their experiments were carried out and by the intellectual traditions that went with this context. And so, before describing Thorndike's work, this chapter looks at the way experimental psychology grew within the German and American university systems.

German science and psychology

In the nineteenth century Germany's scientific activity expanded and became organized in a way that had no previous parallel. A major factor in this growth was the reform of universities carried out early in the century, following Prussia's defeat at the hands of Napoleon. In 1809 a new kind of university was founded in Berlin and this provided a model for higher education elsewhere in the many German-speaking states and principalities of Central Europe.

The organization of German universities was guided by a particular set of beliefs about the general function of university education. Many of these beliefs were summed up by the phrase 'academic freedom'. Those who taught in these institutions, the professors, were free to run their own affairs and decide what to teach, with a minimum of state or religious interference. Unlike their fellow citizens in many of the German states, they were guaranteed freedom of speech. They were also free, and encouraged, to engage in scholarly work, in addition to their relatively light teaching duties. For students, academic freedom meant the opportunity to select lectures and courses and to transfer from one university to another.

This new system was not planned to further the development of empirical science. The intention was that universities should provide a broad education, whose various parts would be integrated by the kind of idealist philosophy, particularly that of Hegel and of Fichte, then dominant in Prussia. In such a philosophy the use of logical insight was not confined to the task of formulating metaphysical principles. Explanations of natural phenomena were to be obtained either by logical deduction from metaphysics or on the basis of intuitive knowledge. This tradition of *Natürphilosophie* was openly hostile to the outlook and the methods of empirical science. Within the new universities scholarly work was to be concerned primarily with the study of cultural development. The idea that the purpose of a university should be as much the accumulation, as the transmission, of knowledge first bore fruit in the fields of history and in the comparative study of language. By 1830 the work of German historians and linguists was unrivalled.

In 1829 experimental science first obtained a foothold within the system when a chemistry laboratory was opened at the University of Giessen. From then on a gradual expansion occurred, notably in chemistry and in physiology. The earlier reforms turned out to have provided an excellent framework for scientific research. A professor would both give lectures and, in his role as director of an institute, supervise the experimental work of students receiving their training in research methods. Thus, science became both a group activity and a profession in a way that had not happened before.

Perhaps even more important than academic organization was the existence of a great number of universities of comparable quality, competing with each other to employ anyone with a good academic reputation. Continued expansion of the universities meant that, once a student had finished his research training and had shown some promise, a variety of academic positions were open which would allow him to continue with a scientific career.

This situation in German-speaking countries was in stark contrast to that in England where, until well into the nineteenth century, only two universities existed and those served as a combination of finishing schools for the upper class and training establishments for the Church of England. To a very large extent intellectual and scientific activity took place outside of the educational system. Until the 1880s a young Englishman could engage in scientific work only if, like Darwin, Romanes or Galton, he could afford such a hobby, or if he was prepared, like Wallace or Spalding, to face financial hardship in the hope that eventually a scientific reputation might aid the sale of books and indirectly provide a reasonable income. Where men like Huxley and Morgan held academic posts research was not seen as part of their duties, but was carried out despite teaching commitments. There were many good scientists in mid-Victorian England, but little means for transmitting their techniques and methods. Towards the end of the century, and decades after this had occurred in Germany, English universities began to provide instruction in science and facilities for laboratory research, and to accept that they should become centres of intellectual activity. The change was as much prompted by fears of the growing industrial and military power of the newly united German empire as based on any new concept of the function of a university.[1]

By 1871 Prussia had within a few years defeated first Austria and then France. Military success and attendant economic growth were in part attributed to the merits of the Prussian educational system. German universities continued to receive generous financial support, while academic freedom and the emphasis on scientific training were maintained. By this time students from the rest of Europe and from North America were arriving in increasing numbers to study at one or another renowned German university. Attraction was particularly strong in physiology, where a series of important discoveries, many of them to do with the workings of the nervous system, had been made since the 1830s in the new research institutes.

Hostility towards the kind of speculative philosophy which had initially guided university development was most pronounced among the physiologists. In 1845 a group of young scientists who were later to make outstanding contributions to the study of neurophysiology and of the senses – including the most brilliant physiologist of his generation, Hermann von Helmholtz – solemnly pledged to keep their science free from metaphysical speculation and swore to uphold the principle that 'no other forces than common physical–chemical ones are active in the organism'.[2] In studying the physiological basis of human perception science came closest to intruding upon the traditional domain of philosophers. The animosity was mutual. As Helmholtz noted later, in discussing Hegel: 'His system of nature seemed, at least to natural philosophers, absolutely crazy . . . The philosophers accused the scientific men of narrowness; the scientific men retorted that the philosophers were crazy. And so it came about that men of science began to lay some stress on the banishment of all philosophic influences from their work.'[3]

By the 1860s there were a number of young scientists concerned with the measurement of human sensory abilities and the problem of relating such data to ever increasing knowledge about the structure and function of sense organs. Various methods for determining, for example, what level of intensity of a sound could just be detected by the human ear or what was the minimum change that could be felt when pressure was applied to some area of the skin had been systematized by Gustav Fechner. Fechner had also introduced the term 'psychophysics' for the general approach of discovering the quantitative relationships between perceptual phenomena and physical events. At the same time methods developed for measuring very precisely the speed of transmission of neural events were first applied to the timing of mental processes.

A major step in the transformation of this cluster of intellectual interests into an academic discipline was the appointment of Wilhelm Wundt to the chair of philosophy at Leipzig in 1875. Wundt had worked as a research assistant to Helmholtz and had carried out some pioneer research on human reaction times; he had become one of the foremost proponents of 'physiological psychology', by which he meant the application of the experimental methods developed in physiology to many of the psychological issues formerly regarded as solely within the domain of speculative philosophy. The unusual situation of a man with these interests and with training in science occupying one of the most prestigious chairs of philosophy in Germany provided a firm base for the development of this new kind of psychology. With the opening of a laboratory in 1879 students could receive training in psychological research at Leipzig, in the same way as elsewhere in Germany they could learn to be physiologists, chemists or physicists. The idealist philosophy dominant a generation earlier enjoyed less and less respect and Wundt's aim of re-establishing philosophy as a science was in general regarded sympathetically by his fellow academics.[4]

The growth of experimental psychology within the German university system coincided with the growth of Darwinism, but was surprisingly little influenced by evolutionary ideas. The reasons for this seem to lie partly in the way that the new psychology remained closely tied to the physiological study of sensory processes from which it stemmed, and partly in the way that Darwinian theory was absorbed into German intellectual traditions.

In promoting evolutionary theory in Germany the part played by Ernst Haeckel was initially similar to that of Huxley in England.[5] Haeckel's scientific reputation, like that of Huxley, was first based on work in marine biology. The book describing this work gained him in 1862 the chair of zoology at the University of Jena where he remained for the rest of his academic career. At about the time of this appointment he first read Darwin's *The Origin of Species* and became an instant convert. From 1863 he wrote a series of papers and books on evolution that were extraordinarily popular and established Haeckel as a leading evolutionist. His *Natürliche Schöpfungsgeschichte* (Natural History of Creation) of 1868 became one of the main sources of the world's knowledge of evolution. Its account of human descent, which Haeckel regarded as central to evolutionary theory, antedated Darwin's own discussion in *The Descent of Man* by three years.

For Haeckel, Darwinian theory had very definite political implications. His work on marine biology had taken him to Italy during the time of the liberation movement. This inspired in him commitment to the reform of political institutions and to the unification of

Germany and evolutionary theory provided Haeckel with a framework in which to develop his social and political beliefs. As a young man he had regularly attended church and shown devotion to orthodox Christian beliefs and principles. As an evolutionist, Haeckel expressed increasingly fierce hatred for religion, which he identified as a major source of political reaction. From the time he first embraced Darwin's theory, Haeckel was as forthright about his political views as his beliefs on human evolution: evolutionary progress, he proclaimed, is 'a natural law which no human power, neither the weapons of tyrants nor the curses of priests, can ever succeed in repressing'.[6]

The overtones of radical politics and anti-clericalism added to Darwinian theory by Haeckel repelled as many as they attracted. But they ensured that the question of evolution was not neglected. A notable public clash between the champions and the critics of Darwinism took place in Munich in 1877. The occasion was a conference on educational policy. The issue was whether evolutionary theory should be taught in schools. This was opposed by many leading German biologists on the grounds that Darwinism was a socially dangerous hypothesis, with an affinity to socialism; schoolchildren should learn facts, not theories. Haeckel's plea that evolutionary principles should constitute the foundation of education was ineffective. But, although the teaching of evolution was prohibited in schools, by the 1880s Darwinism in general was as widely accepted in Germany as in England.

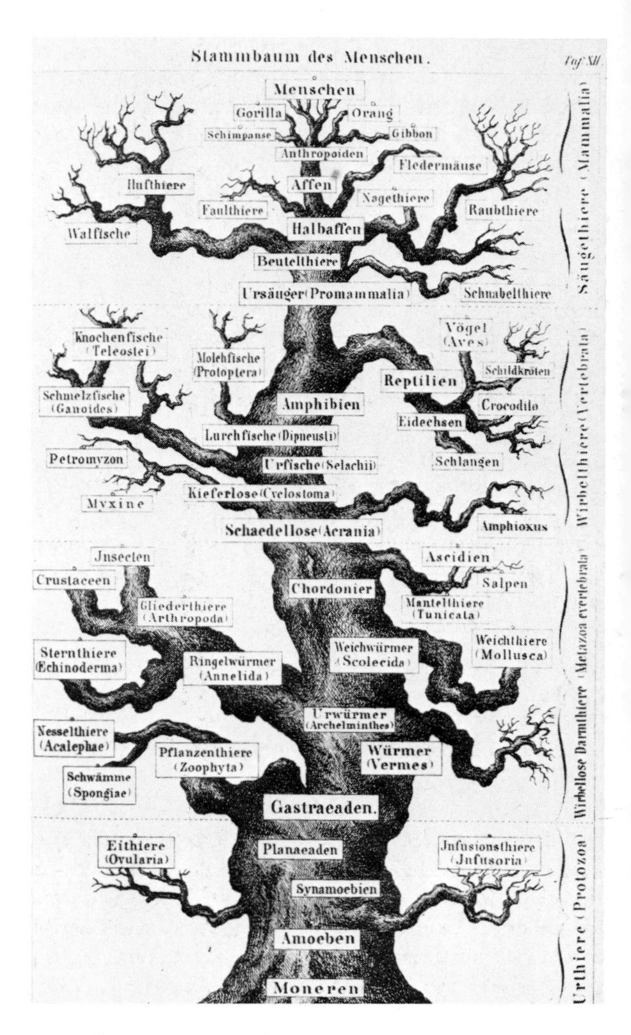

Fig. 3.1. A tree of evolution published by Ernst Haeckel in 1874

From around the time of the Munich debate Haeckel's position in German science became more like that of Romanes or Spencer than that of Huxley. Figure 3.1 reproduces one of Haeckel's evolutionary trees, which graphically illustrates the highly linear, 'progressive' view of evolution that he held; Darwin never sketched any tree like this. The inheritance of acquired characteristics was given major emphasis from the start in Haeckel's writing. In the 1880s he was as vigorous as Romanes in the defence of the Lamarckian principle against the criticisms put forward by Weissman.[7] By this time Haeckel's energy was mainly directed towards the advancement of his philosophical views. His first book on evolution had included some discussion of the relationship between physical and mental phenomena. It had proposed a solution, for which Haeckel coined the term 'monism', and which, as discussed in the previous chapter, Morgan later adopted. Over the years Haeckel's concern to derive a philosophy of life based on evolutionary principles became as grandiose as that of Spencer. His fellow scientists began to find his speculations and generalities as unpalatable as those of Hegel.

The German evolutionists were far less interested in psychology than their British counterparts. The only notable extension of an evolutionary approach to a psychological issue was the work of Wilhelm Preyer, whose main interest was in the psychological development of children. His book of 1882, the *Mind of the Child*, was subsequently very influential in North American psychology, but far less so, it appears, within Germany. Preyer's interest may have been prompted by his position as a colleague of Haeckel at the University of Jena, where he was professor of physiology.

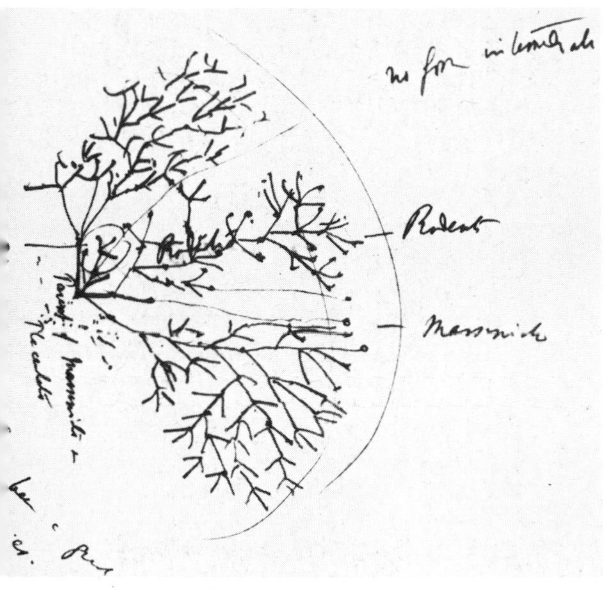

Fig. 3.2. Diagram from one of Darwin's notebooks

In Germany the experimental science most influenced by evolutionary ideas was embryology. A central element in Haeckel's form of Darwinism was the 'biogenetic principle'. Haeckel's claim was that the development of an individual human being, and of every other living creature, repeats in an abbreviated form its past development through evolution. In Haeckel's words, 'the development of the embryo is an abstract of the history of the genus'.[8] The idea had appeared early in the 1800s.[9] It had been discussed briefly, but given some weight, by Darwin in *The Origin of Species*. The crucial importance attached to the biogenetic principle in Haeckel's books on evolution, where he cited facts about human evolution, generated great interest in the subject. It was also one of his more testable principles. Eventually 'recapitulation theory' was found to be an unreliable generalization. Nevertheless, in the study of embryology it proved a good example of the kind of claim where, as an earlier embryologist noted, 'inaccurate but definitely pronounced general results have, through the corrections they call for and the keener observation of all the circumstances which they induce, almost invariably proved more profitable than cautious reserve'.[10]

It was from embryology that, after decades of dominance by a mechanistic approach to natural phenomena, a reaction occurred in the 1890s which became known as vitalism. Showing that the elements of organisms could all be found in inanimate matter seemed to leave untouched the distinction between the living and non-living. All organisms must possess, it was argued, some distinct non-material, vital element. Attempts to create animate forms from purely physical elements were futile. However detailed the analysis of what a fertilized egg might contain, it cannot lead to an understanding of its future development into a complex, living creature.[11]

The conflict between the vitalists and their critics, between emphasis on structure and organization and emphasis on the reduction of processes into component parts, which came to the fore in embryology at this time, later became important in animal psychology. However, during the last twenty years or so of the nineteenth century, experimental psychology continued to thrive within the German university system without paying much attention to the issues which engaged the evolutionists and embryologists. Physiology remained the subject of major interest to German psychologists, particularly the borderland between psychology and physiology in the study of the senses. Considerable advances were made in understanding the nature of, say, colour vision or the way that the ear functions. The observations and theories on the developing child described by Preyer must have seemed pallid in comparison to the achievements of Helmholtz. The latter's analysis of the complex and unconscious processes of inference underlying everyday perception of the world was important to Romanes' account of intelligence.

The evolutionists' relative disinterest in psychology, together with psychology's inclination to share the suspicious attitude towards Haeckel that many physiologists displayed, provide at least part of the explanation as to why in Germany there was no debate over the minds of animals comparable to that in England. Despite the development of an appropriate methodology, there appears to have been no experimental research of the kind undertaken by Spalding. Furthermore, the stories about animal life and the casual observations of behaviour that were for a while taken so seriously by British scientists do not seem to have had the same appeal among German academics. This no doubt reflected the attitude of the now professional scientist towards observations made by amateurs. Darwin, Romanes, and even Morgan paid attention to what the country parson, whose hobby was watching birds, or the colonial officer with time to take an interest in local fauna, had to say about the behaviour of animals. German science had thrived on its rejection of Natürphilosophie, its specialization and its insistence on rigour and on the primacy of evidence obtained in the laboratory. Eventually it was Romanes' *Animal Intelligence* that prompted Wundt to direct his attention to this topic. In his *Lectures on Animal and Human Psychology* of 1892 Wundt remarked that he was appalled by the lack of sophistication in Romanes' work, its simplistic 'popular psychology' and the absence of any logical criticism. He suggested that this unfortunate situation had arisen because of the unjustifiable neglect hitherto shown by psychologists towards the study of animal behaviour.

The views Wundt expressed in these lectures are strikingly similar to those that Morgan discussed at greater length three years later. Wundt argued that, if due attention is paid to the law of parsimony 'which allows recourse to be had to complex principles of explanation when the simpler ones have proved inadequate, it seems that the entire intellectual life of animals can be accounted for on the simple laws of association'.[12] Like Morgan, he was led to this conclusion by informal testing of his pet dog, as well as by a critical reading of Romanes' evidence. Wundt too accepted the general belief that animals possess some ability for learning by imitation. Where he differed from Morgan was in the absence of any attempt to account for the evolution of mind and in his complete acceptance of the Lamarckian principle for the inheritance of previously acquired habits.[13] And, unlike Morgan, Wundt made no experimental observations of animals that might have led to a more detailed development of suggestions such as: 'Intelligence springs from association, and then turns round again to enrich it by new connections which will facilitate the employment of thought in the future.'[14]

There was no sudden beginning of animal psychology in Germany as a result of Wundt's lectures. A young German scientist who wished to study such matters would have found it difficult to do so. By 1892 the expansionary phase of the German universities had long been over and the opportunities for the growth of new research areas were now

Fig. 3.3. Wilhelm Wundt lecturing late in his career

limited. Also, there were no longer many academic posts open to aspiring new graduates. This was not true in the United States where the system of higher education was changing and expanding. Experimental psychology had become firmly established within this system in a way that is described in the next section. Many young Americans who had chosen a career in psychology no longer went to Leipzig, but nonetheless they still read Wundt's books with great respect.

In England there had been considerable theoretical interest, but no institutional support for the study of animal psychology. In Germany appropriate institutions existed and an appropriate methodology was to hand, but there was insufficient interest. The growth of academic psychology in America, together with an intense concern with the psychological implications of Darwin's theory, turned out to provide excellent conditions for laying the foundations for systematic research on the animal mind.

American university reform and Herbert Spencer

In the 1870s the mind was studied in a variety of ways by men in Germany and Britain, but by almost no one in North America. Yet within forty years there were more Americans counting psychology as their profession and more American publications on the subject than in any other country. A number of factors contributed to the rapid expansion of psychology within North America. One was its early association with a new form of university organization, first introduced at the Johns Hopkins University in Baltimore, Maryland. The success of this new university meant that many of its innovations were copied widely during the expansion of American higher education that reached its peak just at the end of the nineteenth century.

The end of the Civil War was the beginning of a period of rapid industrialization and of increasing respect for scientific work, particularly for the technology that it might spawn. What little scientific research there was in America took place, as in England, outside the educational system. Until the 1870s the sole function of the American college was to teach. At most colleges the content of teaching was very restricted and predominantly classical. The idea that a college teacher should devote some time to his own scholarly interests was rarely entertained and, in most colleges, actively discouraged. Very few provided any opportunity for a student to learn the scientific principles and skills that were producing the wonders of the age: the telephone, photography, and electric lighting. Although many colleges had long abandoned the idea that their principal purpose was to train ministers of religion, most were still administered by governing bodies and by presidents whose principal qualifications were theological. Instilling students with a strong sense of morality, in the sense of Christian ethical principles, was understood to be a major purpose of college education. Postgraduate education of a non-professional kind was almost unobtainable in the United States until a new university was founded in Baltimore in 1876.

In the middle of the century one source of very profitable investment was in railroads. And one rail system that had an early financial success was the Baltimore and Ohio Railroad. A portion of its profits accrued to a Baltimore businessman, Johns Hopkins. On his death he left an unprecedented sum of four and a half million dollars for a university whose design was handed over to the man selected as the first president, Daniel Gilman.

Gilman was determined that this university would be a new kind of educational institution: entirely secular, a centre of scholarship and research, and with a major emphasis on science. He crossed the Atlantic to study the various university systems of Europe. Like Matthew Arnold, who had made a similar comparison a few years earlier, Gilman found that the English university had no science, the French university no freedom, but the German university had both. He already knew that American colleges had neither.[1]

One of the problems faced by Gilman was to ensure that an institution which was devoted to the acquisition of knowledge and to the provision of training in research for the few students likely to benefit, could also provide a satisfactory general education to the many with either no desire, or no ability, to extend the frontiers of knowledge. It was not clear that the German system had resolved this problem. Academic freedom meant that a German professor might lecture only on the highly particular topic that happened to engage his interest at the time. A von Helmholtz might be an inspired and devoted teacher to the few students chosen to work in his laboratory, but he put little time or effort into general lectures. The university education received by a German student could be very general at first, but then became increasingly narrow, even if there was no prospect of this specialized knowledge being of any subsequent use.

Gilman's solution to the problem was one that has subsequently become widespread throughout the English-speaking world. He created a two-tier system, whereby the essentials of the German university were superimposed upon the traditional pattern of the Anglo-American college. An undergraduate followed a general education and was awarded a bachelor's degree. Subsequent education could be either of a professional kind, mainly in law or medicine, or further specialized academic work. The latter consisted of teaching by means of seminars, in the German manner, and participation in research. The second degree, the doctorate in philosophy or Ph.D., was awarded for an original contribution to knowledge.

In order to ensure that there would be candidates for this Ph.D. degree at Johns Hopkins, Gilman established research fellowships. The advertisements for this unprecedented offer of paying people to study what was of interest to them were received with near unbelief. The response was so overwhelming and the quality of applications so high that Gilman promptly doubled the original number of fellowships.

Another innovation at Johns Hopkins was its departmental system. In addition to a full professor, a department also contained a number of university teachers who were paid reasonable salaries and enjoyed fair security of tenure. Gilman may have been aware of the potential problems of the more despotic German system. During its expansionary phase a young German scientist, whose ideas or personal conduct had offended his professor, could easily move to another university. When expansion ceased, the same professor might wield for decades immense power over those that studied with him, controlling their livelihood, the allocation of research facilities and often the opportunity to publish their work.

A student of psychology at Leipzig almost inevitably became a Wundtian. The departmental system of Johns Hopkins meant that a Ph.D. student might be strongly influenced by any of a number of more senior members of his department, or by none of them. The tradition was soon established that, in general, a student benefited most if he studied for a Ph.D. at some university other than the one where he had obtained a B.A. This also ensured a cross-fertilization of ideas that was rare elsewhere.

The Johns Hopkins University was opened in 1876. As a symbol of its secular nature and its devotion to scientific knowledge – and as a tribute to his work on behalf of English education – Thomas Huxley was invited as a principal speaker at the opening ceremony. His speech was unobjectionable. But it was only two years since his Belfast address and in the local Christian community Huxley's presence at the celebration, and the absence of a religious service, were not well received. There were the first signs of future discord between the university and the city.

Despite its rich endowment, the Johns Hopkins University depended on the goodwill of the citizens of Maryland. It needed, for example, the fees paid by these citizens for their offspring to obtain an undergraduate education. Providing a general education meant that teaching appointments could not be guided by research criteria alone. Philosophy was a particularly difficult problem. It was a subject that undergraduates wished to learn. And yet a vigorous, scholarly pursuit of philosophical problems seemed much more likely to create antagonism in the city of Baltimore than, say, research in chemistry.

For a few years this dilemma was resolved by employing a number of part-time lecturers in philosophy. Then, in 1884, a small department of philosophy was started and one of these lecturers, G. Stanley Hall, was appointed as professor of psychology and pedagogy. The reasons for Gilman's choice of Hall were varied. Hall had studied at Harvard with William James and had then spent some time in Germany. He had returned to America as an enthusiastic propagandist for the new experimental psychology. Hall's interests could be seen perhaps as a new kind of philosophy, scientific in spirit, yet not necessarily threatening to religious belief. In addition, Hall's personal life and attitude to religion were less likely to cause offence than those of the other candidates. Finally, his interest in applying psychology to educational issues might be useful in demonstrating to Baltimore the advantage of having a major university in its midst.[2]

Hall proved to be very energetic. He set up a psychological laboratory, in which he carried out a number of collaborative studies with his graduate students. These were concerned with much the same kind of problem as Wundt assigned his students in Leipzig. For example, one student, Henry Donaldson, discovered the separation of warm and cold receptors in the skin and Hall later took part in this work.[3] In 1887 Hall founded the first professional journal of psychology outside of Europe, the *American Journal of Psychology*.[4] By this time Hall had gained wide respect in academic circles, although Gilman was less impressed. When plans were made for a new university in Worcester, Massachusetts to be named 'Clark' after its benefactor, Hall was invited to become president. Gilman made no great effort to deter Hall from leaving Johns Hopkins. Its philosophy department, together with the psychological laboratory, were wound down after Hall's departure.

By the time Clark University opened in 1889, Hall

had attracted there some of the best scientists in America. He made sure that psychology was one of the major departments. Many of its members, including Donaldson, came with Hall from Johns Hopkins. Hall decided to abandon Gilman's two-tier system and establish Clark as a purely graduate university, offering no undergraduate teaching whatsoever. But Hall was not as successful as Gilman. Clark University's endowment was small compared with that of Johns Hopkins, its benefactor was alive and very much concerned with various aspects of his university, and locally there was little support for an institution that offered only graduate education. Hall's autocratic and sometimes dishonest relationship with the faculty caused increasing resentment.[5]

Shortly after Clark's inauguration, all previous benefactions to universities were outclassed by John Rockefeller's contribution to the founding of the University of Chicago. Its enterprising first president travelled eastwards with a large cheque book to raid other universities. In 1892 he arrived at Clark and a majority of its faculty left for the higher salaries and ample opportunities for research now available in Chicago.

In the last decade of the century the profits from the Baltimore and Ohio Railroad were small. The consequent financial problems for the Johns Hopkins University meant that it was no longer pre-eminent among American universities. However, by then most of the principles that Gilman had established in Baltimore had been adopted both by the older universities with their vaster and more stable wealth, and by the many new universities, founded, like Chicago, on later industrial fortunes or on state land grants. Along with the emphasis on science, the provision of time and facilities for research, generous salaries and the two-tier degree system, came the founding of laboratories of psychology as a symbol of the modern university, a symbol that Gilman himself might well not have chosen.

Changes in American education were not confined to the universities. In the last decades of the century, Americans moved increasingly from the land to the cities, whose population was further swollen by accelerating immigration from Europe. Rapid urban expansion made it very difficult to maintain universal education and began to pose urgent questions about the organization and teaching methods of elementary and high schools.

At the same time another social change created a further opportunity for applying the results of research in psychology, in this case at a far more personal level than that of formal education. The

Fig. 3.4. G. Stanley Hall

beginning of a steady decline in infant mortality rates meant that, on a scale not previously known, parents could be concerned with the psychological development of a child and not simply with his or her physical survival and state in what might be a very imminent after-life.

Hall was both the first person to establish an institutional base for psychology within the new university system, and the first American psychologist to encourage the idea that his subject could provide important practical benefits to pedagogy and to the raising of children. The research carried out during the first few active years of the Johns Hopkins laboratory mainly followed Wundt's lead and was concerned with the senses. The possible fruits of such research were not likely to enrich educational practice. Hall's principal interest, and one that came to dominate research at Clark University, was in child development. Evolutionary theory, rather than Wundt's psychology, provided the framework for these studies. Preyer's book on the *Mind of the Child* became well known in the United States. It inspired dozens of studies that looked at various aspects of child behaviour at different ages. Much of this loose, observational work took place outside any academic context and what became known as the 'Child Study Movement' was encouraged, rather than directed, by

Hall.[6] A major theme in Hall's work, as in that of Preyer, was Haeckel's biogenetical principle: with reference to psychology, the idea that the mind of the child passes through various stages that closely correspond to the developmental history of mental functions in evolution. The incorporation of this idea into Romanes' theory of mental evolution, which was contemporary with Preyer's book, was noted in the previous chapter.

An evolutionary view of psychology was already very familiar in America from the volumes of Herbert Spencer's *Synthetic Philosophy*. The introduction of psychology at Johns Hopkins was one reason for the subsequent growth of the subject in the United States; Spencer's work was a further major factor. During the period of rapid change that followed the Civil War, his books were bought by the thousands. His philosophy of progress appeared to provide a much needed guide to the age, and an important part of this philosophy was faith in the benefits to be derived from advances in the social sciences. Hall can be seen as making more specific Spencer's general claims about the utility of psychology. With the Civil War in progress, no equivalent to the encounter between Huxley and Bishop Wilberforce could have attracted public attention to the issue of evolution in the years immediately following Darwin's publication of *The Origin of Species*. When, in the 1870s, general interest turned to such matters, it was Spencer's view of evolution rather than Darwin's that generated most excitement. His books enjoyed extraordinary popularity both among the elite of that society, the philosopher as well as the industrial baron, and among the thousands of people, scattered among the countless small towns of America, who thought seriously about their lives, but were no longer so content with the answers provided by their religion. Spencer became 'the metaphysician of the homemade intellectual and the prophet of the cracker-barrel agnostic'.[7]

Although Spencer's books were also very popular in Britain, his influence on social and political opinion was not as strong as in America.[8] Industrialization in Britain had already advanced to the stage where the social evils created by the form of unbridled *laissez-faire* system that Spencer advocated were already apparent. To a large extent he was preaching a gospel that had already been found wanting. In the United States rapid industrialization on a large scale was a more recent phenomenon. Spencer's slogan, 'the survival of the fittest', was eagerly accepted by business, where it was quite clear that fitness was to be measured in terms of wealth. As a historian has noted: 'with its rapid expansion, its exploitative methods, its desperate competition, and its peremptory rejection of failure, post-bellum America was like a vast human caricature of the Darwinian struggle for existence'.[9]

North America also differed from England in the variety of its social conditions. In the summer of 1876 Huxley arrived in New York, on his way to the opening of Johns Hopkins University, and when viewing the harbour felt so exuberant over the scenes of bustling, modern life that he exclaimed: 'If I were not a man, I think I should like to be a tug'.[10] Just a few months earlier and two hundred miles up the coast, the operation of a telephone system had been demonstrated for the first time in Boston. And yet, during Huxley's visit, two thousand miles westward a combined force of Sioux and Comanche Indians defeated the U.S. 7th Cavalry at the Little Bighorn River. It was the last major check to complete domination of the continent by peoples of European descent. Although Spencer personally abhorred violence, his philosophy could be used to justify the decimation of one race and culture by another, as well as unrestricted competition between individuals.

The widespread acceptance of Spencer did not mark a totally new departure. The extreme individualism fitted in with what was already a well-established national tradition. The progressive outlook, the belief in a glorious future to be reached with the aid of science, also suited another national tradition. And as for the strong religious elements of American life, Spencer's philosophy was conveniently ambiguous. By the 1880s his form of evolutionary theory had been endorsed by most religious leaders. American scientists, especially the biologists, might be persuaded by Huxley's blunt, uncompromising arguments, but most people concerned with the implications of evolutionary theory read Spencer.

The vogue for Spencer's books lasted for twenty years or more. The inevitable reaction occurred in a variety of contexts. There was increasing labour unrest. An early event was a strike by employees of the Baltimore and Ohio Railroad, which hurt the fortunes of the Johns Hopkins University. A working man's doubts as to whether to join in a strike for the sake of other workers at some time in the future, and at the immediate cost of prolonged hardship for himself and his family, were not easily solved by turning to traditional principles. Spencer's answer, that the well-being of the individual and of his immediate relatives should be the principal criterion in reaching such a decision, increasingly appeared to represent an ethical system that favoured only the rich.

Concern with such ethical problems became part of a movement in American philosophy known as

Fig. 3.5. Herbert Spencer at the time his popularity in the U.S.A. reached its height

pragmatism. In academic circles, the earliest reaction to Spencer occured among philosophers. Pragmatism started with the work of C. S. Pierce, whose application for the chair at Johns Hopkins had, for reasons to do with his personal life, been passed over in favour of Stanley Hall. Its impact came later with the work of John Dewey and of William James.

The new university system and the attitude towards social sciences encouraged by Spencer's philosophy provided ideal conditions for psychology's growth in America. This might still not have occurred. Fortunately for psychology the main critics of Spencer shared at least the latter's high regard for the subject. Even more important, one of them, James, published in 1890 two volumes on psychology whose power and clarity conveyed prestige and inspired commitment for many years after its publication.

William James

Poor health played a major part in William James' education and subsequent career. He grew up in a wealthy and remarkably intellectual household. By the age of eighteen, family debates and extensive travel in Europe, which was undertaken as a form of convalescence from almost any malady, had produced in James a sharp and broadly educated mind and an ability to express himself with exceptional clarity. They did not give him any clear ambition.

After a year devoted to painting, an interest in natural science led him to study chemistry at Harvard University, one of the very few places in the United States where formal scientific instruction could then be obtained. These first two years at Harvard did not produce a devotion to chemistry. They did, however, allow close contact with Charles Eliot, the professor of chemistry. Eliot was later to play almost as important a part in changing the American university system as Gilman at Johns Hopkins. Eliot may have been disappointed by James' lack of enthusiasm for chemistry, but nevertheless was impressed by his wide knowledge and general promise.

James remained at Harvard, but now switched to the study of medicine. The childhood pattern of frequent illness, leading to yet further excursions to Europe, continued. This meant that his voracious reading in a variety of languages was maintained and that he was able to become closely familiar with intellectual developments in Europe. It also meant that six years passed before James was medically qualified.

A life dedicated to the practice of medicine had no more appeal than one devoted to art or to chemistry. In 1870, at the age of twenty-eight, James was affected by profound depression. Eventual recovery, and a growing interest in psychology, began with an unusually intellectual event: some essays he read convinced him of the reality of free will, and produced what today might be called an existential commitment. His new belief in the mind's power over the state of the body was followed by a steady improvement in health. In 1872 he obtained his first job: Eliot offered him a teaching position in physiology at Harvard. Three years later he began to give a course to medical students on the new psychology, the first to be given outside Germany. Although it included the recent discoveries on the senses made by German physiologists and pyschologists, Spencer's *Synthetic Philosophy* served as the basic textbook. In the following year this course was available as an 'elective' for all Harvard students.

The elective system was Eliot's solution to the problem of incorporating new academic subjects within the traditional undergraduate syllabus. It came to serve as another factor that favoured the growth of psychology and other social sciences within American universities. Instead of establishing additional specialized degree programmes, as happened generally in European universities, Harvard students were allowed to choose from a number of additional courses, electives, in new subjects like psychology. Initially the time available for electives was small in comparison to that devoted to the required courses for a bachelor's degree. But the degree of choice steadily increased and other American universities followed Harvard's lead, thus producing what today is the most marked distinction between higher education in America and in Europe. An American bachelor's degree is awarded upon the accumulation of a specified number of units, obtained from a series of self-contained courses. Within certain constraints, the combination of courses taken by individual students varies widely. This contrasts with the situation common in European universities where the degree of specialization is much greater and the degree of choice open to a student is much more limited.

As a result of the elective system, justification for the existence of a psychology laboratory within an American philosophy department, and later for the independent existence of a psychology department, could be based simply on the argument that students wanted, and should be able, to study psychology during their undergraduate years. The much longer, and sometimes unsuccessful, struggles of the social sciences to establish themselves within the less flexible system of European universities depended on the more highly resistible argument that they should be regarded on an equal footing with disciplines already well established.

The elective course in psychology given by James proved very popular. In addition to the attraction of his style of lecturing, there was considerable interest stemming from his use of Spencer's *Principles of Psychology* as a text. The course began just after Huxley's visit, which had further fuelled interest in Darwinism in North America. For discussion of evolutionary theory to be included within a university course was still very much a novelty. Years earlier James had been 'carried away with enthusiasm by the intellectual perspectives' that Spencer's philosophy had seemed to open.[1] Well before teaching at Harvard James had become increasingly repelled by it. Later he was to comment that Spencer's name was 'linked to one virtue and a thousand crimes': the virtue was Spencer's belief in the universality of evolution, while the thousand crimes were 'his 5000 pages of absolute incompetence to work it out in detail'. Spencer's philosophy was, James came to decide, 'almost a museum of blundering reasoning'.[2] Yet for twenty years Spencer's *Psychology* remained the main text for James' course and, although the book served mainly as an intellectual punch bag, it profoundly affected his views.

In 1876 James' position at Harvard became firmly established when he was appointed as an assistant professor in the philosophy department. A sign of the approval for psychology won by James was the university's allocation of space and money for setting up a small laboratory to be used for experimental demonstrations. James became a leading figure in the intellectual life of Cambridge and Boston. Wider fame developed from a series of papers on psychology that he began at this time and continued throughout the 1880s. Many of these papers became incorporated as chapters in his two volume book, *Principles of Psychology*, which was eventually published in 1890. This was James' major contribution to psychology.

The book succeeded brilliantly in achieving a synthesis of three different approaches to the study of the mind. One was an analytic approach in the tradition of British empirical philosophy. As is often the case, those predecessors receiving the most criticism were the most important influences. James rejected many of Spencer's ideas and also, in a more careful and respectful manner, detailed what he saw was wrong with the analysis of mental processes provided by Bain and earlier British associationists. In particular, he criticized the idea that subjective experience could be decomposed into discrete mental atoms, 'ideas', and substituted an emphasis on continuity, introducing the term, the 'stream of consciousness'.[3] James also dismissed what he saw as the misguided environmentalism of empiricist philosophy. Concepts of time and of space could not be based on experience alone. And, citing Romanes, Spalding and Preyer, a chapter was devoted to the important part played by instinct in human behaviour. Compared to that of Morgan, James' concept of instinct seems a very muddled one. His examples ranged from the relatively simple – *sneezing, coughing, keeping time to music* – to traits that are arguably direct reflections of James' specific cultural heritage. Thus, to *imitation, anger, fear, curiosity, sympathy* and *love* are added *secretiveness* – 'there is unquestionably a native impulse in everyone to conceal love affairs' – *acquisitiveness, cleanliness, modesty* and an 'anti-sexual instinct': 'the instinct of personal isolation, the actual repulsiveness to us of the idea of intimate contact with most of the persons we meet, especially those of our own sex'.[4] Nevertheless, despite this nativism and the stream of consciousness, the basis of James' psychology clearly owes most to Spencer and to Bain.

The second major element in James' *Principles* was its clear account of various current developments in European psychology. James' linguistic ability and wide scholarship enabled him to present the best of

Fig. 3.6. William James when writing *The Principles of Psychology*

French and German psychology, as well as work within the British tradition. In particular, it was one of the first books in English to provide a thorough review of the new physiological psychology. Time and again James expressed his distaste for Wundt's ideas, but just as often discussed at length the results obtained in the Leipzig laboratory and in those of other German universities. 'The experimental method has quite changed the face of the science so far as the latter is a record of mere work done'; with this new approach 'the Mind must submit to a regular siege' by assailants with 'little of the grand style'; experimentalists 'mean business, not chivalry'.[5]

The third major element in James' book is the one that has maintained its freshness over the years. Its chapters on memory, attention or the will consist largely of his own sensitive reflections on the experiences of everyday life; or perhaps, more narrowly, on the conscious mental activity of a leisured, intelligent and well-educated Western male.

In looking through James' two volumes and picking out the themes that were later emphasized by animal psychologists in America, what stands out are its view of the brain as a mass of potential connections, and the great stress laid by James on the importance of habit in human affairs.

An early chapter discussed the current state of knowledge about the anatomy and physiology of the brain. By then certain areas of the cerebral hemispheres had been identified as intimately related either to specific senses or to the control of bodily movements. The functions of the extensive remaining areas of the cerebral cortex were obscure. As was common practice at the time, James referred to these as the 'association areas'. Also common was James' belief that in these areas new connections or pathways were formed and 'stamped in' during an individual's life time. Less common was James' attempt to describe in detail how such connections might be formed. He provided a series of diagrams to illustrate the changes in neural pathways that could occur, for example, in the classic example of a child learning to avoid fire.[6] The scheme is similar to Morgan's theory of how a chick learns to reject cinnabar caterpillars, as described in the previous chapter. Unlike Morgan, James in general gave little weight to the Spencer–Bain principle and argued that the importance of pleasure and pain had been exaggerated.

The formation of new pathways connecting areas of the cerebrum was emphasized because of James' belief that the acquisition of habits was a central problem in human psychology. For one thing, James argued, habits are of prime importance in practical matters. They simplify 'movements required to achieve a given result, make them more accurate and diminish fatigue'.[7] Habit diminishes 'the conscious attention with which our acts are performed' and thus releases the mind from pre-occupation with the mundane aspects of life.[8] James described with approval the successful effort made by Houdini to perfect the skill of juggling four balls to an extent that allowed him to read a book at the same time.

The study of habits was also of great importance to James because of their varied ethical implications. One of these was to do with the stability of human societies. It is worth quoting James at length on this topic, since it illustrates well the more rhetorical side to the *Principles*.

'Habit is thus the enormous fly-wheel of society, its most precious conservative agent. It alone is what keeps us all within the bounds of ordinance, and saves the children of fortune from the envious uprisings of the poor. It alone prevents the hardest and most repulsive walks of life from being deserted by those brought up to tread therein . . . It dooms us all to fight out the battle of life upon the lines of our nurture or our early choice, and to make the best of a pursuit that disagrees, because there is no other for which we are fitted, and it is too late to begin again. It keeps different social strata from mixing . . . You see the little lines of cleavage running in the character, the tricks of thought, the prejudices . . . from which a man can by-and-by no more escape than his coat-sleeve can suddenly fall into a new set of folds. On the whole it is best he should not escape. It is well for the world that in most of us, by the age of thirty, the character has set like plaster, and will never soften again.'[9]

Habits, James argued, are also central to the moral life of an individual. We should 'make our nervous system our ally instead of our enemy . . . For this we must make automatic and habitual, as early as possible, as many useful actions as we can'. It is a miserable human being who deliberates over 'the lighting of every cigar, the drinking of every cup, the time of rising and going to bed every day, and the beginning of every bit of work'.[10] Half such a person's life is wasted over deciding, or regretting, such matters. For James there was 'no more contemptible type of human character than that of the nerveless sentimentalist and dreamer, who spends his life in a weltering sea of sensibility and emotion, but who never does a manly concrete deed'.[11]

This stress on action and on the importance of habits based on neural connections foreshadowed the pre-occupations of later animal psychologists in America. It is also seen in one of James' most original contributions, his theory of emotion. Emotions do not act as causes. We do not strike out *because* we are angry, blush *because* we feel shame, or flee *because* we are afraid. Rather, James claimed, certain situations directly elicit certain specific and powerful reactions, made up of abrupt changes in our outward behaviour, in our expressions and in our bodily state. It is these reactions that give rise to our subjective experience of emotion. 'Commonsense says, we lose our fortune, are sorry and weep; we meet a bear, are frightened and run; we are insulted by a rival, are angry and strike . . . The more rational statement is that we feel sorry because we cry, angry because we strike, afraid because we tremble.'[12] In essence, James is agreeing, in the case of emotion, with Huxley's view of consciousness as an epiphenomenon.

As with later behaviourists, stress on action and habit was accompanied by a tendency to underestimate the complexities of perception. In contrast, say, to Romanes, James argued against the view proposed by Helmholtz and adopted by Wundt, that perception involves unconscious processes of inference analogous to those employed in conscious reasoning.

On the subject of the animal mind James was very much in agreement with Wundt. He might state that 'perception and thinking are only there for behaviour's sake' and add that 'this result is one of the fundamental conclusions to which the entire drift of modern physiological investigation sweeps us';[13] and yet he was quite adamant that the way we see the world is quite apart from that of any other animal and true thought a unique characteristic of the human soul. Like Wundt, but without the disparaging comments on amateurs, James set out to show 'by taking the best stories of animal sagacity, that the mental processes involved may as a rule be perfectly accounted for by mere contiguous association, based on experience'.[14] Like Morgan, although without the latter's detail, James considered it 'proven that the most elementary single difference between the human mind and that of brutes lies in this deficiency on the brute's part to associate ideas by similarity',[15] where by 'similarity' James meant much the same as Morgan did by 'perception of relations'.

James' views on the mind–body problem were uncertain and to some extent inconsistent. In early years he had accepted the arguments for viewing man and other animals as 'conscious automata'. His theory of emotion can be seen as a reflection of this youthful belief. But during his depressive illness James, with his new belief in free will, came to reject the idea that the thoughts and feelings of which we are conscious can never affect our actions, and that they have an elusive character that prevents the application of scientific methods. The pain from a swollen foot is just as real, if not more so, than the shoe that encases it.[16]

The overall philosophy of mind in the *Principles* was in general a dualistic one. James' dismissal of automaton theory took the form of pointing out the weaknesses of previous arguments in its support. On mental continuity, James opposed the kind of argument put forward by Huxley in the Belfast address, and suggested that instead one could justifiably infer from human experience that an animal's behaviour was influenced by purely mental events. This was just as logical as the claim that, because in principle what an animal does can be explained entirely with reference to physical events in its brain, a science of human behaviour must adopt the same stance. On the difficulty of conceiving how purely mental events could act as causes of material changes, James simply pointed out that the nature of *physical* causation was equally hard to understand. 'As in the night all cats are grey, so in the darkness of metaphysical criticism all causes are obscure.'[17]

The adoption of a dualist position and the rejection of a materialistic one was provisional. For the moment psychology should remain metaphysically naive and employ the language of commonsense. James saw a future need for some kind of 'metaphysical reconstruction', but, until that was achieved, 'to urge the automaton theory upon us, as it is now urged, on purely *a priori* and *quasi* metaphysical grounds, is an unwarrantable impertinence in the present state of psychology'.[18]

James' *Principles of Psychology* greatly helped the general status of this new discipline and persuaded many students to devote themselves to its study. But by the time it appeared in 1890 there were already a number of established American psychologists and in the main they were disappointed by James' two volumes. Brilliantly written, but impressionistic; not a solidly scientific work; these were common reactions. To men busy at their various universities in searching for money, space and approval for new laboratories of psychology, James' sceptical and almost disdainful attitude towards much of the recent experimental work in psychology must have seemed close to sabotage. The dualism, the discussion of the human soul, the interest in spiritualism and the stress on the problems of establishing a firm conceptual basis for psychology were not what was expected from a leading figure of a thrustful new science.

James himself had no desire to lead American psychology. After the *Principles* was finished, his main interests became philosophy and the nature of religious belief. The change was marked by the reversion of his academic title from professor of psychology to professor of philosophy.

The other early pioneer of American psychology, G. Stanley Hall, very much desired to be seen as its leading figure. The research on child development that Hall encouraged was attractive in its promise of demonstrating the utility of psychology, where James chose to say very little about the applications of psychological research. Yet Hall's methods and theories were distressingly loose and vague for most of his fellow psychologists. Moreover, Hall generated a great deal of distrust, while on personal grounds James was universally admired.

In October 1895, Hall published an extraordinary editorial in the *American Journal of Psychology* which set out his claim to be personally responsible for almost every institutional development in American psychology over the previous fourteen years. A speedy and fierce reaction came from a number of younger psychologists, who pointed out the variety of factual errors in Hall's editorial.[19]

Two men were particularly angry. One was James Cattell. Cattell had been the first American student to work with Wundt. In the face of the Leipzig insistence on studying general psychological phenomena, Cattell's research there was concerned with psychological differences between individuals. This interest was strengthened when, after his studies in Germany, Cattell visited England and met Galton. Cattell was greatly impressed and came to regard Galton as the world's greatest psychologist. On returning to America, Cattell founded a psychological laboratory first at the University of Pennsylvania and then, in 1891, at Columbia University in New York. He turned out to possess a considerable organizational ability and his arrival at Columbia was timely. This was another older institution where extensive reforms were being introduced and very considerable expansion was taking place. By the time Hall's editorial appeared, Cattell had created a large and energetic department at Columbia and had firmly established in North America a Galtonian tradition that stressed mental measurement, the use of statistics and a highly hereditarian attitude towards human differences.[20]

Another angry objection to Hall's editorial came from James Mark Baldwin. Like Cattell, Baldwin had also spent his year at Germany, had come into contact with Wundt and the new psychology, and on his return to North America had founded psychological laboratories, first at Toronto, and then in 1893, at his *alma mater*, Princeton. Baldwin was much more of a theorist, and less of an experimenter, than Cattell. In his general outlook on psychology he was very close to James, and in his involvement in evolutionary theory close to Morgan. Baldwin was not interested in measuring individual differences. He was far less of a hereditarian than Cattell and shared Morgan's interest in the cultural transmission of psychological abilities. His major interest in psychology was child development, a study that Baldwin regarded as having been perverted by the theories and methods of Stanley Hall.[21]

These two men, Cattell and Baldwin, had been in conflict with Hall a year or so before the exchange over Hall's editorial. The *American Journal of Psychology*, with Hall as its founder and only editor, had remained for many years the only American journal devoted to psychology. By the early 1890s psychologists who were *not* former students of Hall or members of Clark University felt that editorial policy grossly favoured those who were. Outside of Clark, there was general approval when in 1894 Cattell and Baldwin became joint founders of a second American journal, the *Psychological Review*. This rapidly became the most prestigious psychological journal in the world, a status it has maintained ever since.

There were many young Americans who received their bachelor's degree in the summer of 1895 and who, upon reading James' *Principles* as undergraduates, had decided that they wished to become psychologists. Research degrees could now be obtained from a variety of universities and many of these contained active laboratories of psychology. There was also no need for these young men to be wealthy. Most universities now followed Johns Hopkins' lead in providing fellowships for students with likely research ability. And, if they completed sufficient research of an adequate standard to obtain a Ph.D., there were now two journals in which they might hope to publish an account of their work.

One such young man was Edward Thorndike. During his four years as an undergraduate, James' *Principles* was the only book, apart from a few required texts, that he ever bought. In 1895, his final year at college, he successfully applied for entrance to Harvard, choosing to remain in New England and to study with James, rather than to work with Cattell at Columbia, with Baldwin at Princeton, or in the other comparable laboratories now spread across the continent.

Edward Thorndike's puzzle boxes and doctoral thesis

Edward Thorndike's father was a Methodist minister, who moved every two or three years from one small New England town to another, as was the custom in this church. Methodism was on the defensive in this part of America. The size of his congregations in the area around Boston declined over the years as the population became steadily more Catholic. In reaction Methodism became increasingly austere, displaying a pinched moralism suspicious of any form of pleasure. Thorndike's mother regarded such things as dancing and Sunday newspapers as wicked, and even felt that her husband's ability to enjoy a ball game was unworthy. One should be diligent and serious, avoiding lightness and gaiety.

Edward Thorndike was diligent and serious, and very studious. He was an academic success at Wesleyan College, the progressive Methodist institution to which he had been sent, but there he lost his faith in religion. It was replaced by devotion to a scientific outlook that rejected the beliefs and the emotionality of his parents' Methodism, but retained many of its other features. Thorndike's science was a sober, demanding and methodical activity. His background was not unusual. A large number of American scientists grew up in a clergyman's household during the last part of the nineteenth century and then obtained a general education at a small college where their religious belief was replaced by a passion for science.

As an undergraduate at Wesleyan, Thorndike was taught some psychology, read a great deal of Spencer and became inspired by James' *Principles*. He entered Harvard as a graduate student in the autumn of 1895, at the age of twenty-one, with little clear idea of what he wanted from the future except to learn more about psychology. In his first year Thorndike took an advanced psychology course taught by James and based on Wundt's *Lectures on Human and Animal Psychology*. Within a few months Morgan arrived to deliver the series of lectures in nearby Boston, which were later to form the basis of his book, *Habit and Instinct*. Whether Thorndike attended these lectures or had any direct contact with Morgan is not known. Somewhat curiously Thorndike never made any comment about Morgan's visit. But, in any case, shortly afterwards Thorndike was reading about animal intelligence and starting experiments with chickens that were intended to test for the Lamarckian inheritance of habits.[1] It seems that Morgan's lectures came just at the time when Thorndike was searching for a research topic. Morgan's critical attitude appealed to Thorndike's own iconoclastic feelings towards any approach that was not strictly scientific. The only recorded comment Thorndike ever gave on his reasons for choosing this topic was made years later: 'My first research was in animal psychology, not because I knew animals or cared much for them, but because I thought I could do better than had been done'.[2] His first studies were also partly prompted by failing to obtain permission for a study on children that he had planned.

As first Spalding, then Romanes and Morgan had done, Thorndike began by working with newly hatched chicks. Although he had settled on a research topic, there were still practical problems to solve. Psychology departments had not quite reached the stage of encouraging experiments with animals and the only space available to Thorndike was the cellar of William James' house. Harvard also had little financial support to provide for penniless graduate students. In his second year there, Thorndike successfully applied for a fellowship at Columbia University, which involved teaching duties at Teachers College, the division of the university concerned with graduate training in education. So, in the autumn of 1897, he moved to New York, carrying with him a basket containing some of his prize chickens only to discover that the ample laboratory space he thought he had been promised did not exist. He soon found an attic on the fifth floor of Schermerhorn Hall, which still houses the department of psychology at Columbia University. He acquired some cats and dogs, and began the research for his doctoral thesis which was to make his name appear in almost every book on animal psychology that was written thereafter. The work occupied almost all of his waking hours. A full day of experimenting in the psychology department would be followed by an evening taking care of the cats and dogs, and later even a monkey, housed in his apartment.

Cattell had become Thorndike's supervisor, but by now he was a busy organizer, very much occupied with the development of his psychology department and with the advantages offered by the university's move to its new, large site on Morningside Heights. Cattell had very little direct influence on the research carried out by his many Ph.D. students. But he may have helped Thorndike to realize that the slow breeding rate of chickens made them very unsuitable subjects with which to study the heritability of habits. In any event Thorndike turned instead to some of the other topics Morgan had discussed during his American lecture tour, those of learning by imitation and of memory.

Thorndike felt that Morgan's ideas were sound, but that he had not gone far enough in rejecting Romanes' approach. Thus, Morgan should not have accepted that animals possess imitative abilities and retentive memories, even if of a desultory kind, on the basis of only anecdotal evidence. It showed a lack of commitment to argue for the power of associative principles, as seen in trial-and-error learning, while putting little effort into studying these principles. A few informal studies with a few chicks or a dog were not enough. If the subject really were important, then it deserved systematic experimental investigation using a variety of situations and greater numbers of animals; only then would claims for general validity be justified.

Thorndike's way of testing the intelligence of his cats and dogs was inspired by Romanes' claims about the mechanical abilities of cats. Thorndike took the kind of situation described by Romanes' correspondents and also by Morgan in the latter's account of how the fox terrier learned to operate the latch of a gate, and turned it into an experimental method. A number of wooden crates were each fitted with a door which could be operated by one of a variety of devices. These devices included a simple rotating catch, a loop either by the door or by the opposite wall, which was attached via string to a removable bolt, and a treadle on the floor, similarly connected to a bolt. These *puzzle boxes*, as they became known, were very simple, much cruder than suggested by the schematic figures in most subsequent textbooks. Some of the photographs of them taken by Thorndike are shown in Figure 3.8.

Fig. 3.7. Edward Thorndike on entering Harvard University.

Thorndike's aims were: first, to study trial-and-error learning in a systematic, quantitative manner; second, to test memory by first training an animal and then allowing some period of time to pass before placing it in the situation again in order to assess how well the skill was retained; and finally, having established how long it took a representative animal to learn without any aid how to escape from a particular box, to use this as a norm for assessing the possible beneficial effects of various kinds of tuition. His animals were maintained in a state of what he termed 'utter hunger' and on each trial, once they had escaped from the box, they could reach a dish containing a small amount of meat or fish placed just outside. Thorndike's procedure was to put a dog or cat into the box and observe its behaviour, while timing the *latency* of each response, that is, the time between insertion into the box and the occurrence of the appropriate

Fig. 3.8. Four of the puzzle boxes used by Thorndike in the research for his doctoral thesis

movement which opened the door. By plotting the response latency on each trial as a function of the trial number, he provided a graphical representation of the results for each animal, the first of a multitude of learning curves to come. Some examples are shown in Figure 3.9.

The initial achievement of the work was to show that Morgan's observations could be readily replicated in a standardized situation and with a wide variety of responses. In each case there was the same progression from the initial trials, in which an animal made a long series of varied movements ending with the apparently chance production of the appropriate response, to trials in which the animal consistently made the appropriate response in a smooth and rapid manner. The graphs in Figure 3.9 also show the results of the 'memory tests', obtained by replacing an animal in the same box after many days had intervened since the last training trial. Thorndike noted that a considerable lapse of time produced little increase in latency and so confirmed experimentally the belief Morgan had based on informal evidence, that associative learning of this kind is not subject to very much forgetting.

The overall shape of a learning curve could be used as a measure of speed of learning, and this allowed comparisons to be made between cats and dogs. At first rather tentatively, Thorndike suggested that this could serve as a direct index of animal intelligence, both within and across species. This was never a very promising idea. The comparison between cats and dogs showed that the speed with which they solved a problem was not a direct measure of some underlying associative process, but was greatly affected by a range of factors. For example, during early stages of training the cats reacted far more vigorously than the dogs, clawing frantically at and by the door, when placed in a puzzle box. Consequently, cats learned much faster than dogs to operate a device placed close to the door, but much more slowly a

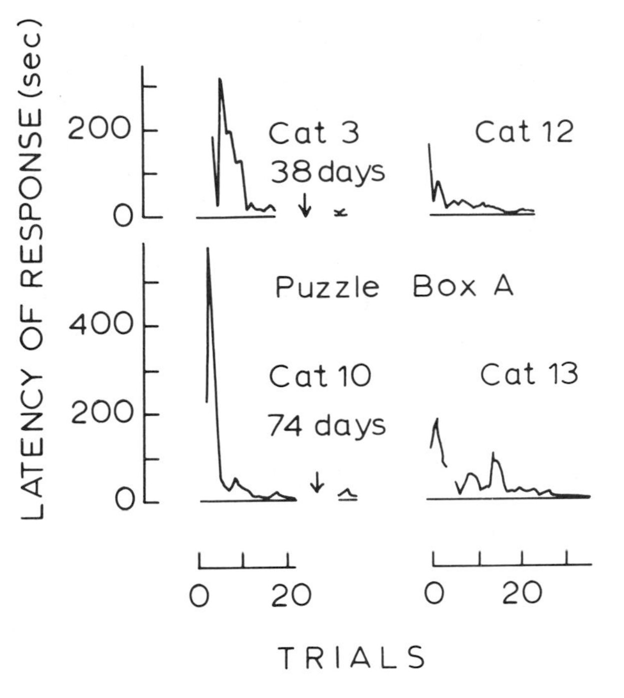

Fig. 3.9. The first learning curves: the results from four cats who learned to escape from Thorndike's Box A, showing the general decrease in the latency of the correct response as training was continued and, in the case of cats 3 and 10, good retention of the response after intervals of 38 and 74 days respectively

device at the opposite end of the box. In addition, Thorndike suggested that rates of learning depend on the ability of the animal to perceive critical features of the situation, that is, on its particular sensory abilities. It is clearly nonsense to claim that one species is more intelligent than another simply because it learns how to solve some particular problem more rapidly. Nevertheless, in later years Thorndike was less tentative in asserting that the speed with which a connection is made increases with the evolutionary development of the species.

At the beginning of his research Thorndike did not set out to make some new discovery; he viewed his experiments as an exercise in method, a demonstration of how one could verify with experimental data analyses based on informal evidence. His results on memory had confirmed Morgan's conclusions on this matter. Thorndike expected that his further experiments on learning by imitation and on the effects of passive tuition would similarly produce hard evidence in support of beliefs that were already widely held. The results did not turn out this way.

The procedure he used to study imitation was to attach to the side of a puzzle box a smaller compartment, which allowed an inexperienced cat confined inside to watch the performance of a trained cat in the puzzle box itself. Subsequently the inexperienced cat was placed in the puzzle box and Thorndike studied how it learned to open the door, looking for evidence of a beneficial effect of its prior observational experience. To his surprise Thorndike found no such evidence. He repeated the tests with dogs and again consistently failed to obtain positive results.

In a similar manner he attempted to find out whether cats or dogs could learn when the appropriate movement was made passively. Thorndike would place an animal's paw on the catch or loop, and make the appropriate movement a number of times, so that on each occasion the door opened and the animal could escape to get at the food. Then the animal was left in the box on its own and the time it took to learn to operate the door was recorded. Again rates of learning were no faster than those of animals which had never been trained in this manner. Suspecting that he might have omitted some crucial factor, Thorndike sent questionnaires to a number of well-known animal trainers to see if they could throw any light on the matter. Their replies were disappointing and failed to suggest any faults in his own methods.

These unexpected failures to confirm the almost universal belief that animals such as cats and dogs were quite able to learn by imitation or by passive movement appear to have had a marked effect on Thorndike's ideas on animal intelligence. As a consequence, they changed his thesis from a solid, original, but somewhat unexciting, demonstration of the use of experimental method in studying animals into a highly provocative piece of work.

Thorndike decided that a simpler analysis of trial-and-error learning than Morgan's might well be correct. The latter had suggested that the improvement shown by the duckling in escaping from the newspaper pen or by the fox terrier in opening the garden gate was based on the progressive establishment of an association between two ideas: the idea of a particular action and the idea of its consequence. Thorndike proposed that trial-and-error learning be explained in terms of a connection between one idea, perception of the situation, and what he called a *motor impulse*. The implication of this 'stimulus–response', or *S-R*, view of learning is that an animal learns what to do in a situation without in any sense knowing the consequence of the action. The sole effect of a reward in such a process is to 'stamp in' the S-R connection.

Such a theory leaves an animal with a very

impoverished understanding of its world. Nevertheless, in addition to the precise results reported in the thesis, a host of informal observations from the long hours of experimental work had also convinced Thorndike that this was a correct view of the mind of animals, at least of those less complicated than primates. By watching for some time, he explained, one gets 'a fairly definite idea of what the intellectual life of a cat or dog feels like. It is most like what we feel when consciousness contains little thought about anything, when we feel the sense-impulses in their first intention, so to speak, when we feel our own body, and the impulses we give to it. Sometimes we get this animal consciousness while in swimming, for example. One feels the water, the sky, the birds above, but with no thoughts *about* them or memories of how they looked at other times, or aesthetic judgements about their beauty; one feels no *ideas* about what movements he will make, but feels himself to make them, feels his body throughout. Self-consciousness dies away. Social consciousness dies away. The meanings, and values, and connections of things die away. One feels sense-impressions, has impulses, feels the movements he makes; that is all.'[3]

By the end of February 1898 his experiments were finished and the same energy and long hours were devoted to the writing of his thesis. Thorndike could now challenge established views on the nature of animal psychology, as well as castigate all previous investigators for their failure to use proper experimental methods. He described the situation and his attitude in a letter to his fiancée: 'On the floor and the book-case are lots of little piles of thesis. On my chair is also thesis. I walk and sit on thesis. I haven't yet reached the stage when the bed has also to serve, but expect by next week to sleep on thesis. It is fun to write all the stuff up and smite all the hoary scientists hip and thigh. I shall be jumped on unmercifully when the thing gets printed, if I ever raise the cash to print it.'[4]

Within ten months Thorndike was awarded his doctorate and the thesis was published in the *Psychological Review*. Thorndike was 'jumped on', as he had expected. The whole tone of his paper, that of the cocksure youngster challenging his elders, guaranteed this. Where Morgan had shown sympathy, as well as polite scepticism, towards Romanes, Thorndike substituted cheerful insults. For example, he compared the efforts of the 'anecdotalist school', to someone trying to base the study of anatomy on dime-show freaks. Even Morgan received this treatment at times, as, when commenting on Morgan's suggestions about language, Thorndike suggested that 'when one says language has been the cause of the change from brute to man, . . . he is talking as foolishly as one who would say that a proboscis added to a cow would make it an elephant'.[5] The provocative presentation was a deliberate attempt to generate a reaction. Thorndike admitted that his theory seemed 'even to me, too radical, too novel'. He cordially invited denial of the theory – as long as the criticism was constructive and accompanied by further experimental evidence.

The immediate public response by other psychologists was one of dismay and rejection. Thorndike was taken to task for confining his animals within small boxes and for keeping them in a state of 'utter hunger', a term whose use he later regretted since it suggested near starvation, whereas in fact the condition had been far less drastic than this. His test situations were rejected as being unnatural and therefore producing misleading results which totally underestimated the intellectual capacities of cats and dogs. A few of those who could not believe his conclusions reacted in the way that he had hoped. They attempted to devise testing methods which they felt would be much more likely to show that animals could learn by imitation or from passive tuition, could solve problems in a manner that indicated some understanding of the situation and could associate an action with its consequences.

The Law of Effect and S-R bonds

By the summer of 1898 Thorndike's student days were over and he needed a job. He was lucky to find a position at a small college in Ohio, where he was required to teach pedagogy. There were no facilities for studying animals and no opportunity for any kind of research. Within his first year he was persuaded to return to Columbia University as a faculty member, not of the psychology department, but of Teachers College.

He had always been an ambitious and highly competitive man. The success of his thesis simply increased his confidence. 'I've decided to get to the top of the psychology heap in five years, teach ten more and then quit', he announced to his fiancée at the time of his return to New York. Later, just before their marriage, he wrote: 'I can go ahead and do something in the world now and you will find looking after me and the world of science lots more worthwhile than anything else you could do'.[1] The route to the top appeared to be via educational psychology, and not by way of further research on animals. In the United States school teaching was being turned into a professional activity; teachers were to be viewed as people possessing special skills and methods that had been scientifically proven. The nature of training courses for teachers was being changed and Teachers College at Columbia attempted to lead these reforms. Thorndike became deeply involved in a wide variety of topics related to education.

For a short period Thorndike decided to work with monkeys again. At the time of writing his thesis Thorndike believed that monkeys and apes were capable of learning in a more sophisticated fashion than other animals and that it should be possible to demonstrate that they possessed at least a rudimentary form of the human capacity for 'free ideas'. Unfortunately the monkey he had kept as a student had been hopeless. He now acquired three Cebus monkeys – the species, as he explained, commonly associated with organ grinders – and adapted his puzzle box methods in a suitable fashion. This time he found it more convenient to place the food inside the box and leave the monkey outside. The learning curves he obtained were in general much steeper than those from cats and dogs and often displayed the abrupt decreases in latency that had been rare before. He was unsure how to interpret these findings. In his thesis research the gradual slope of the learning curves had implied for him the absence of reasoning and had seemed to 'represent the wearing smooth of a path in the brain'. The abrupt curves obtained from monkeys might therefore indicate the presence of reasoning. Thorndike preferred to see them as indicating the rapid establishment of S-R connections in the monkey brain.

He also repeated with the monkeys the tests of imitation and passive learning. Occasionally there were what seemed to be positive results, but the monkeys were difficult to work with and Thorndike was dissatisfied with the whole project. In the report on this work his comment that 'the results are as a whole on their face value a trifle ambiguous' was an understatement. Exasperated by the three monkeys he decided nonetheless that 'given ten or twenty monkeys that can be handled without difficulty and it could be settled within a month'.[2] It turned out to be a much more lengthy business than he anticipated to settle the issues that he had raised.

Over thirty years passed before Thorndike became directly involved again in animal psychology. A student who was to continue the work with cats and dogs never did so. The puzzle boxes were put on the shelf and no laboratory for the study of animal behaviour was established at Columbia University for many years.

Even if the results of these experiments with monkeys had been more encouraging, it is doubtful whether Thorndike would have continued with such research. He continued to carry out some experimental work with human subjects. One study, performed in collaboration with a colleague, was a particularly influential one for educational policy in the United States. They trained people extensively on one particular kind of learning task and tested to see whether this produced any beneficial effect when a second, very different, task had to be learned. No such beneficial effect was found. This negative result was seen as counting against the belief that a thorough education in some subject such as mathematics or classical languages builds up intellectual muscles which can then cope with any other kind of learning.[3]

Most of Thorndike's time, once lectures were out of the way, was spent on writing textbooks, which were largely based on these lectures. Also, under Cattell's influence, he became more and more interested in mental measurement and statistics. He published an introductory textbook on the subject in 1904 and began to take a direct interest in the development of psychological tests. Increasing numbers of his students needed, or at least wanted, to gain expertise in this area for the sake of their future profession as school administrators. As with Cattell and Galton, Thorndike's interest in testing went hand in hand with an increasingly hereditarian attitude, which was reflected in his undertaking the first major study of twins, and with a highly respectful attitude towards mathematics. As illustrated in the following quotation, this respectfulness is in striking contrast to the earlier iconoclasm of his thesis: 'Tables of correlation seem dull, dry, unimpressive things besides the insight of poets and proverb-makers – but only to those who miss their meaning. In the end they will contribute tenfold more to man's mastery of himself. History records no career, war or revolution that can compare in significance with the fact that the correlation between intellect and morality is approximately 0.3, a fact to which perhaps a fourth of the world's progress is due.'[4] This confidence about applying numbers to concepts such as morality and progress, as well as intellect, recalls that of Galton.

The new interest in mental measurement and in the influence of heredity was not as inconsistent with his earlier work on learning as it might at first seem. In 1898 he lay emphasis on the gulf between the animal mind and the human mind. The only forms of animal intelligence, and the only process by which experience could modify an animal's behaviour, consisted of the development of S-R connections. At best one could find slight traces of 'free ideas' in the animal kingdom.

In contrast, it was quite clear to Thorndike that, by the age of three, children were capable of reasoning. In his thesis he compared the way his subjects learned to open the doors of puzzle boxes with the development of human skills such as swimming, playing tennis or billiards, or juggling, and suggested that S-R connections might form the substrate of human mental life. However, in general, since human intelligence was seen as very different from animal intelligence, there was no reason to study it in anything like the same manner or from the same theoretical standpoint.

As the years passed he began to show a renewed interest in the theory of S-R connections. The report on the experiments with monkeys of 1901 included the argument that current views of human psychology be turned upside down and that the study of habit formation should become the central issue. Only then would psychological theories begin to be explanatory and would it become possible to relate them to the physiology of the brain 'without arousing a sneer from the logician or a grin from the neurologist'.[5] Various studies of rote learning by human subjects that were carried out for reasons to do with educational issues appear to have convinced him that the principles of S-R learning were of much greater importance for understanding the human mind than he had suggested in 1898.

It was perhaps this growing conviction that led Thorndike in 1911 to publish yet a further book. In *Animal Intelligence* he put together the various papers on animal psychology that he had written over ten years earlier and added two fresh chapters. There was no strong tradition in this period for authors of books bearing titles such as 'Animal Intelligence' or 'Comparative Psychology' to feel obliged to present a scholarly review of evidence, or of theories, on the topic designated by the title. Instead they tended to air their own viewpoint and describe in detail only the relevant evidence that they themselves had collected, mentioning in passing or in footnotes evidence collected by others, and in detail only rival theories that they wished to rebut. The books by Morgan, and those by Hobhouse and by Koehler discussed in a later chapter, were like this. In the case of *Animal Intelligence* Thorndike did not even include footnotes to indicate that there had been several experiments reported in the intervening period, which, as he had originally invited, were concerned with the issues raised by his 1898 paper.

The tone of the established professor is much more dogmatic than that of the anti-establishment young man. *Animal Intelligence* is notable chiefly for its restatement in one of the new chapters of what up to now has been called here the 'Spencer–Bain' principle and for the use of a new label, which the principle has retained ever since: the Law of Effect. In a section entitled 'Provisional laws of acquired behaviour and learning' he stated two primary laws of psychology. The first, familiar for some time in discussions on habit in human psychology, was termed the Law of Exercise. The second was the Law of Effect and is worth quoting here in full:

'Of several responses made to the same situation, those which are accompanied or closely followed by satisfaction to the animal will, other things being equal, be more firmly connected with the situation, so that, when it recurs, they will be more likely to recur; those which are accompanied or closely followed by discomfort to the animal will, other things being equal, have their connections with that situation weakened, so that, when it recurs, they will be less likely to occur. The greater the satisfaction or discomfort, the greater the strengthening or weakening of the bond.'[6]

Although very similar in many ways to the statements made by Spencer and then by Morgan, which were described in the previous chapter, it deserves some comment.

Morgan had distinguished between a set of empirical observations, which he had termed 'trial-and-error' learning with accidental success, and possible explanations of these observations, for which he suggested there might be two different kinds of associative learning involved. In his statement of the Law of Effect, Thorndike obscured this distinction by mixing together a number of empirical generalizations and a theoretical statement about the basis of these generalizations. Regarded as a set of empirical laws, it contains three parts. The positive part – that responses become more probable in a situation if they are immediately followed by a satisfying event, or *positive reinforcement*, in that situation – was supported by a great deal of evidence by 1911. The negative part, describing the outcome of what became known as a *punishment* procedure – that responses become less probable if they are immediately followed by a discomforting, or *aversive*, event – could have been based only on Morgan's work with cinnabar caterpillars and the research by Yerkes, described in a later chapter, which had been carried out shortly before Thorndike published his book. It is not clear on what he based the third empirical claim, that expressed in the final sentence of the statement: the effect of the magnitude of reinforcement on learning was not seriously investigated until very much later.

The theory of learning contained within Thorn-

Fig. 3.10. Edward Thorndike in 1910, just prior to christening the Law of Effect

dike's statement is the same as that provisionally proposed in his 1898 paper: the formation of S-R connections which are strengthened or weakened by immediately subsequent events. What had started as an explanation for the manner in which animals learned to escape from his puzzle boxes had by now become a general law of behaviour. In a way that was entirely inconsistent with his earlier emphasis on the gulf between the human and animal mind, he now insisted that human behaviour was also to be analysed in terms of S-R connections alone. In 1911 he stated his belief that 'the higher animals, including man, manifest no behaviour beyond expectation from the laws of instinct, exercise, and effect; the human mind [does] no more than connect in accord with original bonds, use and disuse, and the satisfaction and discomfort resulting to the neurones'.[7]

This is an extraordinary claim. A clue to one of the important reasons it was made is provided by the reference to 'neurones'. Thorndike wanted psychology to be respected by other sciences. This seemed likely only if psychological theories were couched in neurological terms. He took from James the belief in the huge importance of habits, but rejected – or, in fact, simply ignored – all those aspects of the human mind that James considered to be irreducible to some physiological process. For the more complex aspects of human psychology it was difficult to see even what form an explanatory theory could take. At least for the study of habit there existed a concrete model which seemed to provide a suitable basis for theorizing. Lurking in Thorndike's theories is a human invention which is never directly mentioned. The telephone had been invented just after Thorndike's birth and by the time he began to study psychology its use was becoming widespread in North America. The first person to draw an analogy between the brain and a telephone exchange was Karl Pearson in a book, the *Grammar of Science*, of 1892, which Thorndike read as a student.[8]

Thorndike's theories very strongly suggest a view of the brain as an exchange in which lines are connected and disconnected, not by some internal homunculus, but by some process analogous to Darwin's theory of natural selection. Thorndike wrote of 'the struggle for existence among neurone connections' and, following the recent discoveries in neurology of the synaptic junctions between nerve cells, speculated that the physiological basis of S-R connections might be changes in conductivity of individual synapses. This conceptual scheme had the added appeal of explaining why habits have the characteristics of innate reflexes; the then current conception of the reflex arc was that of a relatively direct connection between an input stimulus and an output response and so S-R theory provided added meaning to the old observation that habits become 'second nature'. In contrast, Thorndike's critics, and also the later opponents of S-R theories, were unable to support their claims that many learning processes were more complex by producing alternative models of any substance or by suggesting a generally plausible neurophysiological basis. Only with the advent of the computer a half century later was there a human artefact that seemed more suitable an analogy of the human brain than the telephone.

The idea that the human brain consists of myriad S-R bonds of various and varying strengths, and little else, became widely accepted by American psychologists within the first quarter of the twentieth century. Thorndike's further claim in the Law of Effect that the strengths of such bonds vary according to the hedonic value of events following the response had little immediate influence. S-R theories became very popular, but for many years they did not incorporate Thorndike's reinforcement principle.

Although the idea of an S-R bond might be acceptably scientific, the behaviourists who followed Thorndike could not see that it made sense to imply that nerve cells experience satisfaction and discomfort. How do we know whether a particular event is satisfying to an animal? It seemed that either one had to decide this by extrapolating from human experience in the fashion of Romanes, or conclude that the Law of Effect is circular. In Thorndike's statement, and in all 'weak' versions of the Law of Effect, the meaning of 'satisfying event', or of 'reinforcement', is given by the Law itself. Such weak versions appear to be empty of any explanatory or predictive power and thus merely tautologous: a response increases in probability if it is immediately followed by positive reinforcement, and positive reinforcement is an event that increases the probability of a response that immediately precedes it.

It was some years before a clear argument was presented that the circularity of the weak law does not necessarily imply that it is empty.[9] The circularity is the same as that of, say, Newton's second law, *Force* = *Mass* × *Acceleration*, where the meaning of 'Mass' obtains from the law and has no external reference. What gave empirical substance to Newton's law was the assumption that the mass of an object remains constant in any situation. In a similar manner the substance of a weak version of the Law of Effect comes from the assumed 'trans-situational' property of reinforcement, whereby, if some event is found to be an effective reinforcer in one situation, the same event is assumed to act as a reinforcer when it follows a different event in some other situation.[10]

This appears to be how Thorndike understood the matter – he had provided evidence for the trans-situational assumption by demonstrating the effectiveness of a given reinforcer with a wide range of responses – but he never clearly answered his critics with such an argument. He was very explicit about the improbability of a strong version of the Law of Effect, one in which, for example, reinforcement was defined in terms of events of direct biological significance to the animal. Following Spencer, Thorndike argued that, through evolution, reinforcers would in general be events favourable or unfavourable to the life of the individual or of the species, but that, for example, there are obvious examples of positive reinforcement which are injurious and, possibly, beneficial events which do not function as reinforcement. Whether or not an event is reinforcing to an animal in a given state is to be determined only by behavioural criteria.

The first major statement of the behaviourist position was made only two years after *Animal Intelligence* was published. In a number of ways Thorndike was obviously the behaviourists' most important predecessor. His influence persisted throughout the long era, when psychology was dominated by a conceptual approach based on S-R associations, in a major way that has been generally recognized and also in more minor ones that are less familiar. His comments on language learning provided the first statement of an important attempt to understand the acquisition of language by a child.[11] When in the mid-1960s the Gardners set out for the first time to teach a chimpanzee sign language, they initially strongly resisted the suggestion that tuition based on passive movement of the animal's hands be employed. The reason appears to come from Thorndike's failure seventy years earlier to obtain any positive results from using this method. The Gardners subsequently found it to be highly effective.[12]

Nevertheless the early behaviourists did not consider Thorndike to be a fellow spirit. To a large degree they found the same faults in him that he had found in Morgan. Although Thorndike had no sympathy for Morgan's belief that the single ultimate goal of psychology was the analysis of human consciousness, he was content that psychology should be the study both of behaviour and of subjective experience: as a matter of personal taste he was happier in studying behaviour. He had no hesitation in deciding whether or not his animal subjects were aware of some stimulus or of having made some impulsive action. Just as he had pointed at Morgan's discussion of memory and imitation, so his own use of terms like 'attention' was found unacceptably lax by his successors. Allowing the effectiveness of a satisfying event to depend on an animal's level of attention appeared to add another unsatisfactory feature to the Law of Effect.

Following the publication of *Animal Intelligence* Thorndike continued with his work in educational psychology and two years later produced a three volume textbook on the subject. By the 1920s he was widely regarded as the most important American psychologist. According to his own lights, he could be said to have realized his earlier ambition and reached the top of the heap, although it took more than the five years of his original estimate. His career was startlingly successful in financial terms. He must have been the first person to make a fortune from being a psychologist. In 1924, for example, his total income from salary, lectures and, above all, royalties from his books and his mental tests was $68 000.[13]

Thorndike left himself little time to enjoy his wealth. He continued to work long hours and to carry out empirical work throughout an uninterruptedly productive life. He retained the attitudes of his youth: the scientist's scepticism towards accepted opinion and the faith in objective data obtained by using systematic experimental methods. In these respects he was, as he prided himself, 'a scientist'. For Thorndike science consisted of the industrious accumulation of empirical facts, a 'coral reef' – as one critic put it – 'which is built up little bit by little bit by little bit'.[14] The whole enterprise was somehow very organized and business-like in marked contrast, for example, to that of Morgan. The insistence on empirical fact and on straightforward S-R bonds had a solid, no-nonsense air, with the implication that more complex forms of speculation are indulgently aery-faery.

What impressed later animal psychologists was the fertility of Thorndike's ideas. As well as the experimental work described in the previous section, his thesis contained a number of other intriguing ideas and observations, which, however, were not taken any further. It is worth looking briefly at some of those which are both interesting in their own right and which later became central issues in the development of theories of learning.

One such issue is whether all types of behaviour are equally sensitive to reinforcement. The range of responses Thorndike used in the early puzzle box experiments included grooming activities. The door was opened if the cat licked itself, or, in another case, scratched itself. Thorndike found that in some respects these responses were learned in a different manner from those that involved the manipulation of some object in the external environment. He found it curious and stopped using grooming responses. He did not enquire whether there might be classes of responses that are differentially affected by a given form of reinforcement.[15]

Another issue is concerned with how animals learn to discriminate between two similar events. In the first and very informal experiment of this kind, Thorndike trained his cats to climb up some wire netting to receive a fish when he said: 'I must feed those cats', but they received no reward when he said: 'I will not feed those cats'. He later tried out a more complicated form of discrimination training with the monkeys.[16] But the work remained fragmentary.

At one time during his thesis research Thorndike attempted to devise a direct test of whether his animals were capable of forming associations between 'free ideas'; in doing so he came very close to using what subsequently became known as a *sensory pre-conditioning* procedure. For example, suppose we wish to discover whether a dog can learn that a light is always followed by a certain sound, even if neither of these events is related to food reward or anything else of much significance. One solution is to arrange that the animal experiences appropriate pairings of the two events and afterwards to teach the animal to make a specific response to the sound so that we can then test whether the same response is made to the light. Thorndike carried out an experiment like this which was not designed correctly. Although he subsequently realized how it should have been done, he did not try again.[17]

A final example comes from a problem that was first explored in Thorndike's animal studies, and also later in his work with human subjects, that of transfer of training. In some experiments his cats and dogs showed considerable positive transfer. For example, a cat first learned to open the door of one puzzle box by pulling a small loop covered with bluish thread, which hung close to the door. Subsequently the cat was placed in a second box where a much larger loop covered with a black rubber compound hung at the back. The cat learned to open the door of this second box much more rapidly than any cat without previous training. Such positive transfer may be viewed as an instance of what was later termed the principle of *stimulus generalization*: once an animal has learned to make a response to some specific stimulus, then the same response will also tend to occur to similar stimuli. What is more interesting than the observation, is the explanation that Thorndike proposed. This was very different from that adopted a few years later by Pavlov, who investigated this kind of phenonemon much more thoroughly. Thorndike emphasized that these results did not demonstrate that the animal had perceived the similarity of the two situations. He argued that such stimulus generalization, or positive transfer, indicated that the cat failed to discriminate the difference between the two loops and that various details of the first situation had not become associated with the response. The result suggested to Thorndike that the cat's perception of the situation was as ill-defined as the sea to a man who is thrown into it half-asleep.[18]

These four examples alone could have provided the basis for a rich programme of experimental work, but it was a long time before there was very much research within Western psychology on such problems as constraints on response learning or sensory preconditioning. As we have seen, Thorndike's professional situation after he had obtained his Ph.D. did not make it easy for him to continue with animal research. Even if he had really wished to develop the ideas for research contained in his thesis, or inspire students to do so, it would have been very difficult to solve the practical problems that stood in the way. By 1900 psychology had become established in North America to an extent which gave Thorndike a much better opportunity to study animals than that of his British predecessors. Nevertheless the resources for research at the possible disposal of a psychologist of that generation who wished to study animal behaviour – in terms of time, money, housing for animals, equipment, and research assistants – were still negligible.

The study of how people learn and recall lists of words is relatively inexpensive and far less demanding of time and effort than the study of animal behaviour. Add to this advantage a belief that work on human memory is more likely to be of direct help in the solution of the urgent problems of education and it is easy to understand why Thorndike did not produce a solid body of experiments in animal psychology. The surprise that remains is that the first doctoral thesis in psychology to report research on the animal mind should have been so original and so immensely influential.

Oskar Pfungst and Clever Hans

In terms of method Thorndike's work with puzzle boxes marked a considerable improvement over the studies of animal behaviour that have been described in the first two chapters. He used comparatively large numbers of subjects, attempted to standardize the conditions of testing, employed quantitative measures of performance and made systematic comparisons between subjects given different kinds of training. The experiments were impressive and innovative, but not elegant. Compared to the techniques being used at the time in some of the research on human psychology, Thorndike's work was very crude.

As described earlier in this chapter, experimental methods appropriate to psychology were initially devised for the study of perceptual phenomena and were developed primarily in Wundt's laboratory at Leipzig. Subsequently their use was extended to such topics in human psychology as the study of reaction times, decision processes, and memory. With the spread of psychology to other German universities, the pre-eminence of Leipzig was, by the turn of the century, being challenged, especially by the psychology laboratory at Berlin University. It was in Berlin, rather than Leipzig, that the first application of sophisticated methods to the study of animal behaviour was carried out.

Just after 1900 increasingly frequent reports appeared in the Berlin press of a horse which could perform calculations, read and spell, and understand musical intervals. The horse's means of replying to its questioners was to tap with a hoof a number of times or to point with its head towards an appropriate object or card. This attracted intense professional, as well as public, interest. An African explorer, initially sceptical, became convinced that it was no fraud and made the interest respectable. In September 1904 a report was issued by a commission of thirteen men that included a circus manager, professional horse trainers, Dr Heinroth, who was Director of the Berlin Zoo, another zoologist, and Professor Stumpf, Director of the Psychological Insitute at the University of Berlin. The commission's careful examination strongly suggested that the performance of the horse, now widely known as 'Clever Hans', did not depend on intentional help from his questioner. The commission also could not discern, despite most attentive observation, any form of involuntary cueing. The horse was found to answer correctly to a number of relatively unfamiliar questioners, and not just to its owner, Herr von Osten. There were some cases of correct replies when the questioner himself did not know the correct solution or was mistaken about it.

Perhaps Clever Hans would provide an understanding of the apparent missing links of mental evolution. By no means everyone felt as sceptical as Morgan towards Romanes' claim of having taught a chimpanzee to count. Romanes had spent a relatively brief time with the chimpanzee, while von Osten had devoted considerable effort over a period of years to the education of his horse. Professor Stumpf suggested that one of his students, Oskar Pfungst, work with him on the problem. The series of tests that Pfungst devised read, even now, like a textbook illustration of how to apply experimental methods to a psychological problem.

Having first established the reality of the phenomenon by systematically checking that Hans could perform at a high level even in the absence of von Osten, Pfungst carried out properly the crucial test of intermixing trials in which the questioner knew the answer – the 'with knowledge' condition – and trials in which he did not – the 'without knowledge' condition. Hans was found to answer correctly to over ninety per cent of the questions in the 'with knowledge' condition, but at most ten per cent of the questions in the 'without knowledge' condition. Thus, in some man-

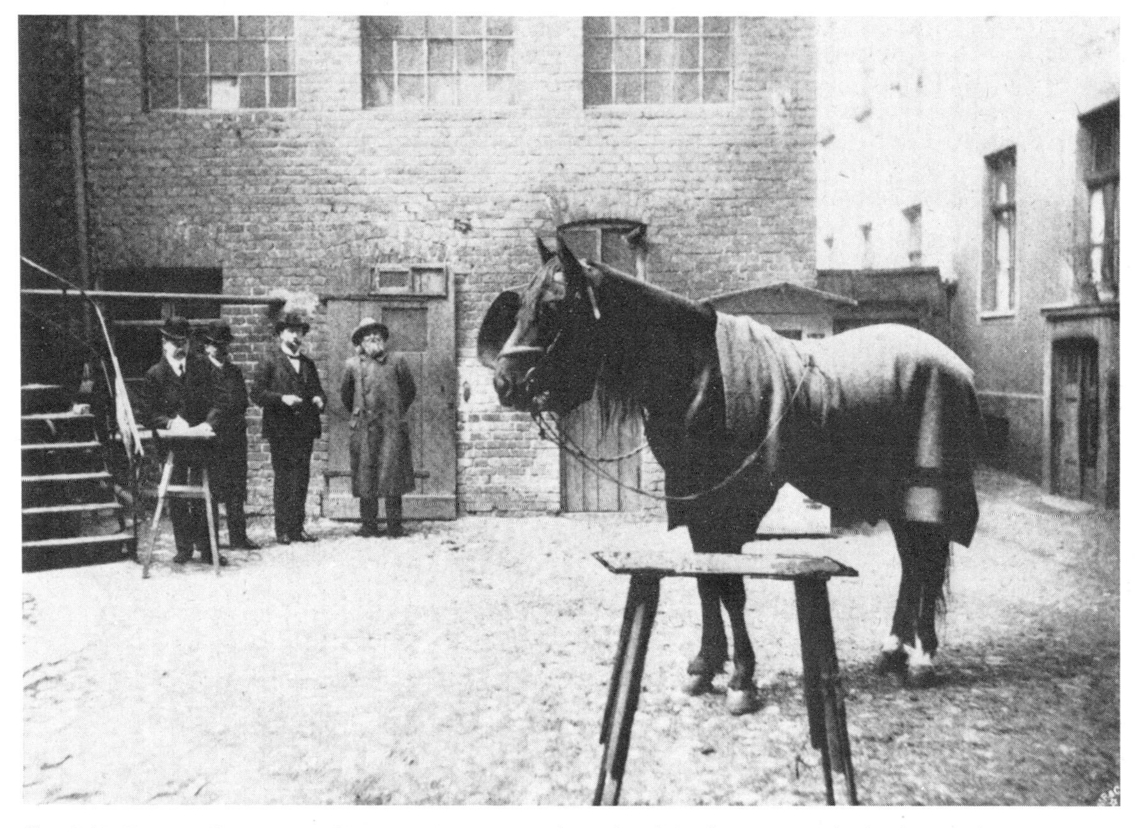

Fig. 3.11. Testing Clever Hans: Herr von Osten is on the right of the four men in the background

ner the horse was gaining information about the right answer from his questioner. From that point it was clear that Hans had no knowledge of German, let alone of mathematics or music. The manner in which his behaviour was guided by signals from the questioner was still a mystery.

The next experiment established that the signals were transmitted while Hans was replying and not while being questioned. Subsequently, it was found that the signals must be primarily visual ones and that auditory stimuli, to von Osten's amazement, were relatively unimportant. At this point Pfungst noticed what appeared to be the crucial events: von Osten would make a barely discriminable downward movement when the horse began tapping and an equally small upward jerk when the correct number had been reached. When this last movement occurred, the horse returned its hoof to the resting position, making what Pfungst called the 'back step' response. Independent timing of the two events showed that the head-jerks almost always preceded the back step response by a small, and relatively constant, interval of about 0.3 s, which was comparable to a person's reaction time in a similar situation.

By making the crucial movements intentionally Pfungst was able to get a more precise idea of the critical stimuli for eliciting a back step. As well as upward movements of the questioner's head or trunk, minimal movements of the eye-brows or even dilation of the nostrils were found to be effective. In a parametric study Pfungst systematically varied the angle of inclination of his body and found that the rate of tapping increased with this angle. Previously the tendency of Hans to tap rapidly when the answer was a large number had been seen as additional evidence of his remarkable intelligence. A further set of experiments showed that Hans' ability to reply to non-numerical questions by moving his head towards an appropriate position was similarly controlled by unintentional and barely discriminable movements on the part of his questioner.

Following this experimental analysis, the case was made conclusive by the demonstration that the effect could be synthesized in a laboratory situation, with Pfungst playing the role of horse. The subjects were of a variety of ages and nationalities. They included a psychologist trained in introspection and a student of psychology named Kurt Koffka. In one study, a subject was simply to think of a number on each trial and then Pfungst would tap with his hand until he saw the kind of signal that he had found Hans to use. The laboratory situation made it possible to monitor the head-movements and breathing rates of subjects and thus obtain objective measurements of the signals. It proved to be a very effective mind reading technique and, just as with the commission of thirteen that had examined Hans, no subject was at all aware of the way in which he signalled the answers. Many were convinced that they were witnessing a demonstration of telepathy.

Herr von Osten, a retired schoolmaster, had begun the training of his horse as a way of demonstrating his own, very environmentalist, belief that the reason why the intelligence of animals appeared to be so inferior to that of human beings is because animals rarely receive an appropriate education. From the account of Hans' training that he provided, it was possible to make a plausible reconstruction of the development of Hans' abilities entirely in terms of reinforcement principles. Unwittingly, it appeared, his trainer had established a chain of stimuli and responses of the kind that both James and Thorndike had described. If, when von Osten leaned forward, Hans began to tap, and subsequently if, when von Osten's back-jerk occurred, Hans then made the back step response, the horse would be rewarded with a piece of carrot or a lump of sugar. The fact that the majority of questioners made the same movements allowed Hans to give correct replies to questioners other than von Osten.

As training was continued von Osten no longer gave Hans a reward for each correct reply, but for a gradually decreasing proportion of correct replies. Pfungst did not pay much attention to this. It was over thirty years later before research began on partial reinforcement effects, that is, on the consequences of providing a reward on only a proportion of the occasions on which a response or chain of responses is made. Such studies have shown that, if reward is made gradually more intermittent, an animal may continue to perform some response in a prompt and vigorous manner even when the response is rewarded very infrequently. It seems that this is what happened in von Osten's training programme. That Hans received a reward only occasionally must have added to the mystification felt by the commission.

There was one final puzzle that Pfungst managed to solve with his laboratory experiments. The solution had very important implications, which Pfungst himself did not emphasize. These experiments appear to be the first direct study of what later came to be termed 'conditioning without awareness'.

Hans' way of providing the answers 'No' or 'Zero' was to move his head first to the left and then to the right. The stimuli controlling this response pattern were not the small 'natural' movements that most unfamiliar questioners would make when expecting such an answer. It appeared that with experienced questioners of Hans the natural movements had been displaced by others that were effective in controlling the horse's reaction. Pfungst tried to discover if a similar phenomenon could be reproduced in an experimental situation. Giving a fictitious reason for the procedure, he instructed a subject to think of 'left' and 'right' in any order, while Pfungst tried to guess the content of the subject's mind, without uttering a word. If Pfungst thought that the subject was thinking 'right', then Pfungst *lowered* his arm; if 'left', then Pfungst *raised* his arm.

For the first few trials Pfungst could base his guesses on whether the subject was thinking of 'left' or 'right' by whether the subject's eyes made a slight movement to left or to right. After six or seven trials the subject's sideways eye-movements began to be replaced by a regular upward eye-movement at the thought of 'left' and a downward one for 'right'. It seemed that the up- or downwards movements that had previously occurred as responses to Pfungst's gestures were now occurring in anticipation of a specific gesture; when the subject was thinking 'left', it was as if he now unconsciously predicted that Pfungst would move his hand upwards and, as a result, his eyes started to move in this direction.

Similar results were obtained from further experiments in which subjects were instructed to think about one or other of a pair of objects. For example, Pfungst might raise his hand if be believed a subject to be thinking of a carriage, and lower his hand for a bowl of fruit. Once again Pfungst could guess with considerable accuracy on the basis of vertical eye movements made by a subject. Afterwards the people taking part in these experiments reported that they had simply tried to imagine the objects; they claimed that they never thought about the related arm movement of the experimenter and were completely unaware of the changes that had occurred in their own behaviour.

The study of Clever Hans demonstrated that a highly complex performance could be the unwitting results of the conditioning of specific responses to

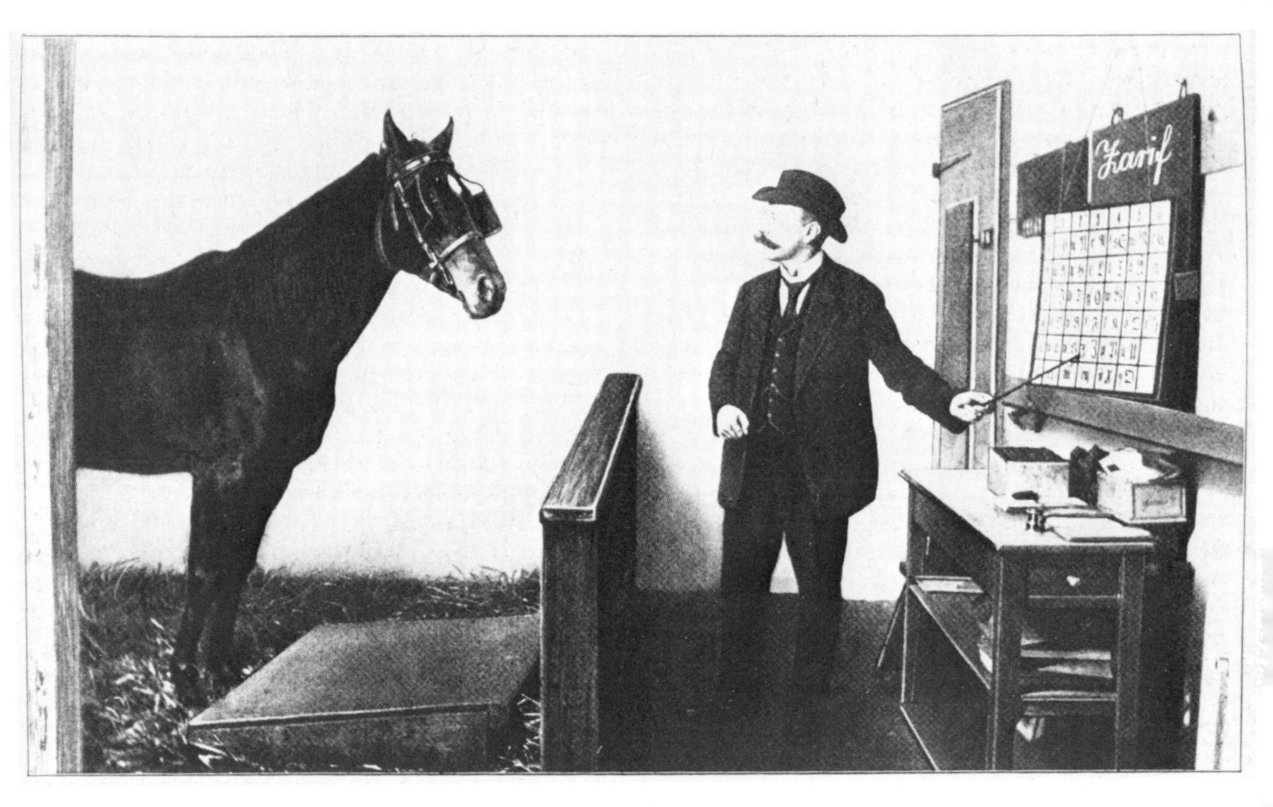

Fig. 3.12 An admirer of von Osten, Karl Krall, who remained unconvinced by Pfungst's evidence, training a successor to Clever Hans

stimuli, which even the most critical and observant among Hans' questioners were unaware of producing. It also showed that the human interrogators' own behaviour was conditioned by that of the horse, again without any realization on their part. Pfungst's series of analytic and synthetic experiments provided graphic support for Morgan's speculations about the counting abilities of chimpanzees and for comments Thorndike had made in passing on a counting horse at a circus, whose performance had been nowhere as impressive as that of Clever Hans.

The intrinsic interest of the case, the elegance of Pfungst's work and the obvious implications for claims supporting the existence of various psychic phenomena attracted wide interest to the report that Stumpf and Pfungst issued in December 1904. The report was very favourably reviewed by a young instructor at the University of Chicago, John Watson.[1] This was followed by publication of the first English translation in 1911, which inspired a number of investigations of mind reading in the United States. One of the first of these studies included, as a collaborator, Edward Tolman, another psychologist who was to contribute to the development of animal psychology.

Concluding discussion

As psychology developed within the German university system it acquired many of the features that had already become common throughout the German scientific tradition and that were particularly strong in physiology. A highly empirical outlook, which stressed the value of the results from well-controlled experiments and viewed other sorts of evidence with suspicion, had its origins in the conflict between laboratory science and idealist philosophy earlier in the century. It was seen later in the unreceptive attitude towards Darwinian theory. Another feature was the deliberate emphasis on pure science, a disinclination to study problems because of their practical relevance or their bearing on everyday life. After all, chemistry had not sought to be an applied science and yet the search after knowledge for its own sake in this subject had happened to convey enormous advantages to the German chemical industry; arguably more so than if the university chemists had aimed directly, say, at the goal of synthesizing a range of dyes.

It is curious that the two Americans who first publicized the new psychology and initiated a similar growth within their own university system had little sympathy for these aspects of German science. Both G. Stanley Hall and William James were enthusiastic evolutionists, liked theory and were impatient with the drudgery and detail of experimental work. James somehow managed to convey both an exciting account of the experimental approach in psychology and a disdainful attitude to what he viewed as an obsession with experimental method; he would have enjoyed the comparison, suggested later, between the method-centred scientist and a drunk searching for a lost wallet, not where he left it, but under the street lamp, because the light was better there.[1] James was very much in favour of the development of psychology as a natural science, but did not believe that its scope should be restricted to topics in which experiments were viable.

Hall paid lip-service to the virtues of laboratory work and for a while was marginally involved in the research he had helped to launch at Johns Hopkins. But when, a few years later, he criticized James' *Principles* for its disrespectful attitude towards experimental methodology, this was pure hypocrisy. Hall himself had already abandoned such an approach since he could find no way of fulfilling the promise he had made that the new psychology would be useful, unless it left the laboratory and tolerated the deployment of evidence of a much more ambiguous nature.

America encouraged the growth of experimental psychology, but did not expect it to remain a purely academic exercise. Spencer's call for the development of social sciences was based on the promise that they would make a major contribution to solving the ills of society. Obvious areas in which it seemed psychology could be usefully employed were those of child development and of education. The problem was that psychological research of that time still seemed far away from spinning off a technology in the way that chemistry had done a generation earlier.

When he published the *Principles* in 1890 James had not yet become interested in the application of psychology to problems of society. He was, however, very much concerned with its pertinence to the more private aspects of life. As we have seen, his emphasis on the force of habit stemmed largely from the moral considerations that this appeared to involve.

Thorndike's work represents the point in psychology when the German physiological tradition and the Anglo-American mixture of associationist philosophy and evolutionary thought began to combine. The questions addressed by his thesis work were all grounded in evolutionary theory, reflecting the general background provided by Spencer and the more immediate stimulus of James, Romanes and Morgan. Yet Thorndike's sceptical attitude and refusal to accept anything but quantitative experimental evidence as contributing towards adequate answers to these questions was more in the German tradition.

Although Thorndike's choice of a thesis topic was in no way affected by any consideration of possible practical consequences, once he had obtained his doctorate there were strong pressures upon him to do work of an applied kind. Together with the lack of suitable facilities for animal work, these pressures, and enticements, led him to confine his subsequent research to human psychology. Unlike Hall, Thorndike believed that psychology could both remain a laboratory-based science and make direct and immediate contributions to the world outside. This belief could be maintained only by assuming that the really important factors in, for example, the functioning of a school classroom were the same, straightforward ones that could be manipulated within a psychological laboratory.

There were two further ways in which Thorndike's psychology seems distinctly un-German. One is its quasi-religious tone. Many young Americans began to hope that psychology might provide answers to practical questions because they had a decreasing faith that religion could do so. Evolutionary theory had weakened traditional beliefs and the authority of traditional rules of conduct. For some, science became a substitute for religion. Thorndike provides a particularly good example of a man whose system of beliefs changed from traditional Christianity to empirical science, while most other aspects of his outlook on life remained unchanged. There was a very strong puritanism in his insistence on clean experimental method and the identification of science with hard work. His attacks on Romanes and the anecdotalist school were like a distorted reflection of the conflict between the Methodism of Thorndike's father and the Catholic encroachment into New England.

The other un-German characteristic of Thorndike's work is its stress on action. He paid little attention to the many complex and subtle mental processes discussed in James' *Principles*, but was enormously impressed by the importance James attached to habit and the moral virtues attributed to action, so unexpectedly from a writer leading an exclusively cerebral and sedentary life. One major reason for the appeal and long influence of Thorndike's 'S-R' theory of learning was that it provided a simple solution to what later became known as the 'performance problem': if, as everyone up until Thorndike assumed, an animal can learn an association between a response and its outcome, one still needs to explain how such learning is translated into performance. No such problem exists if, as in Thorndike's theory, the only function of a reward is to 'stamp in' a connection between an *S* and an *R*, since the occurrence of the appropriate stimulus or situation directly calls up the connected response.

With Thorndike the purpose, as well as the methods, for studying animal psychology began to change. His work as a student in the 1890s was inspired by the same general questions about the evolution of the human mind that Spencer and Romanes had discussed. Because of the unexpected failure to find much sign of imitation or passive learning among the cats, dogs and monkeys that he tested, by the time his thesis was completed Thorndike suggested an even wider separation between human and animal intelligence than Wundt or Morgan had proposed. The implication was that such research was unlikely to enlarge our understanding of mental evolution. By 1911, when Thorndike came to publish a collection of his old papers as a book, he had lowered his estimation of human intelligence and become even more convinced of the importance of habits in human affairs. Consequently the ways in which his cats and dogs had learned to escape from his puzzle boxes over ten years earlier now seemed of considerable interest to the study of human learning.

Animal psychology, that began as an attempt to understand mental evolution and was changed by Morgan into a de-mystification of the animal mind, now became a way of removing the mystery from the human mind. If Thorndike was at all right, the experimental study of principles of reinforcement should enable psychology to deliver the practical benefits which were expected of it; in particular to help teachers to teach more effectively.

The results that came from Pfungst's study of Clever Hans were consistent with Thorndike's claim. There were two main messages. The obvious, and well-remembered, one was that apparently complex behaviour may have a very simple basis – and that it may sometimes require a great deal of careful observation to show that this is so. The much less obvious one was that a human being's behaviour may change in a manner of which he is completely unaware: introspective reports of conscious experience may be irrelevant, or may even hinder, an understanding of how habits are acquired.

> Our study of psychology must be conducted according to the fundamental principle of every young branch of science, namely, to proceed from the simple to the complex . . . The psychical phenomena of animals, and not those of men, should be used as the primary material for studying psychical phenomena . . . Physiology will begin with a detailed study of the simpler aspects of psychical life and will not rush at once into the sphere of the highest psychological phenomena. Its progress will thereby lose in rapidity, but it will gain in reliability.
>
> Ivan Sechenov: *Who must Investigate the Problems of Psychology, and How* (1871)

4 Reflex action and the nervous system

The theories and research discussed in previous chapters took very little account of the nervous system. The subject of this chapter is a scientific tradition concerned with understanding how the human body functions, and some of the attempts within this tradition to explain the actions of living creatures in terms of neural processes. Towards the end of the nineteenth century experiments on learning arose from the study of the neural control of certain bodily functions. The general approach of those carrying out this research was different in many respects from that of Thorndike and his predecessors such as Morgan. In comparison with that of the evolutionists, this tradition had been aggressively experimental for many decades, was entirely concerned with mechanism and immediate causes, and was supported by an institutional structure whose main justification was to provide training for medical practitioners. It took the reflex as its basic conceptual unit for understanding behaviour.

Until the beginning of the modern era in science the vast majority of Europeans, and probably most members of all other cultures, explained many aspects of the world around them in an animistic way. The kind of sensitivity and desires experienced by human beings were attributed to all things that move, non-living as well as living. Even in the late 1500s when magnetism, for example, was first studied, it was common to describe magnets as moving together in 'voluntary union' and terms like 'attraction' and 'repulsion' were not intended merely as colourful metaphors.

Early in the seventeenth century this general outlook was challenged by a new, mechanistic view of the world. This first sought to abolish animistic explanations of inorganic phenomena and to substitute for them the search for universal laws of nature whose application to specific issues could then be worked out in detail. It was a momentous departure to apply this new world view to living systems. A major event in this development was Harvey's discovery of the circulation of blood; principles of mechanics and hydraulics obtained from the study of inorganic matter and of man's own artefacts were now successfully applied to understanding so symbolic an organ as the heart. Rene Descartes (1596–1650) was one of the people who appreciated the full significance of such discoveries and became a leading proponent of the view that an understanding of nature would come only through application of the laws of physics.

Descartes has had an enormous influence on our views of the mind in two different ways that ultimately came into conflict. One was the absolute distinction he made between the human mind and the body; his doctrine of dualism claimed that the body alone was subject to the natural laws of matter. With this went a second, and equally clear-cut distinction: between the activities of human beings, which could be guided by an immaterial mind, and the activities of animals which, being mindless, could be explained in terms of principles gained from the study of the non-living world. Descartes also suggested a general approach to provide such an explanation, one that has become known as the concept of reflex action.

The influence of Descartes' ideas on animal movement was for a long time largely confined to the study of physiology. Although the specific suggestions he made about the way that nerve fibres function and the nervous system is organized were later rejected, the general idea of reflex action became incorporated within an increasingly experimental tradition of neurophysiology.

There were developments of Descartes' general views. A notable one was the extension by Julien

Offray de la Mettrie (1709–1751) of the Cartesian view of animals to man himself, in urging an entirely physiological account of the human mind. Another consisted of an attempt to understand the learned behaviour of animals in terms of nervous processes; detailed suggestions on this were made by David Hartley (1705–1757) in a work that was the first to link in a systematic manner philosophers' ideas on associations with what was known about the brain.

Until well into the nineteenth century, the work of neurophysiologists did very little to change general views of the mind in the directions pointed out by la Mettrie and by Hartley. In the late eighteenth and early nineteenth centuries there was a vigorous reaction to Cartesian physiology; a vitalist belief in the need to appeal to non-material causes to explain the properties of living matter became prevalent among many scientists, particularly the *Natürphilosophen* of Germany. A factor in inclining many to reject the reflex as an adequate concept to explain animal movement was the continuing inability to understand the basis of neural activity.

By the middle of the nineteenth century this problem had begun to yield and a vigorous school of German physiologists led the study of the nervous system. As noted in the previous chapter, Hermann von Helmholtz (1821–1894) and his young friends committed themselves to the Cartesian programme of explaining every function of the body in terms of the forces of chemistry and physics. But they did not extend their ideas and attitudes to the study of either animal or human behaviour. This finally occurred only when experimental physiology took root in Russia. The man most responsible for this, Ivan Sechenov (1829–1905), strongly advocated that physiologists take over the job of studying the mind and his theories greatly influenced the next generation of Russian physiologists.

Rene Descartes and the beast-machine

It would be difficult to find any other book that has managed to express so many novel, powerful and profoundly influential ideas in as brief, but beautifully effective, a way as Descartes' *Discourse on Method* of 1637. It is divided into six parts, of which the first four are concerned with the problem of establishing what is true, with the method of doubt, and with arguments for the existence of the human mind, of God and of the physical world. Having divided the world into mind and matter, with mankind representing a mixture of the two, the next question to consider was the position within this scheme of other worldly creatures. The fifth part of the *Discourse* provided the answer to this question.

Fig. 4.1. Rene Descartes

Descartes' basic point was that the actions of animals, unlike those of man, could be explained entirely in terms of physical principles and therefore there was no reason for believing that any animal possessed any kind of immaterial mind or soul. Thus, he began by explaining how various bodily organs work and suggested that any reader familiar with man-made devices would appreciate how bodily movements as a whole could be explained in such terms: 'nor will this appear at all strange to those who are acquainted with the different automata, or moving machines, fabricated by human industry . . . such persons will look upon this body as a machine made by the hands of God, which is incomparably better arranged, and adequate to movements more admirable than in any machine of human invention'.[1] But, say that human invention improved mightily and that it became possible to build automata that appeared, and behaved, just like a particular animal; could we tell the difference? Descartes thought not: if machines were made exactly resembling 'an ape or any other irrational animal, we could have no means of knowing that they were in any respect of a different nature from these animals'.

What about a machine that looked and acted just like a man? In this case Descartes argued that it would be possible to tell the difference: 'if there were machines bearing the image of our bodies, and capable of imitating our actions as far as it is morally possible, there would still remain two most certain tests whereby to know that they were not therefore really men'. These two tests provided the empirical justification for deciding that man, but neither any animal nor the most human-like robot, possesses a mind: one was the use of language and the other the ability to act rationally in a new situation. Meaning by 'language' a use of spoken words, or of other kinds of sign, that allows one individual to convey his thoughts to another, Descartes pointed out that 'even men of the lowest grade of intellect' can speak, whereas there was no evidence for language in any kind of animal, not even in those, like the parrot, capable of vocal mimicry. He wrote very little about the second test, that of adaptive action in a new environment, but simply indicated that he wished to distinguish between 'acting from knowledge' and acting 'solely from a disposition of parts'. Nevertheless, it seems likely that the kind of problem-solving experiment devised early in the twentieth century to assess the intelligence of apes illustrates what Descartes meant by his second test. The absence of even the most fragmentary study of this kind in the 1600s and the near complete ignorance on communication among apes did not lead Descartes to suspend judgement on such matters: failure on these two tests, he asserted, 'proves not only that the brutes have less reason than man, but that they have none at all'.[2]

Descartes gave strong ethical, if not empirical, reasons for refusing to attribute a mind or soul to any animal. To accept continuity between animals and man could lead directly to the belief that the only rational basis for human conduct was that of seeking worldly pleasures. Why should we lead a good life to attain Heaven, if it can also be reached by other creatures who are far less capable of truly virtuous behaviour? For Descartes such a belief was almost as evil as atheism: 'For after the error of those who deny the existence of God . . . there is none that is more powerful in leading feeble minds astray from the straight path of virtue than the supposition that the soul of the brutes is of the same nature with our own; and consequently that after this life we have nothing to hope for or fear, more than flies and ants.'[3]

Another ethical reason for viewing animals as simply machines was that it sanctioned the use of animals in experiments. Descartes is regarded as the founder of the rationalist tradition in analysing the foundations of knowledge because of his method of doubt and his belief in the self-evident truth of certain mathematical statements. Nevertheless, he held a great deal of respect for experimental work and most of his ideas on physiology were based on his own experience from many years of dissecting various bodily organs. Following his discussion on animals in the fifth part of the *Discourse*, the next and final part is concerned with the practical advantages of science and the need for experimentation.

It begins by explaining why, despite the recent condemnation of Galileo for his heretical views and the possibility of valid disagreement with the conclusions reached in the *Discourse*, Descartes had nonetheless decided to make his ideas public. To have kept silent would have been to offend against the principle by which useful knowledge should be made available to all, a principle which was to become a major part of the scientific ethic, but which in the seventeenth century was not commonly held. Thus, for example, within the Hermetic tradition, which had flourished in the 1500s and to which Descartes was bitterly opposed, scientific knowledge was regarded in a comparable manner to the rituals and symbols of a secret society and so to be made accessible only to the initiated. Descartes was very definite about why his ideas should become known: 'I believed that I could not keep them concealed without sinning grievously against the law by which we are bound to promote, as far as in us lies, the general good of mankind. For by them I perceived it to be possible to arrive at knowledge highly useful in life . . . and thus render ourselves the lords and possessors of nature. And this is a result to be desired, not only in order to promote the invention of an infinity of arts, by which we might be enabled to enjoy without any trouble the fruits of the earth, and all its comforts, but also, and especially for the preservation of health, which is without doubt of all the blessings of this life, the first and fundamental one.'[4]

Descartes believed that the medical knowledge of his time was of little practical value. An enormous amount of careful and laborious experimental work would be needed before men became 'lords and possessors of nature' so that they were no longer the helpless victims of disease and ill-health. He knew the difficulties of carrying out satisfactory research and stressed the need, which was only recognized widely two hundred years later, for society to support medical science in a generous fashion if it wished to obtain improvements in the practice of medicine: 'If there existed anyone whom we assuredly knew to be capable of making discoveries of the highest kind, and of the greatest possible utility to the public; and if all other men were therefore eager by all means to assist him in successfully presenting his designs, I do not see that they could do aught else for him beyond contributing to defray the expenses of the experiments that might be necessary; and for the rest, prevent him being deprived of his leisure by the unseasonable interruptions of any one.'[5]

There is a dilemma facing medical research of a kind that does not arise in the physical sciences. As Descartes argued, to decide between equally plausible theories requires careful experimentation before knowledge relevant to human health can advance; and yet very often a direct test requires experiments on human beings that are unacceptable by almost any set of ethical standards. A possible solution in such situations is to take the less direct route of experimenting upon animals. However, if one believes in a continuum of organic life, then the dilemma remains, in that knowledge of a kind most useful for the promotion of human health is likely to be obtained from those animals that most resemble human beings. Descartes' dualism removes this dilemma: if even the most human-like ape is a mere machine, a physiologist need not hesitate to wield his scalpel in whatever way seems most effective in promoting an understanding of how the organs of some animal function. A beast might seem to suffer excruciating pain during a vivisection experiment of the days before anaesthetics, but this would be a false impression; it might well respond in a manner that resembled the human experience of pain, but it did so without experiencing pain, since it had no mind and was not conscious in the way that a man is conscious.

More than three hundred years later Descartes' dualism with respect to man's body and soul still appears to be pervasive, and yet there is deep unease about his dualism with respect to animal and man. Even after repeated experience very few people would feel no more hesitation in cutting up an alive, yet helpless, monkey than in taking apart some mechanical gadget. And yet there are very few who, when seriously ill, are not reassured to learn that the diagnosis is based on extensive medical knowledge, derived as much from research on animals as from clinical experience, and that the medication has been extensively tested on beasts.

The Cartesian reflex

Descartes' opinions concerning the nature of animals as presented in the *Discourse on Method* summarized a more detailed discussion that should have appeared some years earlier. It was to have been included in a book, *The World*, which he had almost completed when he learned of Galileo's treatment by the Inquisition and decided that it would be dangerous to publish. The part of *The World* concerned mainly with human physiology finally appeared only after his death under the title *Treatise on Man*. In this work Descartes suggested a general principle by which both the involuntary actions of man and the entire behaviour of animals might be understood, namely, the principle of the reflex. This idea represented the very first attempt to comprehend in specific, physical terms how it was that living beings moved. More than two centuries passed before there was any development of the general concept.

Animated statues in the gardens of the palace at St Germain served as the inspiration for Descartes' notion of reflex action. These ingenious contrivances worked by hydraulic action, and their movements were triggered, for example, by some inadvertent pressure on a panel concealed in a pathway, which would open a valve, thus releasing a flow of water and causing the statue to move. A bathing Diana would, when approached, hide in the rosebushes and, if the visitor tried to follow, this would cause Neptune to appear, threatening with his trident.

Instead of pipes and water, Descartes suggested that in the body nerves and 'animal spirits' fulfil comparable functions. As these spirits 'enter more or less into this or that nerve they have the power of changing the form of the muscle into which the nerve is inserted and by this means making the limb move. You may have seen in the grottoes and fountains which are in our royal gardens that the simple force with which the water moves in issuing from its source is sufficient to put into motion various machines . . . And, indeed, one may very well compare the nerves of the machine which I am describing with the tubes of the machines of these fountains.'[1] As illustrated in the famous diagram of the kneeling man, shown in Figure 4.2, Descartes' idea was that when some sensory organ is excited by some external stimulus – as the skin receptors in the foot, B, by heat from the fire, A – delicate threads contained within the nerves are moved. The threads are attached to valves within the brain and such movement leads to the release of animal spirits, which pass through the nerves to appropriate muscles.

In withdrawing a hand from the fire or in fleeing from some frightening sight the body performs, speedily and automatically, actions which the mind could well have chosen. A further example of reflex action provided by Descartes is one where the body's reaction is not consistent with the mind's understanding of the situation: 'If someone suddenly places his hand before our eyes, as if to strike us, even though we know him to be our friend, that he does it only in jest,

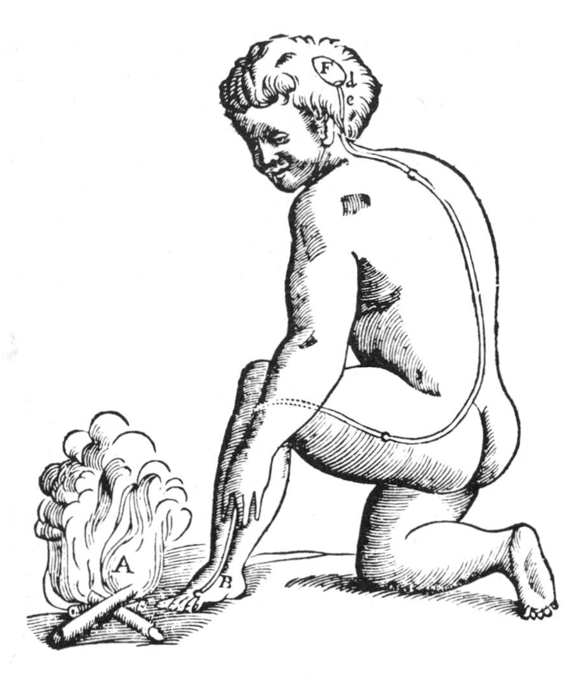

Fig. 4.2. The diagram of the kneeling man, used by Descartes to illustrate his idea of reflex action

and that he will take care to do us no harm, we always have difficulty in not closing them.'[2]

The term 'reflex' derives from the idea that animal spirits can be 'reflected' within the brain in a way that is directly analogous with the reflection of light or of waves on the surface of a liquid.[3] Descartes proposed that the interaction between mind and body was based on the mind's modification of the way that animal spirits are reflected in the pineal gland. Human voluntary action is produced when this gland deflects the animal spirits, or 'very active and pure fire', produced when certain particles of the blood reach the brain, into particular nerve channels. The example of the kneeling man is a case where little 'reflection', in Descartes' terms, takes place. But other examples of involuntary action that he gave involved vision and in these cases a reaction occurs when the animal spirits are reflected by an image formed in the pineal gland; thus a terrifying apparition may cause the spirits to be reflected into muscles that 'dispose the legs for flight' and into those that 'increase or diminish the orifices of the heart'.[4]

Reflex action was a masterly idea, but even among Descartes' contemporaries there were many who suspected the specific assumptions he made about the nervous system in order to give substance to the idea. Descartes' belief that muscular action consisted of the inflation of a muscle by spirits entering from the nerves that joined to it was a very ancient one; in his own lifetime this was shown by experiment to be incorrect. Equally ancient was the tradition that the nerves were hollow, a tradition that stemmed only from the belief that such hollowness was necessary in order to contain the animal spirits. Careful efforts by various Renaissance scientists of the Italian medical schools had failed to find the hollowness and this had become another tradition to be called in question by the time the *Discourse* was published.[5] Furthermore, there had been some progress towards making a distinction between *sensory* nerves, those conveying excitation from receptors to the central nervous system, and *motor* nerves which convey excitation from the central nervous system to the muscles and glands. This, too, was ignored in Descartes' account of reflex action.

There was also the problem posed by decapitation. In this respect the macabre story of what happened to Descartes' corpse is peculiarly fitting. After twenty highly productive years spent in a large number of quiet Dutch towns, after day upon day of peaceful solitude spent mainly in bed, Descartes was persuaded to go to Sweden to instruct young Queen Christina in philosophy. In 1650, within months of his arrival in Stockholm, he died of an illness severely aggravated, it has been said, by having to begin the queen's lessons at five o'clock in the morning. He was buried in the cemetery for distinguished foreigners. Sixteen years later, his body was exhumed, as it had been decided by various friends and disciples that it would be more fitting for his bodily remains to rest in France; perhaps they did not respect as seriously as he might have wished, Descartes' belief in the possibility of a disembodied spirit and the existence of mental processes in the absence of any brain. The French ambassador to Sweden took charge and first cut off Descartes' right forefinger as a personal souvenir. It was then found that the special copper coffin provided for transporting the body was too short. So the neck was severed and skull removed to be shipped separately. The coffin returned safely to Paris and Descartes' headless body was reburied with great pomp. The skull had a more sordid fate: it was stolen by an army captain, passed from one Swedish collector to another, and took 150 years to reach Paris where it was awkwardly shelved in the Academie des Sciences and has apparently remained there ever since.[6]

It was strange that Descartes did not consider the reasonably common knowledge that certain animals may continue to move and to react to some kinds of stimulation for many minutes after their heads have

Fig. 4.3. One of Descartes' diagrams showing the supposed reflection of animal spirits within the pineal gland

been chopped off. Under some conditions animals with their cerebrum removed may show signs of life for some hours. How could this happen if all movement is caused by animal spirits released in the brain ?

He also said almost nothing about how, unlike the automata at St Germain, appropriate movements by animals might be acquired.[7] However, this does not seem to have been considered a central problem either by his immediate followers or by his critics. For the latter, there were ample positive claims about the nervous system to reject without bothering about the lack of any very specific claims on the nature of learning.

Julien Offray de la Mettrie's man-machine

Descartes' views on the beast-machine were not widely accepted. Two major obstacles appeared to exist. One was the problem of perception and whether this can occur without cognition: how can human sensation involve thought, and yet animals, capable of sensation, be incapable of thought? The second was simply reluctance, based on everyday experience, to accept the view that all animals were equally lacking in the power of thought. Various compromise positions were advocated: for example, that animals possess various degrees of cognitive ability, but never any immortality of the soul. Over the hundred years since Descartes' death, there were many men of letters and scientists who were deeply influenced by Descartes' philosophy and yet rejected to some degree or other his views on animals.[1] The least compromising and most notorious of them all was Julien de la Mettrie.

La Mettrie was a Breton, born in 1709, who in the course of an education designed by his father to lead to a career in the church, became passionately fond of poetry and literature. The intervention of a local doctor, who pointed out to the father that even a mediocre physician earned more than the most inspiring village priest, changed the direction of the son's education towards medicine. La Mettrie qualified as a doctor but, before beginning the life of a general practitioner in his native Brittany, spent some time at the University of Leyden in Holland which was the leading medical centre of the day. Here, he studied under Hermann Boerhaave who, although he made no major scientific contributions himself, appears to have been an unusually inspiring teacher.

La Mettrie returned to France with considerable enthusiasm for the scientific study of medicine. His practice provided an income and allowed time to write a number of monographs and publish some translations on medical topics; for example, on venereal disease, vertigo and smallpox. While in his early thirties, he was becoming widely known as an energetic young man with a great deal to say and also, it appears, widely resented since he had a facility for antagonizing people, a taste for satirizing the medical profession, and a very quick temper. In 1742 he left Brittany for Paris and came under the protective wing of an aristocratic patron, the Duc de Gramont, who obtained for La Mettrie a commission as physician to a regiment of guards in time for the war between France and Austria.

La Mettrie's military experience prompted a new interest in metaphysics. At the siege of Freiburg he contracted a violent fever; the resultant disorder of his thoughts and feelings made him meditate during his convalescence on the relationship between mind and body and decide that this was much more intimate than Descartes had believed it to be; as a biographer wrote: 'For a philosopher illness is a school of physiology; he believed that he could clearly see that

Fig. 4.4. Julien Offray de la Mettrie

thought is but a consequence of the organization of the machine, and that the disturbance of the springs has considerable influence on that part of us which the metaphysicians call soul.'[2] Immediately on recovering la Mettrie published a monograph, entitled *The Natural History of the Soul*, to explain his ideas on the mind–body problem. These ideas were seen as heretical and aroused a great deal of hostility.

The next major event in the war was the French victory at Fontenoy, where the Duc de Gramont was killed by a cannon shot. With the loss of his patron la Mettrie was advised to flee from the French army and he returned to Leyden. Here he continued his philosophical speculations. In *The Natural History of the Soul* he had expressed his belief in the continuity between animal and man and rejected the Cartesian idea of animals as machines without souls. Three years later he retained his belief in continuity, but had come to agree with Descartes on the notion of beasts as machines; consequently, it followed that man was also a machine. He expressed this view in an appropriately titled monograph of 1748, *Man a Machine* (*L'homme machine*).

The monograph contained three major themes. The first was the close and direct relationship between physical factors and mental states such as beliefs, judgements and feelings. La Mettrie expressed a very strong version of what may be termed the medical approach to psychological matters. The views of philosophers, who are not at the same time physicians, deserve no respect; the mistaken dualism of a philosopher like Descartes stems from ignoring the kind of evidence that is so overwhelming in clinical practice. In Descartes' system the interaction between mind and body is a one-way affair: the mind influences the body's actions via the pineal gland, but no effects occur in the opposite direction. For la Mettrie such a view makes incomprehensible the everyday effects of wine, coffee, or a good meal – the English, he believed, were literally bloody-minded because they ate meat that was insufficiently cooked – as well as the less frequently experienced effects of the kind of fever he had contracted at Freiburg or of taking a drug like opium. Such effects are usually transient, but la Mettrie believed that more permanent changes in mental states might also have simple, direct physical causes. He mentioned disorders such as hypochondria, delusional beliefs and insomnia. Can a Cartesian system explain how the introduction of some substance into a man's body may lead him to believe that he is a wolf-man or that his nose is made of glass?[3]

Far less compelling is the other kind of evidence la Mettrie cited in support of the view that 'the diverse states of the soul are always correlative with those of the body'. He suggested that philosophers should take note of comparative anatomy and observe how the mental characteristics of different species are related to the size and structure of their brains. However, comparative anatomy of this kind had made little progress by the mid-seventeenth century. From what was then known he simply drew the surprising conclusion that the fiercer animals have less brain,[4] and then moved on to the obvious question raised by such comparisons: given the close physical resemblance between man and ape, why do their mental abilities appear to differ so greatly? His answer provides the second major theme in *Man a Machine*.

'The transition from animals to man is not violent.'[5] The superiority of man over ape comes from education and from the use of language which enables symbolic thought to occur. By 'education' La Mettrie meant a lot more than explicit instruction. He stressed the way that the mind needs constant nourishment from experience or, to use his metaphor, the brain contains muscles for thinking which require exercise as much as the leg muscles for walking. A good mental condition needs to be maintained by good company since 'in the society of the unintelligent, the mind grows rusty for lack of exercise, as at tennis a ball that is served badly is badly returned'. From our social environment 'we catch everything from those with whom we come in contact; their gestures, their accent, etc.; just as the eyelid is instinctively lowered when a blow is foreseen, or as (for the same reason) the body of the spectator mechanically imitates, in spite of himself, all the motions of a good mimic'.[6] Thus, according to la Mettrie, learning, even learning by imitation, could be as involuntary and unconscious as the kind of reflexive movement discussed by Descartes.

Pre-linguistic man differed from the ape only in that he looked wiser; in no sense was he king over the other animals. But once language had been invented, and education ensured that it continued from generation to generation, man's pre-eminence was established. It followed that, despite Descartes' claims to the contrary, an ape should be able to acquire a language and, having done so, become capable of solving problems in a way that would overcome the second barrier erected by Descartes between animals and man. La Mettrie wrote that because of the structural similarities of the ape to man he had 'very little doubt that if this animal were properly trained he might at last be taught to pronounce, and consequently to know, a language. Then he would no longer be a wild man, nor a defective man, but would be a perfect man, a little gentleman, with as much matter or muscle as we have for thinking and profiting by his education.'[7]

The third and final theme of *Man a Machine* concerned morality and religious belief. Here again, la Mettrie implied that philosophers who stay in bed and shun society see too little of the seamier side of life. Better acquaintance with cases of cannibalism, infanticide and wars, in which 'our compatriots fight, Swiss against Swiss, brother against brother, recognize each other, and yet capture and kill each other without remorse, because a prince pays for the murder', would shake their confidence that man possesses a unique capacity for distinguishing between good and evil. If only man can feel remorse for his actions, and if remorse is so strong a force, why is the threat of hell-fire needed to deter man from sin? A clear comparison between human behaviour and that of animals shows that 'man is not moulded from a costlier clay; nature has used but one dough, has merely varied the leaven'.[8] As for conventional religious belief, la Mettrie considered the existence of a supreme being to be a 'theoretic truth with little practical value . . . since we may say, after such long experience, that religion does not imply exact honesty, we are authorized by the same reasons to think that atheism does not exclude it'.[9]

Descartes' idea of the beast as a reflex machine had been inspired, as we have seen, by moving statues operated by hydraulic action. These were crude in comparison to later mechanical contrivances produced for the French court. One man in particular, Jacques de Vaucanson, appears to have been a genius in producing life-like figures; his flute-player and a duck that could swim, eat and digest were renowned. Thus, by the time of la Mettrie, such devices had become even more plausible as models of living organisms and, for la Mettrie, of man's behaviour as well.

The image of the clock had become more appealing than ever: 'Man is to the ape, and to the most intelligent animals, as the planetary pendulum of Huyghens is to a watch of Julien Leroy. More instruments, more wheels and more springs were necessary to mark the movements of the planets than to mark or strike the hours; and Vaucanson, who needed more skill for making his flute-player than for making his duck, would have needed still more to make a talking man.'[10] But la Mettrie's view of the world was no cold, cruel cynicism; his mechanistic outlook was combined with an almost mystical respect for nature. He saw belief in the uniqueness of man as a primary source of evil; to admit our ignorance, to accept that we, like animals, are machines, will reunite us with nature and make a degree of happiness possible that is not to be attained in a society based on religion and dualism. 'The materialist, convinced that he is but a machine or an animal, will not maltreat his kind, for he will know too well the nature of these actions, whose humanity is always in proportion to the degree of analogy proved above between human beings and other animals; and, following the natural law given to all animals, he will not wish to do to them what he would not wish them to do to him.'[11]

The climate of opinion in Holland of 1748 was not so enlightened as to tolerate such near-blatant advocacy of atheism. As expressed in a eulogy by la Mettrie's next, and final, patron, Frederick the Great of Prussia: 'Calvinists, Catholics and Lutherans forgot for the time that consubstantiation, free will, mass for the dead, and the infallibility of the Pope, divided them: they all united again to persecute a philosopher who had the additional misfortune of being French, at a time when that monarchy was waging a successful war against their High Powers'.[12] La Mettrie was forced again into flight, this time to Berlin where, as a member of the Royal Academy of Science, he resumed his work on medical topics, writing on dysentery and on asthma. In less than four years he was dead, at the early age of forty-nine. According to the official eulogy, he died of a fever which, like that at the siege of Freiburg, attacked his brain. In private, Frederick the Great admitted the truth of the popular legend that la Mettrie died of indigestion caused by over-indulgence in a pasty of pheasant and truffles. A theological issue of great interest at the time was whether an atheist could die in peace. The circumstances of la Mettrie's death seemed fitting for someone so impious.[13]

La Mettrie was a brave man, fully aware of the danger of writing honestly about what he believed. In introducing a collection of his essays he advised: 'So write as if thou wert alone in the universe, and hadst nothing to fear from the jealousy and prejudices of men, or – thou wilt fail thy end'.[14] He wrote in praise of sensual pleasure and discussed sexual matters in an explicit manner. His publications on such subjects, seen by many as cynically justifying lust; the scandal surrounding his major work – the very title, *Man a Machine*, was enough to offend deeply; the legend of his death; all of these combined to provide him with a monstrous reputation. Although la Mettrie appears to have made a great impression on many writers of the late eighteenth century, his works were almost never cited and his name rarely mentioned.

The ideas on equality and on the importance of an individual's physical and social environment became commonplace among the philosophers of the French Enlightenment. In France they grew to inspire and then decline with the revolution. Before that occurred these ideas had crossed the Atlantic and taken a deep hold in the minds of many young Americans, like the backwoods lawyer, Thomas Jefferson, in Virginia. The way that human skills, knowledge and institutions were beginning to flourish in the novel surroundings provided by the thirteen colonies brought strong conviction to such beliefs. Jefferson had probably never heard of la Mettrie nor read about the system of ethics described in la Mettrie's *Discourse on Happiness*. In Jefferson's *Declaration of Independence* of 1776 the appeal to self-evident truths looks back to Descartes. But, the inclusion as one of those truths that all men are created equal owes more to la Mettrie, as does the claim that one of the unalienable rights which a government has the duty to safeguard is the pursuit of happiness.[15]

Man a Machine seems to have been equally unknown to the English evolutionists who discussed similar issues over a century later. As was described in an earlier chapter, in his Belfast address of 1874, Huxley paid full tribute to Descartes, to the latter's method of doubt and to his hypothesis that animals are automata, but in modifying Descartes' doctrine, Huxley was unaware that almost exactly the same arguments had been made by la Mettrie. Perhaps more striking is the way that la Mettrie's beliefs in the continuity between animal and man and in the power of 'education', plus the conviction that widespread adoption of these beliefs will reduce human misery, anticipate behaviourism; but the early American behaviourists also knew nothing about his work.

The tradition which remembered La Mettrie was that of physiologists concerned to improve upon the picture of the body's machinery which Descartes had proposed.

David Hartley's 'Observations'

For all the emphasis he laid on physiology, on the need for a philosopher of mind to have a good understanding of scientific knowledge, la Mettrie said almost nothing in detail about how the nervous system might work or about what kind of machine man might be. There was compensation for this neglect by two books that appeared shortly after *Man a Machine*.

In 1749, David Hartley published his *Observations on Man, his Frame, his Duty and his Expectations*. It mixed together scientific ardour with religious certainty in a way that was much more common in England than in continental Europe. Hartley was the son of a clergyman and had completed his training as a minister in the Church of England before turning to the study of medicine.[1] Where la Mettrie, the atheist, defended the value of worldly pleasures and saw the search for scientific knowledge as a way of escaping fetters of prejudice imposed by institutional religion, Hartley believed that the deep purpose of attempting to understand natural phenomena was to increase one's faith in God. One of the two main points that Hartley wished to emphasize in the Preface to his book was that its ultimate aim was to deter from wickedness: 'I do most firmly believe, upon the authority of the scriptures, that the future punishment of the wicked will be exceedingly great both in degree and duration . . . And were I able to urge anything upon a profane careless world, which might convince them of the infinite hazard to which they expose themselves, I would not fail to do it'.[2] La Mettrie and Hartley were opposites in general outlook and in personality – Hartley was a calm, complacent man; assured, benevolent and tolerant – and yet their views on the mind were remarkably similar. This is seen very clearly in a chapter Hartley devoted to 'the intellectual faculties of brutes'.[3]

It is remarkable that such a religious man should have found so little difference between man and brutes and should have traced it to entirely natural origins. Hartley rejected as firmly as la Mettrie the deep Cartesian divide between man and animals. Even at a detailed level the similarity in the views of these two contemporaries is striking. Hartley finds five major reasons for the intellectual inferiority of animals. The first is from comparative anatomy and is based on brain size. Anticipating later views much more closely than la Mettrie's odd comments on the relative ferocity of different species, Hartley argued that the smaller brains of animals contain less cerebral matter available for the formation of associations, and also less that is devoted to vision and hearing, the senses crucial to man's unique intellectual development. The second suggested reason is not one that has received subsequent support; Hartley thought that the quality of human nerves might be unusually good and that it appeared 'probable that the texture of the nervous system in brutes should tend more to callosity and fixedness in its dispositions to vibrate, than in men'.

The third reason was again one that la Mettrie had discussed. Hartley thought that animals' 'want of words and such-like symbols' was of general importance, and particularly critical for the intellectual differences between man and ape. However, he did not discuss whether teaching an ape to speak might be possible or whether, if successful, this would transform him into a little gentleman. Hartley probably did not think this likely, since the fourth reason he gave for the intellectual failings of animals was that brutes – including apes, one assumes – are able to do very well by instinct things for which man needs intellect. The final reason was a strong echo of la Mettrie's conviction that a stimulating environment is important for sustaining intellectual abilities; Hartley proposed that, because of their coarser powers of sight and hearing, animals 'converse with far fewer objects than men' and this produces a narrowness of understanding.

Hartley's view on the intellect of brutes make an interesting comparison to those of la Mettrie, but they are not particularly remarkable. He was a key figure in the development of psychology for other reasons. He took the various principles of association discussed by earlier writers and formulated a systematic theory of mind based on the doctrine that all mental activity derives from associations between sensations and ideas. This psychological theory was combined with

Fig. 4.5. David Hartley

an account of how the nervous system works, to produce the first major theoretical step after Descartes. Following an earlier suggestion by Isaac Newton, Hartley supposed that nervous action consisted of vibrations in the Aether contained within nerve fibres, together with vibrations of smaller amplitude, 'vibratiuncles', within the central masses of the brain. Such vibratiuncles represent external events and, if two such events occur close together in time, then the subsequent recurrence of one event will cause central vibrations corresponding to the other.

According to Hartley, three major characteristics mark off the animal from the vegetable world. The first, the capacity for sensation, and the second, that for perceiving ideas, he believed to have been adequately explained by his theory of vibrations. The third characteristic, the 'locomotive faculty' or ability to produce 'muscular motions', should be amenable to explanation along the same lines. He was clearly well acquainted with Descartes' idea of the reflex, and also with subsequent evidence indicating that muscular activity is controlled entirely by the nervous system. Hartley extended the Cartesian reflex to cover human voluntary, as well as involuntary, actions and made explicit Descartes' hint that learned habits – or 'automatic motions of the secondary kind' as Hartley termed them – might have the same physiological basis as an inborn reflex. In doing so he briefly described what was to become known as the conditioned reflex.[4]

Hartley wanted to show how voluntary behaviour could develop from automatic or reflexive behaviour. He gave the example of the grasping reflex in a young child. Initially, a child closes its fists to grip an object only when the object actually makes contact with its palm. Hartley supposed this to occur because of vibrations transmitted along sensory nerves from palm to brain, which then pass to 'motory' nerves and cause the contraction of specific muscles. When a favourite plaything is repeatedly given to the child, a vibratiuncle is generated which corresponds to the sight of the plaything and this becomes linked with the vibratiuncle corresponding to pressure on the palm. Consequently, the child will start to close its fist at the sight of the toy. 'By pursuing the same method of reasoning we may see how, after a sufficient repetititon of the proper associations, the sound of the words *grasp, take, hold*, etc., the sight of the nurse's hand in a state of contraction, the idea of a hand, and particularly of the child's own hand, in that state, and innumerable other associated circumstances, i.e. sensations, ideas, and motions, will put the child upon grasping, till, at last, that idea, or state of mind which we may call the will to grasp, is generated, and sufficiently associated with the action to produce it instantaneously. It is therefore perfectly voluntary in this case.'[5]

Hartley went on to outline how such a process could account for the manner in which we learn to speak. He also claimed that it applied to the development of powerful voluntary control over actions such as 'swallowing, breathing, coughing and expelling the urine and faeces', as well as the 'feeble and imperfect power over sneezing, hiccoughing and vomiting'. He was modest and undogmatic about his analysis of voluntary and involuntary actions. This section of his book ended simply with the conclusion that 'thus, we are enabled to account for all the motions of the human body, upon principles which, tho' they may be fictitious, are, at least, clear and intelligible'.[6]

The *Observations* presented a major theory of mind and the first comprehensive attempt to explain mental events in terms of natural principles. Unfortunately, the book was as ponderous as its full title and it was not widely read. Many of the ideas it contained were presented in an obsessively exhaustive manner; to read the complete work, including the lengthy theological section, took a great deal of determination. As the above quotations may suggest, the style compares very unfavourably with the powerful, elegant way in which Descartes wrote, or with the spontaneity and liveliness of la Mettrie. In *Man a Machine* there is a sense of urgency and transience; a feeling that, unless the work is brisk and to the point, a cannon ball, angry mob or pheasant pasty might prevent its completion. Hartley's presentation of his thoughts was as unexciting as his life; an honest, comfortable life of a Yorkshireman who had early decided to devote his leisure to philosophical enquiry and who, upon a second marriage to a wife with a respectable income, succeeded in finding a good share of leisure in the pleasant town of Bath.

The main impact of Hartley's work came many years later, after James Mill had made his young son study very carefully what to the father was the 'master-production in the philosophy of mind'. The *Observations* made a deep and lasting impression on John Stuart Mill. Later, as a young man developing his own philosophy of mind, with his friends he pored once again over Hartley's detailed claims.[7]

The spinal cord and nervous energy

By the 1750s purely speculative accounts of nervous action evoked little respect from the leaders of the various medical schools of Europe. To an increasing extent a person's views on such matters required support from the authority conferred by personal experience of appropriate experimental work. Hartley's interest in the nervous system did not extend to the testing of his theories. However, at the same time as he was writing the *Observations* in Bath, a physiologist in Edinburgh named Robert Whytt was carrying out a series of experiments that produced the most substantial increase in empirical knowledge of the nervous system since the time of Descartes. Whytt's results were first extensively reported in the *Essay on the Vital and other Involuntary Motions of Animals* of 1751, just two years after Hartley's book. His work belonged to what was now a thriving tradition of physiological studies with bases throughout Western Europe.

As noted earlier, one of the problems ignored in Descartes' description of the nervous system was that movements can occur in animals that have been beheaded. In his account the spinal cord is viewed simply as a large cable conveying nerves to the brain. The movements of decerebrate animals suggested to many that the spinal cord might be more complicated than represented by Descartes. A major problem in neurophysiology until well into the nineteenth century was that of understanding the functions of the spinal cord.

It is interesting that Hartley in his *Observations* already describes connections between 'sensory' and 'motory' nerves occurring in the spinal cord. He was probably familiar with the experiments some twenty years earlier of the Rev. Stephen Hales, one of those amazing eighteenth-century clergymen who must have been the despair of their parishioners. Hales decapitated frogs and found that reactions could still be obtained to various kinds of stimulation as long as the spinal marrow was not destroyed. This finding was confirmed and extended in a careful and systematic series of experiments by Whytt. For example, he showed that in special cases only a small segment of the spinal cord needed to remain intact in order for a specific reflex to function.[1]

Whytt identified a whole number of involuntary functions as being based on reflexive action. These included digestion, coughing, sneezing, erection of the penis, the pupillary reaction and the secretion of saliva. His *Essay* of 1751 also included a discussion of psychological issues that was very close to that of Hartley. Without attempting to tie his ideas on actions into a general theory of mind as Hartley had done, Whytt nonetheless made very similar suggestions about the relationship between innate reflexes, voluntary actions and habits. He was also just as explicit about the possibility of stimuli *acquiring* the ability to elicit a reflex action; using what was to be the highly significant example of salivation: 'Thus, the sight, or even the recalled *idea* of grateful food, causes an uncommon flow of spittle into the mouth of a hungry person; and the seeing of a lemon cut produces the same effect in many people'.[2]

The combination of Hartley's philosophy with its principle of association and the detailed physiological discoveries of Whytt provided a system of considerable power for understanding the activity of animals. However, the full significance of such a combination was not recognized until over a century later when in 1855 Herbert Spencer used the reflex as basic unit in his evolutionary theory of mind, as described in an earlier chapter.[3] The delay was partly due to the unappealing way in which Hartley's ideas were presented; another reason was that the importance of studies like those of Whytt was masked by the continued failure to understand how nerves function. Whytt was convinced that all muscular action was due to a 'power or influence lodged in the brain, spinal marrow and nerves', but he did not understand the nature of this influence. He referred to it as the *vis nervosa* and was very candid about his lack of any plausible ideas about what kind of energy it was.

We have seen that Descartes' view of a nerve was that it contained two major elements: fibres that connected sensory receptors to the brain, and a hollow tube allowing the passage of animal spirits to the muscles. By the end of the eighteenth century this view had been long abandoned and it had become common to distinguish, as Hartley did, between different kinds of peripheral nerves: sensory ones whose specialized function was to convey information to the central nervous system, and motor ones that served primarily to activate the muscles and glands. Together with the idea of reflex action, this distinction suggests what is termed the 'sensori-motor' theory of the nervous system: the principle that every part of the nervous system can be categorized as being either sensory or motor in function.

This sensori-motor view is an ancient one. Although it was held by some physiologists at the beginning of the nineteenth century, it was widely regarded as overly speculative and by no means universally accepted. Then some experiments were carried out which were seen as providing such important support for the sensori-motor view that the findings were elevated to the status of a 'law'. In 1811 Charles Bell in London reported what he considered the most important discovery in the history of neuro-anatomy. It had been known for at least sixty years that where the peripheral nerves, consisting of bundles of hundreds of individual nerve fibres, join the spinal cord they divide in two, so that a double connection is made: for each nerve an anterior and a posterior root. Bell opened up the spine of stunned rabbits and found that when he pricked the anterior roots convulsive movements occurred, but he obtained no effect at all from pricking the posterior roots. He was convinced that the posterior roots contained only sensory nerves, but he had no direct evidence for this. Despite his belief in the importance of this finding, he circulated his report only among friends. As a surgeon and anatomist skilled in the dissection of dead tissue, Bell had little taste for experiments involving vivisection and never pursued this research further.

Eleven years later Francois Magendie undertook the same kind of study and carried it through in a more complete fashion. Magendie succeeded in keeping a six-weeks-old puppy alive following exposure of its spinal cord; after repeating Bell's discovery that stimulation of the anterior roots produced specific movements, he found that severing the anterior roots left intact sensitivity to stimulation of the appropriate part of the skin. He was therefore able to conclude that the posterior roots contained the sensory nerves. This joint discovery was seen as a solid, factual demonstration that the distinction between sensory and motor was a fundamental principle on which the organization of the nervous system was based. It was termed the Bell–Magendie Law.[4]

By the time that Magendie had settled the question of the spinal roots in 1822 a little progress had at last been made in understanding how nerves function. Electrical phenomena had been of great interest throughout the eighteenth century. In 1751, the same year as Whytt's *Essay*, the remarkable Benjamin Franklin produced an account of the experiments and observations he had made in Philadelphia on static electricity, condensers and electrical discharges; his book was at first widely regarded as a hoax – since it was not believed that anyone outside of Western Europe could have any understanding of such matters, let alone display such scientific originality – but then Franklin's explanation of lightning and the introduction of lightning conductors was appreciated as a striking illustration of the experimental route to an understanding and control of even the more frightening and hitherto unpredictable of natural events.[5]

General interest in electricity was also stimulated by theatrical demonstrations of the conduction of electricity over the surface of the human body. By the middle of the century most of the royal courts of Europe had witnessed the way that electricity, generated by friction in an 'electrical machine', could pass from the feet of a page-boy suspended on ropes to activate an electroscope held close to his nose. The kind of arrangement used is shown by the contemporary illustration in Figure 4.6.

Earlier in the century Hales had suggested that electricity might be the elusive form of energy responsible for nervous action. However, there appeared to be incontrovertible reasons for rejecting this idea. The major one was the problem of insulation: it was known that electricity could be conducted from one end of a metal wire to the other, say, only as long as the wire was well insulated along its length. No appropriate insulation seemed to be present in the nervous system. To the contrary, the wet tissues inside a living organism appeared to provide optimal conditions for immediately dissipating any electrical charge that might be generated. Even towards the end of the eighteenth century the grounds for rejecting an electrical hypothesis were seen to be just as strong as those for rejecting any other theory, including Descartes' spirits and Hartley's vibrations. Having turned down all previous suggestions as to the nature of the *vis nervosa*, the physiologists could now find no new theory to take their place. The breakthrough eventually came from the work of a number of scientists in Northern Italy.

By 1780 Luigi Galvani had begun experiments in Bologna with frogs and electrostatic machines which convinced him of the reality of 'animal electricity', the idea that electrical energy was generated within the nervous system of all living creatures. Since none of his published work actually demonstrated this, his conclusions were justifiably challenged by Alessandro Volta of Pavia who, on repeating the experiments, decided that they demonstrated only 'metallic electricity'. The bitter arguments turned out to be enormously productive ones, leading on the one hand to an understanding of bimetallic electricity and the development of Voltaic piles and batteries, and on the other, to the beginning of electrophysiology.[6]

Galvani's work was mainly continued by his nephew, Giovanni Aldini, who mixed serious research with showmanship. One of the more gruesome of Aldini's displays, designed to emphasize the effectiveness of electrical stimulation for obtaining spasmodic movements from muscles, involved using the recently severed heads of two criminals, as shown in Figure 4.7. Despite such publicity most research on electricity followed Volta's lead. This led to the development of various kinds of apparatus, including galvanometers which could be used to measure relatively small amounts of electricity. When interest returned to animal electricity in the 1830s, technical innovations in the generation and measurement of electrical energy made it possible to employ new methods of stimulating and recording from live tissue. The lead in this work was at first taken by Carlo Matteucci who, starting in Florence in 1834, concentrated on electrical activity accompanying muscular contractions in frogs' legs. There was still no understanding, even in very general terms, of the nature of electrical activity in the body. Even though Matteucci gave wrong and inconsistent interpretations to many of the effects he detected, his findings were a considerable step towards discovering what were later termed 'action potentials'.[7]

Electrical instruments were just some of the many technical developments in the first half of the nineteenth century that accelerated the study of the nervous system and gave birth to what is now termed 'neuroscience'. Other developments included the use of anaesthetics like ether for vivisection experiments; the compound microscope, which provided much greater resolving power than previous optical instru-

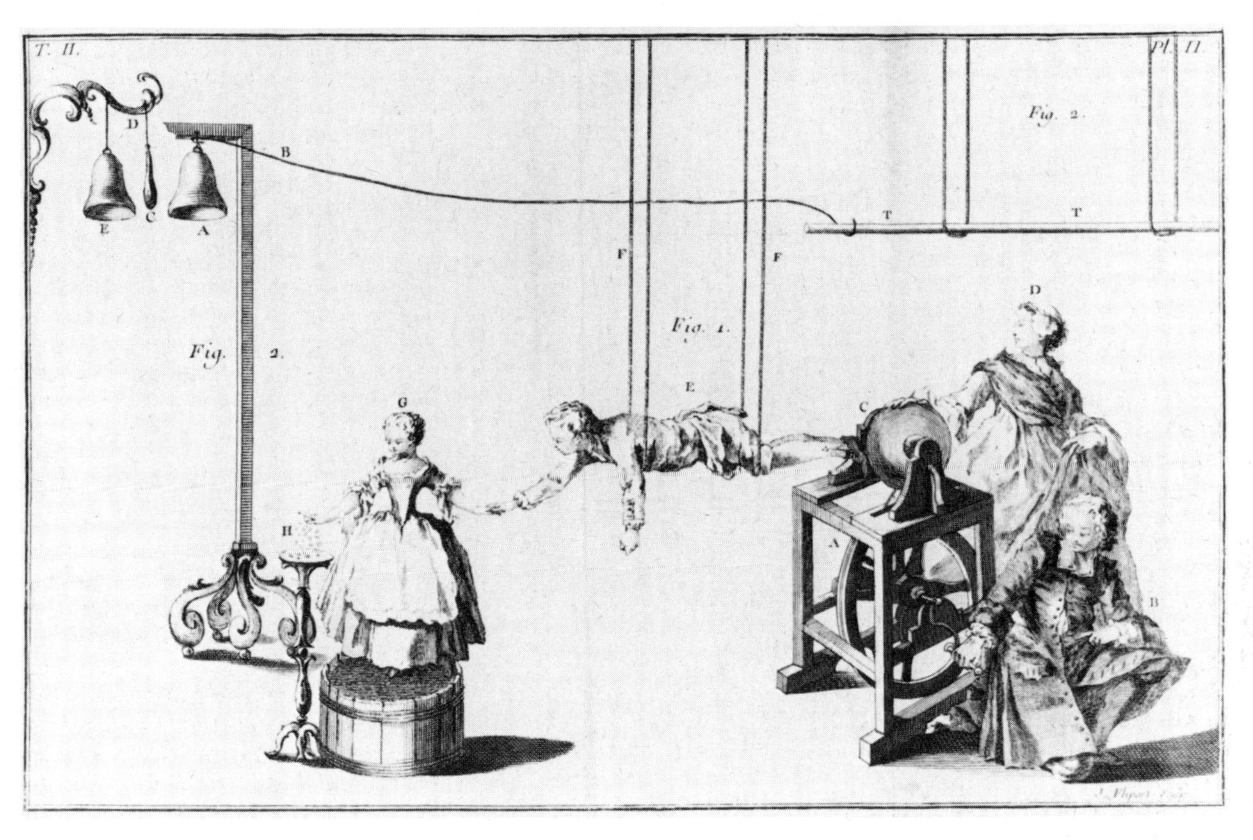

Fig. 4.6. An electrical demonstration of the mid eighteenth century

ments; and various techniques for staining nervous tissues in a way that made structures visible. The consequent advances in histology made it possible to learn about the fine anatomical detail of the nervous system.[8]

Institutional changes were at least as important. Neuroscience became established at any early date within the reformed university system of Germany which, as described in the previous chapter, made possible a new kind of professional science and a continuity of research traditions. There was a succession of notable French physiologists working in the academies of Paris throughout the century, but from the 1830s most of the major discoveries about the nervous system for at least fifty years were made in the research institutes of German universities.

A key event in German physiology was the appointment of Johannes Mueller as professor of physiology at the University of Berlin in 1833. Mueller was a man of wide interests in medicine and biology. He was phenomenally productive; it has been calculated that on average he published a scientific paper every seven weeks from the age of nineteen until his death at the age of fifty-seven. Quantity did not displace quality; many of these papers reported major experimental findings. He also wrote a definitive textbook on human physiology which remained a major influence for at least thirty years. Nevertheless, his major role in developing German physiology, his considerable scientific achievement and his textbook are not remembered as well as his abilities as a teacher. In a way that is not well understood, he managed to inspire a large number of students who became outstanding scientists themselves. Mueller's students included Hermann von Helmholtz, Rudolf Virchow, Theodor Schwann, Emil du Bois-Reymond, Ernst von Bruecke, and Carl Ludwig.[9]

The first of these men to make a major contribution to the understanding of nervous function was du Bois-Reymond. In 1841 he was working as a research assistant to Mueller when the latter gave him a copy of Matteucci's book. Du Bois-Reymond became fascinated by the topic of animal electricity, began his own experiments and seven years later, published his own first volume on the subject. The preface to this volume contained a scathing criticism of Matteucci's ideas and

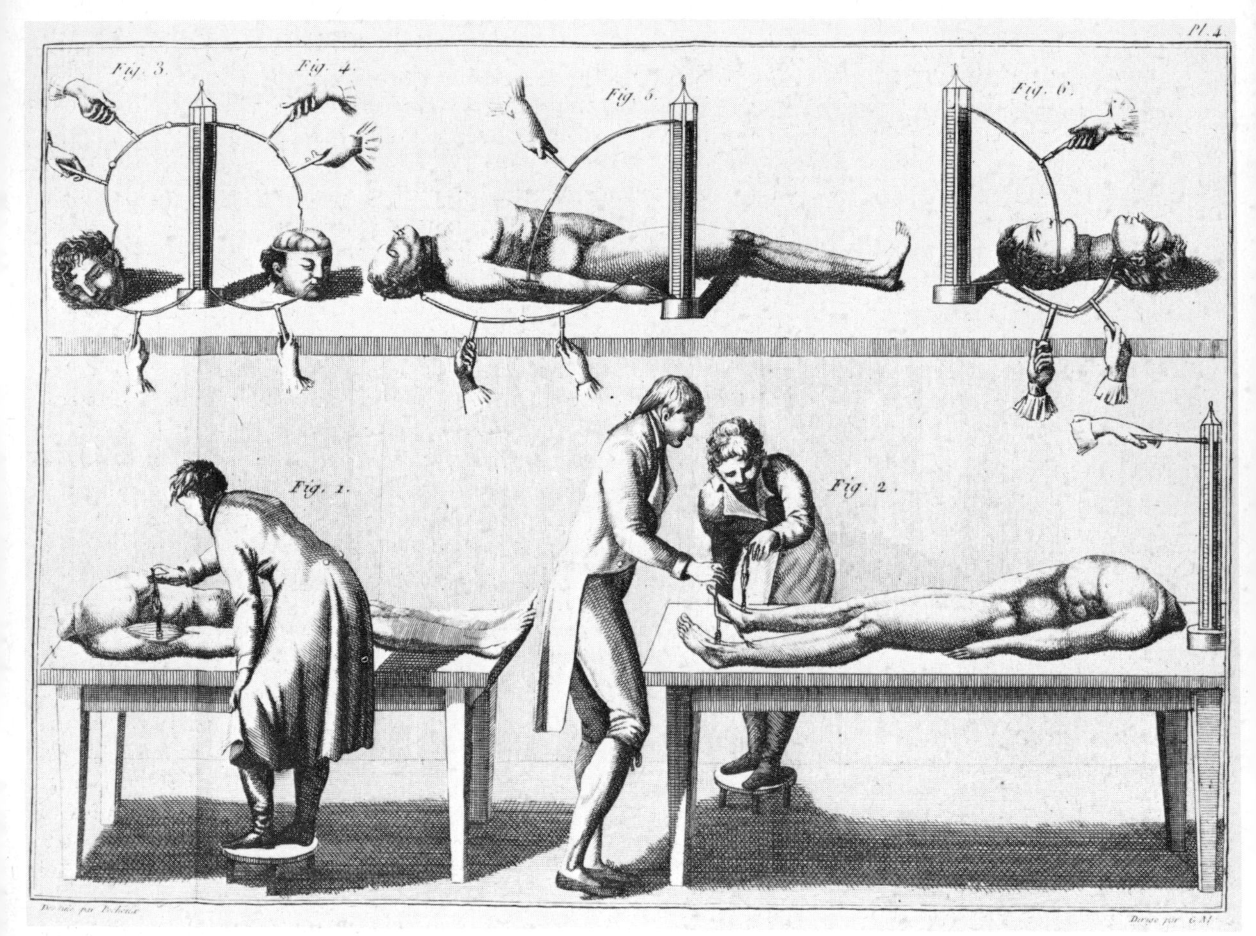

Fig. 4.7. Various examples of electrical stimulation of the human body carried out by Giovanni Aldini

work, and marked the transference from Italy to Germany of the lead in electrophysiological research.

The way that nervous conduction occurs is very complicated indeed and there is nothing in the inorganic world that is at all similar. Since the degree of understanding of electricity in the middle of the century was still quite limited and the apparatus available for experiments still very crude, it is not surprising that progress in identifying the electrical nature of nervous action remained slow. Matteucci discovered electrical activity only in muscles. In his textbook, Mueller firmly rejected the idea that electricity was the elusive *vis nervosa*. Contrary to his master's conclusion, du Bois-Reymond gathered increasing evidence for the electrical basis of neural action and reached the crucial insight that every bit of nervous tissue contains an electromotive force or 'resting potential'. Even then, another of Mueller's students, Carl Ludwig, continued strenuously to reject the idea on the now familiar grounds that resistance of a nerve was too great and its insulation too poor for it to act as a good conductor.[10]

Like many scientists of his time, Ludwig could think of electrical transmission only in terms of a flow of current through a conductor. The speed of such transmission is very close to that of light. Our everyday experience suggests that transmission in the nervous system is just as immediate; we are not aware of any delay between the decision to wiggle a toe and the time that the toe begins to move, despite the metre or so of nervous tissue between the brain and the muscles that move the toe. Whatever the nature of nervous action, Mueller was certain that the speed of transmission was far too rapid for it ever to be measured. This turned out to be another case in which one of Mueller's pupils was to prove his judgement to be mistaken. The experimenter who solved this problem was Helmholtz.

Fig. 4.8. Johannes Mueller

Helmholtz took some large frogs and isolated a motor nerve which could be stimulated five or six centimetres from its junction with a leg muscle. The procedure was trivial in terms of today's technology, but at the time showed an elegant mastery of contemporary physics and instrumentation. He constructed an ingenious switching system which allowed him to time the interval between the onset of electrical stimulation of the nerve and contraction of the muscle, using the total displacement of a ballistic galvanometer as his measure. The apparatus gave readings that were accurate to a thousandth of a second and the experiment produced a good estimate of neural transmission speed, in the range of 25 to 45 metres per second. This was reported in 1850. Two years later, Helmholtz devised another technique, still using frogs, which came up with the same results. Yet a further method was then developed to measure the transmission speed of a human motor nerve. From Hartley's suggestion on the superior texture of the human nervous system a much higher speed could have been expected. In fact, a similar figure, 30 to 35 metres per second, was found to that for the frog.[11]

So it turned out that, although transmission along a nerve was extremely rapid, it was far slower than that of direct electrical transmission through a conductor and only about a tenth of the speed of sound. The experiments were not just of great significance for neurophysiology. These particular studies by Helmholtz were very important for the new science of experimental psychology that started to emerge a few years later, guided by his former research assistant, Wilhelm Wundt. If precise measurements of events in a single nerve could now be obtained, it might also be possible to measure mental events. Helmholtz's experiments were a prelude to the millions of human reaction times that have since been recorded.

The speed of neural transmission was first measured just two hundred years after the death of Descartes. In all that time, the general concept of reflex action stayed the same; the basic idea remained that activity in a sensory receptor would cause the transmission of some kind of force along a nerve to the central nervous system, and this in turn would excite further activity which would be conveyed centrifugally to excite some target muscle. Two centuries of research had made it possible to understand such action in much more detail and had corrected the wrong assumptions that Descartes had made. By the middle of the nineteenth century, the distinction between sensory and motor nerves, and the structure of the peripheral nervous system in general, was well-known.

The function of the spinal cord was also far better understood. Since the time of Whytt, work had continued on this subject. The English physiologist, Marshall Hall, confirmed that the spinal marrow contained numerous connections between sensory and motor nerves and in 1833 he introduced a particularly clear notion of the 'reflex arc', a unit consisting of a sensory nerve, interconnecting nervous tissue which might run through a few segments of the spine, and a motor nerve.[12] Mueller himself performed many experiments on this topic and independently obtained findings that replicated those of Hall.

In 1850 there had been a great deal of progress, but some fundamental aspects of the nervous system remained a mystery. A clear picture of transmission along a nerve was not obtained for many decades. No one knew that the nervous system is made up of individual nerve cells, or neurones, that are separated

Fig. 4.9. Hermann von Helmholtz at about the time he first measured the speed of nervous conduction

by minute gaps, now called synapses. In 1839, one more pupil of Mueller, Theodor Schwann, had proposed his 'cell theory' which stated that 'there is one common principle of development for the most diverse elementary parts of the organism; and that principle is the formation of cells'. However, it was not until the 1870s that many physiologists adopted the belief that the nervous system is made up of individual cells. The controversy between this view and the opposing theory of 'nerve nets' continued until effectively settled in favour of the neurone by the Spanish physiologist, Ramon y Cajal, towards the end of the century.

At mid-century, the most massive concentration of nervous tissues, the brain, remained an enigma. Unlike the spinal cord or peripheral nerves, it was found to be almost completely unresponsive to stimulation by either knife-cut or electrical voltage. No one had detected any electrical activity within it. Apart from its lower part, the *medulla oblongata* which adjoins the spinal cord, the functions of the brain's various parts were completely obscure. Early in the century, the phrenologists had made detailed and dogmatic claims about the psychological functions of different areas of the cerebral hemispheres. Their assertion that variations in these functions were indicated by bumps and indentations on the skull ultimately made their general theories look silly. Phrenology's willingness to exploit public credulity and its promotion as a substitute for palmistry led to its exclusion from the ranks of respectable science.[13]

The physiologists of Europe's medical institutes had no alternative account to offer of the brain's activity. Their science may have advanced in the direction pointed out by Descartes, but none completely shared his view of the animal as simply a mechanism. The most Cartesian of them all, Pierre Flourens in Paris, cut out the cerebral tissue from the brains of birds; he found that pigeons could still fly and chickens still react to certain kinds of stimuli. From the lack of spontaneous or purposive actions in these decerebrate animals, Flourens judged that he had turned them into Cartesian machines. In the intact animal the cerebrum served as the seat of the soul, as in man, and that was that.[14]

The kind of technical ingenuity shown in the measurement of neural transmission speed could perhaps have produced methods for examining the brain. However, it would have been contrary to their philosophical beliefs for the experimenters to look for the physiological bases of the principles of association and of voluntary action described by Hartley, or to test whether these principles were fictitious or not. As one commentator on early nineteenth-century physiology has noted: 'From Haller to Lorry to Legallois, onwards to the best observer of them all, Flourens, and on again to Magendie and everyone else, all were agreed on this – the brain was unresponsive except at the lower and lowest levels. The hemispheres were the seat of the "will"; they excited movement by playing on the motor mechanisms. But how they did so no one knew and no nice man would ask!'[15]

Spontaneous activity and the Berlin physiologists

It has happened frequently that novel findings and basic ideas have arisen within one tradition, while the major scientific advances that followed from them occurred within another. One instance of this was the shift of research on animal electricity from Northern Italy to Germany. Another instance, crucial to the understanding of reflex action, was the way that the initial discovery by German physiologists in the mid-nineteenth century of inhibitory effects provided the starting point for two different lines of outstanding research on reflex activity that flowered outside Germany some fifty years later.

In 1845 a crucial observation was made by Edouard Weber at the University of Leipzig: he found that stimulating a frog's vagus nerve, a major nerve trunk linking the brain to various internal organs, made its heart beat more slowly. It was the first case of inhibition to be reported whereby increased activity in one part of the neuromuscular system produces *decreased* activity in some other part. This kind of effect later became central to the study of the mammalian brain and spinal cord which for a few decades after 1875 became almost the preserve of British physiologists, and which culminated in the superb work of Charles Sherrington in showing how individual spinal reflexes become integrated into adaptive patterns of action. The other tradition in which inhibition became a central concept was Russian physiology; the remainder of this chapter describes how this happened largely as a result of the work of Ivan Sechenov.

An important element in Sechenov's outlook, and in that of the subsequent tradition of Russian psychology that he inspired, was rejection of the possibility that some aspects of an animal's activity might occur in a spontaneous fashion. This has remained a feature which distinguishes Russian analyses of behaviour from major theories in Western psychology. Most Western learning theorists have proposed some interpretation of the Law of Effect – the empirical principle that a rewarded response subsequently increases in frequency – which assumes that actions may occur even when there is no immediately preceding event that can be identified as the stimulus for that action. The assumption has always been rejected by Russian psychologists. This difference in attitudes towards spontaneity has interesting origins in German physiology.

The central doctrine of vitalism is that living matter contains some special creative force and that the principles underlying the properties of inert matter are not enough to explain life. Ever since the late 1600s vitalism has represented an alternative to Descartes' explanation of the movement of animals. The philosophical climate of idealism pervading Germany of the early 1800s made this a period and a place where vitalism was even more widely held than ever. The way German science developed within the university system was as a reaction to idealist philosophy in general and to vitalism in particular. Nevertheless, many of the pioneers of nineteenth-century German science were deeply influenced by vitalist ideas.

The major pioneer in biology, Johannes Mueller, was no exception. As noted earlier, Mueller's research, textbooks and, above all, his students made him the dominant influence in German experimental physiology. His doctoral dissertation had been very mystical in its approach, but later his outlook changed and in old age he tried to destroy every copy of his youthful work that he could find. Yet, throughout his life he continued to believe a special creative force to be the essential condition of life. Unlike earlier forms of vitalism, Mueller's did not regard this force as identical to that of a conscious soul. His outlook was much more like that of Aristotle: all life depends on a creative 'vital' force, but only the higher animals possess the further non-material quality of consciousness.[1]

One consequence of this outlook was that, despite his early research on reflexes, Mueller viewed the movements of animals in a different way from Descartes. The Cartesian view holds that for every action that an animal performs, there must be some immediate cause in the form of a triggering event outside the nervous system; in nineteenth-century terminology, for every response there must be a stimulus. In opposing this view, Mueller maintained that even in quite simple animals movements could be autonomous, occurring in the absence of any external stimulation.

Within German physiology, belief in autonomous activity continued to be seen as a vitalist doctrine. This was by no means inevitable, as is shown by the example of the British tradition of psychology where, because of the work of Alexander Bain, the idea of spontaneity shed its more mystical aspects.

It seems quite certain that Bain picked up from an English translation of Mueller's textbook the idea that activity could be spontaneously generated by the nervous system.[2] Bain's friend, John Stuart Mill, may also have been interested, since one of his main objections to the associationist doctrines of his father and Hartley was that they allowed no place for spontaneity of ideas or creativity of thought. As described earlier, Bain combined the ideas of spontaneous activity with assumptions about the selective effects of pleasure and pain to produce an explanation of voluntary action which was first known as the Spencer–Bain principle and later christened by Thorndike the 'Law of Effect'.

Bain made the idea of spontaneous activity the basis of a theory of learning. In contrast, the vitalist belief that life is the result of some special form of energy is usually accompanied by the view that the behaviour of animals is not greatly modified by their environment and that their skills are inborn rather than learned. This was Mueller's attitude. He claimed that 'all the ideas of animals, which are induced by instinct, are innate and immediate; something presented to the mind, a desire to attain which is at the same time given'. He then continued with an example that may well have induced young Bain to accompany a shepherd and watch the first few hours in the life of a lamb: 'The new-born lamb and foal', stated Mueller, 'have such innate ideas which lead them to follow their mother and suck the teats'.[3] While accepting spontaneous movement, Bain decided on the basis of his own observations that Mueller had a mistakenly exaggerated view of the power of instinct.

Many of Mueller's students regarded the idea of spontaneous activity as one of the many relics of vitalism that physiology needed to shed. One of their striking achievements was Carl Ludwig's account in 1842 of the formation of urine in a way that provided the first detailed and comprehensive account of a complex physiological process to be based entirely on known physical and chemical principles.[4] Three years later Ludwig and three former students of Mueller – Helmholtz, du Bois-Reymond and Bruecke – made their pledge to rid biology of all explanations that appealed to anything but the laws of physics and chemistry. The 'organic creative forces' of Mueller were no longer to be allowed within physiology. These four men, and their later associates, became known as the Berlin School.

As their fame grew, the four radicals became the pillars of German science. Within physiology there was a shift in interests with the growth of its prestige and with its movement towards a central position in the thriving business of medical education. Those involved in physiological research during the second half of the century acquired much more responsibility than their predecessors for educating the future medical practitioners of Germany.

In 1845 a university degree in medicine had uncertain status and declining appeal. Over the next seven years most of the German states enacted laws that gave university-trained physicians a monopoly over medical care that they had not previously even approached. The increased financial value and social status of a medical degree reversed the decline in numbers of medical students attending German universities and ensured a steady growth throughout the 1850s and 1860s. At the same time, the study of physiology based on experimental research was accepted more and more as the most important part of a medical student's training. This was not because of any evidence that such study made him better at healing people than, say, supervised clinical experience, but simply because of the scientific prestige generated by discoveries in physiology.[5]

The close ties with professional medicine reduced the diversity of problems studied by German physiologists. Many of the topics Mueller had included within the scope of physiology in the 1830s were excluded twenty years later. The goal of physiological research became that of obtaining knowledge that would ultimately aid in restoring health to the sick. Although Helmholtz's work on vision and hearing and that of du Bois-Reymond on neural transmission did not have any immediate target of understanding some specific disease or disorder, it was clear that their research would have medical application in the long run. On the other hand, better understanding of some of the other problems that had interested Mueller, and were to become central issues for the early advocates of Darwinism, like Ernst Haeckel – issues that arose in studying evolution, embryology or heredity – were far less likely to be of benefit in caring for the sick.

The change of emphasis within German physiology may have been one reason that Weber's discovery of inhibitory action made less of an impact than it deserved. The reduction in heart rate that occurred when he stimulated the vagal nerve came as a complete surprise to Weber. He was fully aware of the importance of his finding, stating in his report that 'this kind of effect of a nerve upon muscular organs, whereby movements which occur independently are inhibited rather than stimulated, or even prevented completely, is new and startling'.[6] The discovery quickly became well-known, but there appears to have been no appreciation for some time of the possibility of using inhibition to develop more powerful theories of reflex action than that of Descartes.

One person who decided to extend Weber's work was Edouard Pflueger. He was one of a number of German physiologists who explicitly dissociated themselves from the Berlin group. Although he had studied with du Bois-Reymond, Pflueger would have preferred to be known as a student of Mueller. He continued Mueller's earlier work on reflexive action in the frog. In 1853 he put forward in a deliberately provocative fashion his reasons for believing that the laws of physics and chemistry were inadequate to explain the behaviour of even a decerebrate animal. He explained how the high degree of co-ordination and adaptiveness found in the behaviour of a beheaded frog meant that some part of the frog's soul must reside in its spinal cord.[7]

One of the more dramatic findings that Pflueger cited in support of his argument was as follows. A frog's brain was removed and, when the animal had recovered from the effects of ether used as anaesthetic in the operation, a spot of acetic acid was placed on its thigh. This caused the foot of the stimulated leg to move towards, and rub at, the spot of acid. This was then repeated with another frog in which the foot was cut off from the leg that was to be irritated. At first the stump moved to and fro, as if in an unsuccessful attempt to remove the acid. This stopped and then, after a period of general restlessness, the other leg bent towards the spot and the remaining foot rubbed at the acid. It appeared that the spinal frog possessed a number of plans for action and, when one failed, it could choose another. Pflueger concluded that behaviour like this could be explained only in terms of some non-material influence that partly resided in the spine.[8]

Four years later, Pflueger reported the first extension of Weber's discovery of inhibition; it was found that stimulation of a nerve could inhibit activity in a frog's intestine.[9] Yet, although he was so intimately familiar with inhibitory action, he failed to see that this provided the key to understanding co-ordination of movements in a way that made no appeal to a spinal soul. At this point there was no strong reason to believe that stimulating a nerve could reduce activity in another nerve, as well as in muscle tissue.

At about the time Pflueger's report on inhibition appeared, his earlier paper on spinal reflexes in frogs was being carefully studied by a young Russian, Ivan Sechenov, who had just arrived in Berlin and wished to carry out an experiment in what was for him the entirely novel topic of electrophysiology. Using an eel Sechenov repeated some early experiments on reflex action that Pflueger had carried out.[10] Subsequently he was the first to appreciate the general significance of inhibitory action.

Ivan Sechenov and inhibition

Sechenov's life is worth recounting in considerable detail, since it illustrates the state of biological sciences in Russia in the middle of the nineteenth century and the way in which Russian physiology was influenced by earlier developments in Germany. Sechenov came to Berlin just as a new era was starting in German physiology. 1856 was the last year before Mueller retired from the chair of physiology at Berlin, and du Bois-Reymond enjoyed only the insecure position of *professor extraordinarius*. The recent medical reforms had begun to increase the number of students studying physiology, but still very few were interested in studying specialized research topics. Thus, attendance at du Bois-Reymond's course on the physiology of nerves and muscles was not required of medical students and Sechenov that year was puzzled to find that there were only six other students – and one of these, like himself, newly arrived from Russia – listening to du Bois-Reymond's well-prepared lectures on what seemed one of the most exciting developments in contemporary science.

By the 1870s a considerable contrast developed between the well-financed, medically-oriented and experimentally-bound science of physiology and the other biological sciences studied at German universities.[1] There was also a marked difference between the status of experimental physiology in Germany and its status in Russia, where it was still viewed with considerable suspicion. In later years Russian medical training was reformed in a way that closely followed the German model and gave experimental physiology pride of place; the other Russian to attend du Bois-Reymond's course, Sergei Botkin, had a great deal to do with these reforms. However, during the 1860s and 1870s the position of physiology in Russia was very insecure, which partly reflected the state of Russian society. Sechenov's career was very much affected by the major social and political changes occurring during his lifetime. The way in which he became a physiologist in the first place is illuminating about life in Russia at that time.

As a boy Sechenov lived in the country on his father's small estate. An early stroke of good fortune which was to influence his adult life was to have a governess who gave him unusually effective instruction in French and German. When he was ten years old his father died. At the age of fourteen he was sent to a college of military engineering at St Petersburg, since this was known as an inexpensive way of obtaining an education. Sechenov got on well in mathematics and physics, but was bored by instruction on methods of fortification. In his fourth year he managed to fall foul of the college authorities and left with an unfairly low grade. Because of this he was given the unattractive posting of field engineer in a regiment stationed near Kiev.

The turning point of Sechenov's life came in 1849 when, at the age of twenty, he fell in love for the first time. He had become acquainted with a cultivated family of Polish origin living in Kiev and began to visit them regularly in the company of a fellow engineering officer. His 'benefactress', as he later called her, was also twenty and had been widowed six months before their meeting. She had 'serious attitudes towards vital questions', knew about writers in France and Germany, and objected passionately to the constraints

Russian society placed on women. Her great respect for university education went with little respect for the army. The intellectual evenings with his benefactress became the central events of Sechenov's life. Gauche, with a Tartar-like appearance and disfigured by smallpox in his childhood, he had far too little confidence to reveal his feelings either to her or to the small circle of mutual friends. In fact the girl and the friends were fully aware of his infatuation, while he was totally unaware that she was being courted by his fellow engineer. When this friend married the girl a year later, the shock was so great that Sechenov resigned from the army and decided to study medicine. He managed to borrow a little money and set off for Moscow.

In his old age Sechenov wrote about his benefactress: 'I went to her house as a youth, up to this time swimming inertly in the channel into which fate had thrown me, without any clear awareness of where it might lead me; but I left her house with my life's plan prepared, knowing where to go and what to do'.[2] However, the commitment that she had instilled to study medicine and help his fellow man was shaken considerably once he arrived in Moscow.

The standard of teaching was appallingly low. Apart from anatomy lectures, most instruction consisted of tedious recitation, often in Latin, of how to classify illnesses and what remedies should be given. No explanation of any disease or disorder was ever offered and no justification for a particular treatment ever given. Few of the professors betrayed signs of interest in such theoretical matters. An exception was a professor who lectured on physiology. His lectures made no mention of any of the recent findings made in France and Germany, of Weber's discovery of inhibition or of Helmholtz' measurement of transmission speed in a nerve. But they were stimulating enough to persuade Sechenov that physiological research was where his future lay, just as his experience of clinical treatment convinced him that he could never be a good physician.

In 1850, when Sechenov began his medical training, student numbers at Moscow University were being reduced. This was part of a deliberate policy by Czar Nicholas to prevent the spread of Western ideas. It did not matter that Russians were kept in ignorance of technological developments elsewhere, as long as they were also isolated from the pernicious doctrines that had prompted the revolutions of 1848 in many of Russia's western neighbours. Science was believed to be as much a threat as any other foreign import to the stability of Russian society and to traditional religious beliefs.[3] The books for Sechenov's courses were translations of antiquated German texts. Mueller's textbook, written twenty years earlier, was still not available. However, Sechenov managed to find out that this was a key work for physiological research despite the barriers against contact with the outside world. In his final year at the university Sechenov's mother died and he renounced any further claim to a share in the family inheritance in return for the sum of six thousand rubles from his brothers. This money, and the relaxation of restrictions on foreign travel which followed the death of Czar Nicholas, allowed Sechenov to proceed with his plan of studying physiology in Berlin with Mueller. Just before he left, the celebrations announcing the coronation of Czar Alexander II marked the end of an era of severe repression and intellectual stagnation in Russia.

Germany was at first a bitter disappointment. In Russia Sechenov had been unaware that Mueller had long ago given up the study of physiology after finishing his textbook on the subject. Mueller's lectures in 1856 were solely on the comparative anatomy of vertebrate genitals. He was a tired, sick man. Few of the other professors were any more inspiring.

Sechenov was, however, impressed by du Bois-Reymond and he decided to learn first-hand about electrophysiological research. At the time du Bois-Reymond had only a small room in which to perform his experiments and no facilities for an assistant. Sechenov had to be content with a bench in the corridor outside, where he was able to repeat Pflueger's experiment. He received little guidance from du Bois-Reymond who remained coolly distant. In one of his lectures the latter suggested that 'the long-headed race possess all kinds of talents, but the short-headed, in the best instance, only imitation'. It was an interestingly phrenological remark from someone who on purely scientific grounds would have been dismissive of phrenology; the suggestion may have been intended to dissuade the two short-headed Russian members of his small audience from any attempt at original scientific work.

Electrophysiology was only one of Sechenov's interests. During this first trip abroad, and throughout his life, the other topic he concentrated upon was the absorption by the blood of various substances. From Berlin his joint interests took him first to Leipzig and then, after a further year of research, to Vienna. Here he settled upon the topic for his doctoral dissertation, the absorption of alcohol by the blood, and established a life-long friendship with Carl Ludwig. Ludwig was an incomparable teacher and later, after moving to Leipzig, acquired enormous international influence by providing thorough training to far more students than any of his contemporaries.[4] Following a year in Vienna Sechenov went on to Heidelberg to study physiology with Helmholtz and chemistry with Bunsen; he carried out work on topics they suggested, but enjoyed much less contact with them than he had with Ludwig.

Three and half years of study, with long and leisurely walking tours in the Alps and in Italy and with holidays in Paris, cost more than six thousand rubles. By 1859 it was time to return to Russia. Helmholtz gave three copies of his unpublished work on the perception of sound to Sechenov for delivery in Berlin, where du Bois-Reymond was slightly warmer in manner towards the young scientist who now had a couple of papers to his name. From Berlin Sechenov returned to St Petersburg where his dissertation was speedily published. There was now a great demand for lectures on the latest physiological discoveries and Sechenov was the first Russian in many years to return after extensive training abroad. He became an assistant professor at the Military-Medical Academy, which was the part of the University of St Petersburg responsible for the training of both military and civilian doctors.

It was a period of social change in Russia; the most important event was the liberation of the serfs in 1861. There was a strong movement favouring the opening of higher education to women, and Sechenov, remembering his benefactress, became very much involved in this cause. One of the first two women to enlist his aid in their attempts to obtain medical training later became his wife. Sechenov had ample space and funds for his research. However, none of these things led to settled contentment with his new life. As soon as possible he applied for leave, and in the Autumn of 1862 set off to study in Paris with Claude Bernard, who was known as the most skilful vivisectionist in Europe and who had recently made the third discovery of an inhibitory effect of nerve upon muscle.[5]

One aspect of human behaviour that fascinated Sechenov was our ability to suppress involuntary reactions. Although not always successful, we can often suppress, or at least delay, a sneeze or a cough, or resist a temptation to scratch. Over fourteen years earlier, in connection with his discovery of inhibitory action in the frog's vagal nerve, Weber had noted that spinal reflexes were sometimes more sluggish in intact animals than in animals with their cerebrum removed; as an explanation of this, he had suggested that the cerebrum might normally serve as a source of inhibitory influences on reflex activity. This had not

Fig. 4.10. Ivan Sechenov in his laboratory in the Military–Medical Academy with three frogs

prompted any general interest, but Sechenov decided that the idea was an important one and that experimental investigation might provide the key to understanding voluntary control over actions that are normally involuntary.[6]

Most of Sechenov's work during his seven months in Paris was on this topic. He received some help from the old man who assisted in Bernard's laboratory, but little from the master himself; Bernard proved to be even more remote than du Bois-Reymond or Helmholtz had been. Sechenov used as his experimental preparation the reflex withdrawal of a frog's leg when it was dipped in dilute acid; the reaction time between first contact with the acid and the beginning of the leg's movement provided a measure of the intensity of the reaction. Sechenov then stimulated the frog's brain; not electrically, but by placing a salt crystal in various places. Such stimulation could be removed by means of a few drops of water and blotting paper. He found that he could depress reflex withdrawal of the foot by stimulating certain central structures of the frog's brain, but not by stimulation of either the surface of the cerebral hemispheres or the top of the spinal cord. For example, quick reactions by the leg would slow down when the salt was applied to points on the thalamus and become rapid again when the salt was removed. The results provided the first demonstration of the inhibition of neural as well as muscular activity, the first experimental evidence that inhibitory effects could arise within the brain itself, and confirmation of Weber's suggestion that such a process could modulate spinal reflexes.[7]

Sechenov's extension of physiology to mental processes

Justifiably pleased with his discovery and the outcomes of the further control experiments he ran, Sechenov quickly wrote a paper describing the research and travelled in Germany to discuss it with Bruecke, Ludwig and du Bois-Reymond, before heading homewards to St Petersburg once again. One aspect that excited Sechenov was its general implication for psychology: it was now possible to begin the analysis of mental processes in terms of specific physiological mechanisms. He saw the concept of inhibitory action within the central nervous system as a means of removing a major objection to an analysis of all behaviour in terms of reflex action.

Despite the moving statues in the gardens of St Germain-en-Laye that had provided Descartes' original inspiration, it had been generally understood over the next two hundred years that reflex action included as a necessary feature a relationship between the intensity of a stimulus and the intensity of the reaction it produces; for example, a sudden ear-splitting crash normally produces a much more pronounced startle reaction than a softer sound. However, many other aspects of behaviour do not show such a close relationship between the intensity of the perceived event and the reaction that immediately follows. We may think about and subsequently remember some violent event which at the time produced in us no discernible response; on the other hand some very slight event, difficult to distinguish, may produce extremes of terror.

Sechenov's ideas on such issues were expressed in a long article written soon after his return. This was originally titled *An Attempt to Bring Physiological Bases into Mental Processes*, but, on the instructions of the St Petersburg censor, appeared as *Reflexes of the Brain*. The change was intended to convey the impression of a technical work of no general interest. The new title was a misleading one, since the article ranged almost as widely as, for example, Hartley's *Observations* and, although on a much smaller scale, similarly attempted to show the way in which all psychological phenomena might be explained in terms of physiology. Nevertheless it contains little detailed physiology and the term 'reflex' is used in only the very general sense of indicating the principle that for every movement of a muscle there is some immediately preceding event which has acted as the stimulus for that movement.

Mueller and his notion of autonomous activity are never named. However, it is quite clear that Sechenov's principle of the reflex signifies a firm rejection of Mueller's idea that actions can be spontaneous. Even when a thought appears to come from nowhere, there is always an immediate cause to be found, as Sechenov illustrates with an example of thinking about a political figure who – no doubt for the sake of the censor – is described as the Emperor of China. 'I devote my daytime to physiology; but in the evening, while going to bed, it is my habit to think of politics. It happens, of course, that among other political matters I sometimes think of the Emperor of China. This acoustic trace becomes associated with the various sensations (muscular, tactile, thermic, etc.) which I experience when lying in bed. It may happen one day, that owing to fatigue or to the absence of work I lie down on my bed in the daytime; and lo! all of a sudden I notice that I am thinking of the Emperor of China. People usually say that there is no particular cause for such a visitation; but we see that in the given case it was called forth by the sensations of lying in bed.'[1]

A number of aspects of Sechenov's concept of reflex action are particularly interesting. The first is the shift away from reflexes as 'reflections' and back to the idea of the stimulus as a trigger requiring little force compared to the reaction it releases. Sechenov analyses emotions in such terms, using as a model a nineteenth-century version of Descartes' statues, namely the operation by a puff of air of a mercury switch, which in turn activates an electromagnet powerful enough to move a huge mass of iron.[2] The second kind of mismatch between stimulus and response intensity is when contemplative inaction occurs, instead of impulsive response to some important event, or silent endurance, instead of frantic movement to some severe pain. This is where Sechenov believed inhibition to play a major role. He saw the slow establishment of inhibitory control as a major aspect of human development; a very simple case would be the ability to move the little finger without moving the others. In very general terms he claimed that inhibition was also crucial in explaining the adaptive nature of actions seen in quite simple creatures; one example he gave was Pflueger's study of decerebrate frogs.[3]

Inhibition was the major new principle discussed in *Reflexes of the Brain*. Almost as novel was the close relationship Sechenov proposed between a physiological approach to psychology and a view of the human mind that was very environmentalist and based on the laws of association. Hartley had described a similar combination, but few others had done so in the hundred years that followed. Hartley was unknown to Sechenov. Instead he learned his psychology from the philosophers of the French Enlightenment who had followed la Mettrie.[4] Towards the end of *Reflexes of the Brain* there is a distinct echo of la Mettrie in the belief that 'by bringing up a clever Negro, Lapp or Bashkir in European society and in the European fashion, a person will be produced whose mentality hardly differs from that of an educated European'. The sentiment is a long way from that of Sechenov's English contemporary, Francis Galton, as is Sechenov's estimate that 'in the majority of cases 999 thousandths of the contents of the mind depend on education in the broadest sense of the word, and only one thousandth depends on individuality'.[5]

In Czarist Russia, even of the early 1860s, the political implications of such beliefs could be judged subversive. Only by good fortune of its timing was *Reflexes of the Brain* published at all. Although for a long time Russian universities had been greatly influenced by the German system, the latter's concept of academic freedom appeared only faintly and for just a brief period on Russian soil. When permission to publish *Reflexes of the Brain* in a specialized medical journal was given in 1863, it was only three years since lectures on any philosophical topic had again been allowed. In 1866 *Reflexes of the Brain* was reprinted as a small book. There was a great deal of interest in its contents among the intelligentsia of St Petersburg and Moscow, many of whom no doubt had only a slight interest in strictly physiological matters. The book's scientific and, by contemporary standards, frank approach to sexual behaviour, and its account in reflex terminology of the transformation of transient sexual craving into loving friendship, must have added to the excitement it created. The St Petersburg Committee of Censorship attempted to ban it on the grounds that Sechenov's 'materialistic theory . . . militates against the view of Christianity and the claims of the Penal Code; consequently it leads to the corruption of morals'.[6] The move was blocked by the Attorney-General who advised that no action should be taken since it was not strictly illegal to teach materialism, as long as Christianity and the Penal Code were not explicitly criticized; and in any case the ban would simply direct further attention towards the book.[7]

The book could not have appeared any later, for in that same year a new Minister of Education was appointed, Count Tolstoy, who has been described as one of the most bigoted and influential reactionaries of the nineteenth century.[8] His position gave him control over censorship and his policy for higher education was one of emphasizing studies of Roman and Greek history at the expense of science. Elsewhere in Europe and North America university instruction in science became firmly established and a scientific outlook, even one that maintained an appropriately ambiguous form of evolutionary theory, became very respectable by the latter part of the century. In Russia autocratic insistence on the classics inevitably aroused antagonism and among the more disaffected intellectuals contempt for any kind of humanitarian study was accompanied by an increasing worship of science.[9]

The different status of science in Czarist Russia and Imperial Germany of the 1870s may be one reason for the divergence that appeared between Sechenov's attitude to psychology and that of the Berlin physiologists. Du Bois-Reymond now occupied a central position in the German scientific establishment in his capacity as secretary to the Academy of Science. One of his duties was to deliver an annual lecture on a theme of general interest. The lecture for 1872 was one of the most notable out of an interesting series. It was called 'On the limits of natural science'.[10] Its main argument was that there are not only many mysteries of which science can at present only admit its ignorance, but there are also problems which scientific method can in principle never solve, the two major problems being the ultimate structure of matter and the nature of consciousness. The lecture may have been partly a reaction to the heated, but unprofitable, controversy over the 'spinal soul' that had raged ever since Pflueger's report had appeared. It was certainly directed at Haeckel and his followers with their claims that a scientific understanding of evolutionary principles offered a key to all the mysteries of life. Haeckel was enraged by the lecture, particularly by its keynote *ignorabimus* – 'we shall never know' – which he viewed as an assertion of political conservatism, as well as an attack on his own scientific credentials. The lecture can also be read as a warning to German physiologists to stay clear of the troubled waters in which Sechenov had entered.

At the time of the *ignorabimus* lecture Sechenov was working at Odessa University. Developments in

Fig. 4.11. Emil du Bois-Reymond

St Petersburg had been very disturbing to him and in 1870 the final straw came when an important academic appointment at the Military-Medical Academy was made on purely political grounds. Sechenov impulsively resigned and only by good fortune found another job at Odessa a few months later. It was here that he wrote another long article on the relationship between physiology and psychology with the title, unaltered by the censor this time, *Who Must Investigate the Problems of Psychology and How.*[11]

The message it contained was a very straightforward one, but Sechenov was the first to give it. It might be stated as follows. Psychology is a very backward science. The main reason is because psychological phenomena are very complicated. The principle that has been so successful in the analysis of complexity in other fields has been to start with simpler, but related, phenomena. It follows therefore that 'the psychical phenomena of animals, and not those of man, should be used as the primary material for studying psychical phenomena'.[12] At present most psychological phenomena even in animals remain unexplained and, if asked to give an account of them, the 'physiologist prefers to say laconically: "I do not know" '.[13] However the study of reflexes has at least provided a start. If psychology continues to rely on the introspective analysis of subjective experience, then it will indeed continue to be a field in which we remain ignorant, since such analysis is hopelessly contaminated by our false preconceptions about how the mind works. If physiologists turn their attention to psychology, then the brilliant universal theories may well disappear and the illusion of rapid progress will be dispelled; but in their place will come a slow, but real, advance.

Sechenov made no further contribution to such progress. He never proceeded to show in detail, for example, how his claim in *Reflexes of the Brain* that the 'intelligent character of a movement does not exclude the machine-like character of its origin' could be supported; as noted earlier, this kind of analysis of spinal reflexes was eventually provided by Sherrington in England. In Odessa Sechenov continued experimental work on his other major interest, the blood's absorption of gases. He returned to St Petersburg in 1876 and shortly afterwards his theories on this subject were seriously challenged. From then on almost all his research activity was devoted to obtaining further experimental support for these claims. He never became the solid professor enjoying the approval of the authorities and the steady respect of colleagues. In 1888 he again resigned his professorial chair in St Petersburg to become this time a lowly lecturer at Moscow University, where he had to pay for a laboratory out of his own pocket. Three years later, now sixty-two, he became a professor once more, this time for a full ten years. Yet almost the last event in his life was the interruption of a lecture forming part of a course which he gave to factory workers on physiology and anatomy; the course was banned on the grounds that he was not a suitably approved teacher. His autobiography is not very clear about this ban or about the reasons for the various abrupt changes earlier in his career; one suspects that the censor closely examined even this last publication by Sechenov. He died of pneumonia at the age of seventy-six in November, 1905, when the revolutionary change in Russian society that he had hoped for appeared to have just begun.

Concluding discussion

The group of men associated with Johannes Mueller, who have been referred to here as the Berlin physiologists, had a great deal in common with Descartes. They shared his view that there is a great divide between man and all other creatures, one that makes the differences between various species of animals insignificant in comparison. They strongly affirmed his belief that all living processes can be explained in terms of laws derived from the study of inorganic nature and, unlike Mueller himself, adopted the Cartesian view that all the motions that an animal makes are based on reflex action. They were also deeply committed to the view that a firm understanding of the body's function can come only from appropriate experiments; and, as Descartes had said it should, society generously supported them and provided conditions in which they could pursue their research. After all, the society in which they lived shared their faith that this was a most productive way to increase the effectiveness of medical treatment.

The Berlin physiologists knew a great deal more about the nervous system than Descartes. They knew about the fundamental division into sensory and motor components, about reflex arcs within the spinal cord, about the electrical basis of nervous action and even the speed of nervous conduction. But at a general level their concept of how this system produces behaviour was really no different from that of Descartes. This was one of the main points of Thomas Huxley's address at the Belfast meeting of 1874 to a British audience who knew considerably less than he did about recent German research on neurophysiology; Huxley argued that recent research fully supported what he described as Descartes' 'attempt to reduce the endless complexities of animal motion and feeling to law and order' using the reflex as its basic principle.

What Huxley failed to point out to his audience was that the largest part of the nervous system of any vertebrate, the brain, remained a physiological mystery. Descartes' suggestions on this subject had been discarded, but there was little to substitute for them. Furthermore, no one had managed to be very clear about how adaptive behaviour, even of the kind displayed by Pflueger's decerebrate frog, could be explained on the basis of Cartesian reflexes alone. And as for any idea of how the nervous system might acquire new reactions, there was only Hartley's disregarded theory and no firm physiology at all.

It took some time before the significance of inhibitory action within the nervous system was widely appreciated. Sechenov made one of the key initial discoveries and also was the first to realize the power of a system that allows interactions between excitatory and inhibitory effects. Unlike a Cartesian reflex relying only upon excitatory action, one that also includes inhibition might account in a mechanistic fashion for those integrated and apparently purposeful actions that had led others to believe in a spinal soul. Since, however, Sechenov never demonstrated in detail how this could be done, he would not have been a particularly noteworthy figure in this history if it were not for a further important contribution. This was to establish as a central idea within Russian physiology the extension of experimental physiology into the domain of psychology.

Even though his professional career was so unsettled and much of his own research had little to do with the nervous system, Sechenov's integrity, enthusiasm and personal charm deeply influenced the next generation of neurophysiologists in Russia and ensured that the study of reflexes was one of their central concerns. According to Sechenov, the new physiological kind of psychology should take the reflex as its central concept. By this he meant more than 'reflex' in the Cartesian sense, since in the analysis of mental processes the inhibitory action of higher parts of the brain would play a central part. In some contexts the appeal to the reflex simply reflected a general commitment, one shared with the Berlin physiologists, to reject any idea of spontaneous action; it indicated the assumption that for any kind of event to occur within the nervous system there must be an immediate cause to be found outside the system, even if many times this may be a source of stimulation of which the person concerned is completely unaware.

The quotation at the beginning of this chapter represents a novel attitude to animals. Descartes had said that we should experiment upon the bodies of animals in order to understand our own bodies; Sechenov's contemporaries in England suggested that studying the behaviour of animals would help in the understanding of evolution; but Sechenov argued that we should experiment upon animals in order to understand our own minds.

Again Sechenov did not himself carry out any research of this kind, but in the long run he became responsible for a Russian approach to psychology that, in addition to preferring explanations in terms of physiological concepts, has emphasized the role of experimental work with animals. It has also been highly environmentalist in outlook. Despite Sechenov's youthful dedication to medicine, the primary purpose of the kind of psychology he wished to inspire was not that of bringing improvement to the care of the sick. For Sechenov the general point of such studies resembled the concerns of the French philosophers of the eighteenth century more than those of the German physiologists of the nineteenth, more those of Julien de la Mettrie than those of Emil du Bois-Reymond: Sechenov believed that a clear understanding of man's behaviour and potential would release him from the mental bonds that allowed a stiflingly repressive form of society to continue.

> The integrity both of the individual and of its species is ensured first of all by the simplest unconditional reflexes, as well as by the most complex ones which are usually known as instincts . . . But the equilibrium attained by these reflexes is complete only when there is an absolute constancy of the external environment. Since the latter, being highly varied, is always fluctuating, the unconditional, or constant, connections are not sufficient; they must be supplemented by conditional reflexes, or temporary connections . . . The temporary nervous connection is the most universal physiological phenomenon both in the animal world and in ourselves.
>
> Ivan Pavlov, (1934): *Big Medical Encyclopedia*

5
Conditioned reflexes

Ivan Pavlov (1849–1936) has become almost as famous as Darwin. It is very widely known that he performed experiments showing that dogs salivated to a bell which was regularly sounded just before they were given food; and that he called this reaction a 'conditioned reflex'. Nevertheless, most people would find it much harder to say anything about the significance of Pavlov's research or the general nature of its findings than about Darwin's achievements. In fact, many have felt that Pavlov's reputation has been completely undeserved; one person, for example, who was particularly derisive about Pavlov was George Bernard Shaw.

In 1932 Shaw wrote a story called *The Black Girl in Search of God*. As the black girl travels through the African forest she meets a number of odd characters. One of them is a very shortsighted old man in spectacles, sitting on what appears to be a log. He explains that in responding, as she had just done, with terror to the sound of a lion's roar she was acting on a conditioned reflex. 'This remarkable discovery cost me twenty-five years of devoted research, during which I cut out the brains of innumerable dogs, and observed their spittle by making holes in their cheeks for them to salivate through instead of through their tongues. The whole scientific world is prostrate at my feet in admiration of this colossal achievement and gratitude for the light it has shed on the great problems of human conduct.'

'Why didn't you ask me?' said the black girl, 'I could have told you in twenty-five seconds without hurting those poor dogs.'

'Your ignorance and presumption are unspeakable', said the old myop. 'The fact was known of course to every child; but it had never been proved experimentally in the laboratory; and therefore it was not scientifically known at all. It reached me as an unskilled conjecture; I handed it on as science.'[1]

Unlike the old man in the forest Pavlov never claimed as his discovery the fact that a hungry dog produces spittle in anticipation of food; as noted earlier this had been general knowledge since at least the time of Robert Whytt. Moreover Pavlov did not see his experimental work as a process for transforming what is already known to babes into a firm body of scientific facts. His contribution began with the realization that a commonplace observation, and one that had already been familiar within his own laboratory for some years, namely that a dog will salivate when given a signal that food is imminent, could provide the basis for studying two fundamental problems. As he hesitantly suggested in his first paper on conditioning, it seemed possible to devise from this observation a tool for examining the 'seeming chaos of relations' with which the behaviour of an animal comes to adapt to its world and for identifying general laws that govern changes in behaviour.[2] Such research would also grapple with the second problem, that of understanding the basis of such laws in terms of the functioning of the brain. Now, half a century since Pavlov died, the brain mechanisms responsible for learning remain as mysterious as ever; although he always insisted on calling himself a physiologist, Pavlov's major achievement was in the realm of psychology, that of discovering many of the basic principles of learning.

Pavlov had an enormous effect on both Russian physiology and Western psychology because he completely changed general beliefs about the scope of physiological ideas and methods and about the appropriate way to study psychological issues. Although a change of this kind had been advocated by others, many of whom have been discussed in previous chapters, Pavlov was the first major scientist both to argue for the extension of experimental physiology into psychology and actually to demonstrate on a grand scale how it could be done.

Pavlov's career is remarkable for the late age at which he began to study conditioning; his professional interest in psychological issues did not begin until he was already fifty years old. Up to this point Pavlov had spent twenty-five years on physiological work that concentrated on traditional problems. It was also highly productive and widely esteemed. This makes it all the more unusual that the whole direction of his research should have shifted dramatically, so that the problems investigated in his laboratory began to lie well outside what for many decades had been regarded as the bounds of physiology. The fact that his interest in psychology did not develop early in his career was a crucial ingredient of his success, since the work on conditioning would have made little progress without the assets he already possessed: prestige, ample funding, technical facilities, a stream of eager co-workers and superb judgement in planning and supervising experiments.

It has become commonplace in recent years to distinguish between two kinds of scientific activity: normal science and revolutionary science. According to a standard version of this distinction, normal science consists of work which proceeds within a stable conceptual framework widely held by people active in the particular field of study. This kind of activity is occasionally interrupted by an abrupt change in fundamentals, which occurs as the result of the efforts of young rebels within the field and is usually bitterly resisted by their older and better established colleagues. This kind of change has been called a 'paradigm shift' by Kuhn.[4]

One assumption inherent in this point of view is that scientists are very conservative with respect to their basic beliefs. In general this appears to be true. Despite a commitment to scepticism, to maintaining an open mind and to grounding beliefs on empirical evidence, early in their careers most scientists adopt ways of thinking about their subject, beliefs about appropriate methodology and judgements about the relative importance of various problems, all of which remain immune to even major discoveries and theoretical developments within their field. Significant changes in someone's scientific approach usually occur only when there has been a very major shift in their interests; and where this does happen it seems to require a complete change of discipline.

Pavlov's career provides a striking exception to this general rule. Moreover, as well as giving animal psychology the concept of the conditioned reflex and a host of important facts about conditioning, Pavlov also provided a powerful model of what it is to be a successful scientist. Few of his successors have come close to imitating his career, but since Pavlov there have been many students of conditioning who have to some extent adopted a similar style of science, one that involves complete dedication, unrelenting regularity, wariness with respect to theory and confidence in the productiveness of experimental work so long as it is skilfully and thoroughly carried out. These characteristics developed during Pavlov's early work in conventional physiology. Before describing this, some discussion is appropriate of his place in the intellectual currents of late nineteenth-century Russia.

Pavlov is easily seen as a direct successor to Sechenov and as a central figure within the radical tradition which produced a materialist science and the 1917 revolution. Indeed in his old age this is how Pavlov was regarded in Soviet Russia; in 1921 Lenin arranged special privileges for him and later Pavlov was proclaimed a Hero of the Revolution. In a very real sense Pavlov's work on conditioning represents the extension of physiology into psychological issues in just the way Sechenov said it should happen, even if Sechenov himself was not specific about what particular issue physiologists should study or what particular methods they might employ. Thus, it seems quite appropriate to view Pavlov as a follower of Sechenov, even if he was a little slow in developing his interest in psychology. But if the interests and outlook of scientists like Sechenov within the radical tradition are examined more closely, Pavlov's position becomes less obvious.

Sechenov had been a leading figure among the young scientists and writers who had been inspired by the reforms and the new intellectual currents of the 1860s. As the century went by, various forms of government oppression steadily increased opposition to the Czarist system among the intelligentsia. Hostility towards political and bureaucratic organizations went with a rejection of other aspects of Russian culture and with a readiness to embrace Western ideas. Sechenov's belief in the liberating influence of modern science was widespread among his contemporaries and an enthusiasm for experimental physiology was surpassed only by the degree of interest young Russians showed in theories of evolution.

As in Western Europe and America the works of Charles Darwin and Herbert Spencer were immensely popular. Three independent translations into Russian of Darwin's *Descent of Man* appeared within a year of its publication in England and the other major books by Darwin, and by Spencer too, were translated almost as promptly.[5] To disaffected young Russians Darwinism was a new revelation, establishing materialist faith ever more securely. Its reception made Dostoyevsky angrily remark that what others regard as plausible speculation becomes indisputable dogma in Russia.[6] Sechenov was certainly an ardent Darwinian; he was responsible for one of the three translations of the *Descent of Man*.

There was not the same conflict between Darwinism and traditional religion in Russia as there was elsewhere. The Russian Orthodox Church with its greater emphasis on ritual and mysticism and smaller concern with theology than other varieties of Christianity saw little threat from either evolutionary theory or the general spread of a scientific outlook.[7] So, although the radical intelligentsia tended to be both devotees of Darwin and critics of the Orthodox Church, the latter attitude mainly reflected a general desire to reform Russian society.

Pavlov does not fit easily into this pattern. His social background was much closer to the peasantry than that of most of the academics, writers and scientists of his time. In his childhood he was steeped in Orthodox traditions and, although he lost his faith in God, he always retained a sympathetic attitude towards the church. He could praise at times the energy or the efficiency of the Anglo-Saxons, but he was never pro-Western in his general sentiments. On the contrary he displayed an intense form of nationalism which seems to have been the main factor governing his political beliefs; so that, for example, one of his more severe criticisms of the Czarist regime was for the humiliation it had led Russia to suffer in the Russo–Japanese War of 1904–5.[8]

Pavlov did share with his contemporaries an interest in evolution. As a boy he was excited by popular accounts of Darwin's theories and later he became a great admirer of Spencer. But there is little in his work that shows even an indirect effect of evolutionary ideas. What made a more specific and lasting impression on Pavlov was the translation of an English book written at the same time as *The Origin of Species*, but which hardly even mentions evolution. The book was George H. Lewes' *Physiology of Common Life*. Pavlov read it first in his teens and as an old man could still quote long sections from it.[9]

In this book Lewes gave the general English reader of the 1860s with an interest in physiology what thirty years later William James was to provide for the American reader interested in psychology: an intelligent, detailed and very readable review of recent experimental work, mainly that from the research laboratories of the German universities; to this was added generous helpings of the author's own theories and philosophy, plus engaging comments on everyday life.

An unusual aspect of Lewes's book is the emphasis placed on food and digestion. The book appeared in two volumes; while the second of these largely contains a lively and accurate account of the nervous system, the first volume is almost completely devoted to various aspects of feeding. The subject was of great significance for Lewes. 'Hunger is indeed the very fire of life, underlying all impulses to labour', he wrote, 'Look where we may, we see it as a motive power that sets the vast machinery in action . . . Hunger is the invisible overseer of the men who are erecting palaces, prison houses, barracks and villas. Hunger sits at the loom . . . Hunger labours at the furnace and the plough coercing the native indolence of men into strenuous and incessant activity. Let food be abundant and easy to access, and civilization becomes impossible; so indissolubly dependent are our higher efforts on our lower impulses.'[10]

The scientific tradition to which Pavlov kept throughout his career was that of German experimental physiology, whose early achievements were so ably summarized by Lewes. Furthermore the focus of much of Pavlov's research reflects Lewes' interest in food. Except for one early period, all of the many hundreds of experiments carried out by Pavlov and his co-workers measured what was happening at one point or another along the digestive tract.

If in 1900 a gypsy with a crystal ball had told Pavlov that before his death he would become world famous as a scientific revolutionary and be acclaimed hero of a new communist society, his scorn would have been exceeded only by that of his colleagues at the University of St Petersburg. At that point the kind of experimental physiology he pursued so energetically had been entirely within the mainstream of his time; and outside the laboratory his behaviour and attitudes could be characterized as a cautious, uninvolved conservatism, which was remarkable only because of the time and place in which he lived. The beginnings of research on conditioning are interesting in terms of the preconceptions most people hold about the age and character of those responsible for a new scientific paradigm, quite apart from their importance in the development of animal psychology.

Pavlov's early career

As a child Pavlov lived in Ryazan, a small town some two hundred and fifty miles to the south-east of Moscow, where his father was a parish priest.

Russian priest enjoyed a humbler social position than his Western equivalents and was expected to be more or less self-supporting by tilling his land like a peasant. Only the exceptional priest had time and inclination for intellectual work as well. Pavlov's father was one such exception and he succeeded in inspiring at least three of his sons with a love of learning and determination to obtain a university education. From his father Pavlov also acquired a lifelong love of gardening and of hard physical exercise.

At the age of nine Pavlov suffered severely from a fall off a wall and as a consequence there was a delay of two years before he entered the local school. During this period he spent a large amount of time with his godfather who was abbot of a monastery near Ryazan. The abbot was as influential a figure in Pavlov's childhood as his father. He led a simple, spartan existence and devoted himself unceasingly to his monastic duties and studies. As an adult Pavlov believed in science instead of God, but otherwise the way of life he later adopted was very much like that of the abbot: simple, regular to an unusually precise degree and displaying an other-worldly lack of concern for anything but his work.

Pavlov's formal education began at the local Ecclesiastical High School in 1860 and a few years later continued at the Ryazan Theological Seminary. By then the exciting new ideas and discussions frothing in Moscow and St Petersburg of the early 1860s had even reached theological seminaries in towns such as Ryazan. In 1866 a new climate of reaction was triggered by an attempt to assassinate Alexander II and this reaction quickly began to affect education. Count Tolstoy, the Minister of Education, introduced into the state high schools a much more rigid curriculum that excluded the teaching of science. However, as a result of attending a religious institution, Pavlov was able to gain a progressive and stimulating education, in which there was plenty of opportunity for him to follow his own intellectual inclinations and to learn a good deal of science, when this had become impossible in the harshly disciplined setting of the secular high schools.[1]

Pavlov left Ryazan for good at the age of twenty-one to study natural sciences at the University of St Petersburg, where he was to spend almost all of his life. He walked the whole way, a distance of many hundreds of miles. His arrival in 1870 coincided with Sechenov's departure and so Pavlov's first formal introduction to physiology came from the lectures of Sechenov's successor to the chair of physiology, Ilya Cyon.

Cyon was a fine surgeon, a researcher with an international reputation and an inspiring lecturer. Unfortunately a large number of his colleagues resented the way in which he had been appointed and his unpleasant personality subsequently offended many more. A merciless attitude towards students who did poorly in his examinations helped to spread his unpopularity. Four years after Pavlov's arrival Cyon failed so large a proportion of the students taking the course in physiology that there was a riot and this set in train a series of events leading to his resignation and departure from Russia.[2]

Unlike many of his fellow students Pavlov came to respect Cyon highly and by his third year his childhood interest in physiology had turned into a decision to devote his life to research in the subject. Under Cyon's supervision he began an experimental study of the pancreas which delayed his graduation by a year, but enabled him to start to acquire the delicate and rapid surgical techniques that became a key ingredient of his later success. The project also gained him a gold medal and a four year scholarship for post-graduate study.

Cyon must have been equally impressed by Pavlov. He offered the young man a research assistantship when he graduated. This would have provided an excellent supplement to Pavlov's modest scholarship, and made for a convenient arrangement by which Pavlov could both continue to gain training in research and to keep to his decision to study for a second degree in medicine. However, with Cyon's departure from St Petersburg, Pavlov had to look for another post.

He eventually found an assistantship in the Veterinary Institute. Here his research switched to the study of blood circulation and of the innervation of the heart; these remained his main interests for the next twelve years. Meanwhile his medical studies gained him his second degree in 1879. By this time Pavlov had come into contact with the final major figure of his student years, Sergei Botkin.[3]

It was Botkin who had attended du Bois-Reymond's lectures of 1856 in Berlin with Sechenov. Over the next twenty years Botkin had become one of the most influential men in Russian medicine, having been largely responsible for making experimental physiology as central a part of the training of physicians in Russia as it had been for some time in Germany. This was particularly the case for students in St Petersburg where Botkin had become the professor of clinical medicine at the Military-Medical Academy. In keeping with the doctrine of 'experimentalism' that he preached, Botkin maintained an animal laboratory which, however, consisted of

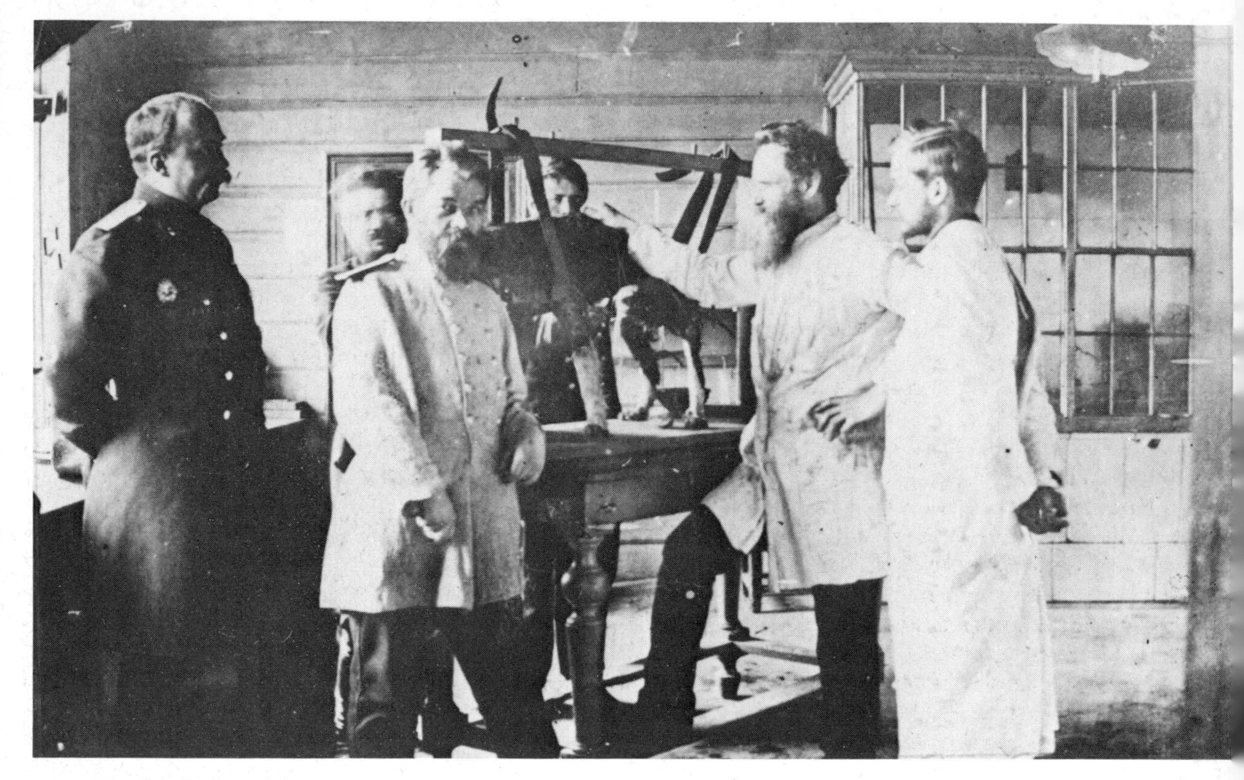

Fig. 5.1. Botkin's laboratory. Pavlov is second from the right with his hand resting on the dog. The dog's harness shown here is essentially identical to that used in Pavlov's conditioning experiments, even though this photograph was taken almost twenty years before Pavlov became interested in the conditioned reflex

little more than a wooden shed in the garden of his clinic. His administrative, teaching and clinical commitments left Botkin no time to become closely involved in research. In 1878 he needed a new director for his small laboratory. Pavlov was highly recommended and, although relatively young and unqualified, was appointed.

The new post provided an unusual degree of independence for someone still enrolled as a student. It also offered plenty of opportunity for learning how to direct other people's research, since Pavlov had to spend a great deal of time and effort, which brought him no immediate gain, in advising the many fellow students who were carrying out research projects under Botkin's nominal supervision. One of the earliest changes imposed upon the universities by Count Tolstoy was an increase in the importance of direct experience of research. Professors of natural science were strongly encouraged to involve their students in experimental projects since, as a highly time-consuming kind of work, such participation would make a student less likely to engage in 'illegal extracurricular activities', a label given to any group activity not explicitly organized by the authorities. From 1870 onwards research projects became a much more important element of a first degree in science in Russia than elsewhere.[4]

One consequence of this policy was that a scientist without the financial means to hire assistants could nonetheless remain highly productive as long as he could first attract and then rapidly train sufficient unpaid labour from students anxious to complete a satisfactory project. Students would compete to work in the laboratory of anyone with a reputation for providing an interesting topic for a project, for giving adequate guidance in how to carry out and report an experiment and, most importantly, for essentially guaranteeing that this particular requirement for the degree would be completed in time. Pavlov later acquired just such a reputation among the medical students of St Petersburg and the bulk of his research came to be based on student projects. This reputation was well deserved; his effectiveness in supervising such work clearly owed a lot to his early experience of helping fellow students who had signed up to work for Botkin.

The experimentalist approach to medicine that Botkin advocated had by the 1870s already become so pervasive in the medical faculties of St Petersburg that Pavlov might well have absorbed such an outlook even if he had never come into personal contact with Botkin. Until quite late in his career Pavlov took it for granted that the experimental study of some problem in physiology, no matter how abstract, was the most beneficial way of contributing to medical practice.

The new scientific medicine of the nineteenth century was anxious to rid itself of any vestiges of traditional beliefs in the importance of 'bodily humours'. Old ideas that various ailments reflected disruption and imbalance among the body's fluids were rejected in favour of new kinds of explanation, notably ones involving invasion of the body by micro-organisms; in this respect the discoveries made by Louis Pasteur were particularly influential. Alternatively, as within the Berlin school of physiologists, the nervous system tended to be seen as an all-pervasive influence in the way that the humoral system had previously been viewed. This doctrine of 'nervism' is still embedded in our language; someone's abnormal state may well be explained in everyday English as due to the state of his or her nerves as an alternative to blaming some hormone, or a virus or germ.[5]

Botkin strongly believed that understanding the nervous system was of paramount importance to medicine. In this case direct acquaintance with Botkin probably was an important factor in Pavlov's similar commitment to nervism.

In 1881, three years after his appointment to Botkin's laboratory, Pavlov married. When he had first arrived in St Petersburg the burden imposed by his complete indifference towards wordly matters – he never so much as bought himself a pair of shoes – had fallen upon his brother, who had taken full responsibility for finding the two of them somewhere to live and enough to eat. Through his brother Pavlov first met his future wife, Sara.[6]

At the time of their meeting Sara was a lively and independent young woman in her early twenties, training as a teacher and with a passion for modern literature. Pavlov was immediately enchanted by her, but it was two years or so before Sara took much notice of this shy man, six years her senior, who talked so earnestly of his current enthusiasm for the works of Herbert Spencer and was so surprisingly lacking in radical views for a student of his generation. He courted her with persistence and with the kind of idealism that remains insensible of the material benefits it is likely to confer.

Fig. 5.2. Sara and Ivan Pavlov at about the time of their wedding

The full extent of Pavlov's lack of concern for the ordinary things of life only became apparent on their wedding day when it turned out that he had no money whatsoever to contribute towards the wedding or to pay the return railway fare to St Petersburg. Sara had soon to take charge of all financial decisions, which were made the more difficult by Pavlov's refusal to let her find a job. They lived in unrelenting poverty. When a group of colleagues managed to raise money to pay Pavlov for giving some lectures so that the young couple's situation would be eased for a while, Pavlov simply used the money to buy extra experimental animals.

Poverty was not the only cause of the grim state of their early married life. During Sara's first pregnancy Pavlov insisted that their regular long walks continue and that she keep pace with his rapid stride; as a result she had a miscarriage. Their lack of a home of their own and his single-minded devotion to his experiments meant that they were often apart; she would stay with a friend while he spent day and night in the laboratory. During her second pregnancy Pavlov was more attentive; a son was born but, partly due to their lack of money, the child fell sick and died within a year.

What sustained Sara was belief in her husband's genius and in the supreme value of his work. In the early years of marriage they agreed upon a pact which both were to keep for the rest of their long life together. If she was to devote herself entirely to his welfare so that there would be nothing to distract him from his scientific work, then he was to regulate his life accordingly; she made him promise to abstain from all forms of alcohol, to avoid card games and to restrict social events to visits from friends on Saturday evenings and entertainment, in the form of concerts or the theatre, to Sunday evenings. His life was to be as ascetic as that of his uncle.

During the 1880s Sara's belief in her husband's future success began to be confirmed, as he continued as a full-time researcher, working partly as director of Botkin's laboratory and partly on his own projects. The results from his medical degree examinations earned him a second gold medal and a further small fellowship. In 1883 he submitted a dissertation on the innervation of the heart and gained his Doctorate of Medicine. This was shortly followed by his appointment as a lecturer in physiology at the Military-Medical Academy. A further fellowship enabled him to study abroad.

Within a year of Sara's miscarriage Pavlov departed alone to work in Germany for two years. Some of this time he spent, like a large number of the physiologists of his generation, at Carl Ludwig's laboratory in Leipzig. The remainder was spent with Rudolf Heidenhain, the professor of physiology at Breslau, whom Pavlov had visited for a couple of months some years earlier. Heidenhain shared Pavlov's interests in both the heart and the digestive system, was an expert on glandular secretion and was famous for his surgical skills.

In subsequent years Pavlov said little about his experience abroad. He appears to have regarded these two years simply as an opportunity to learn more physiology and improve his experimental techniques. It certainly did not leave him with any special love of Germany or of foreign travel. His nationalism was unusual for a scientist of that era. He never became fluent in any language but Russian. It is said that in old age, whenever he returned from a scientific trip abroad, he would bow down to kiss his native soil; and that when he needed to undergo a serious operation he insisted that it be carried out by a Russian surgeon and not by a foreign specialist.

By the end of the 1880s Pavlov had become widely known as one of Russia's most distinguished young physiologists. But at forty years of age he still had no stable position with a respectable salary. His various fellowships and appointments brought him prestige, but little money. A reputation for being outspoken about the lack of scientific merit and reliability of even those in important places and an unwillingness to compromise did not help his applications for the few permanent academic posts that became available.

Pavlov's situation was symptomatic of a crisis in higher education in Russia of the 1880s. Some of the problems were as elsewhere; the end of the expansionary period of the sixties and early seventies left a large number of able young Russian scientists eyeing academic positions likely to be occupied for a very long time by men only a few years older than themselves and in many cases far less distinguished.[7]

This also happened in Germany, but the position of science in Russian universities was made much worse by special political factors. The authorities continued to regard science with ambivalence. It was appreciated that research based in the universities played a vital role in the necessary modernization of the economy; and the material benefits proferred by applied science were much enjoyed. But the critical attitudes that arose with the spread of a scientific outlook were highly suspect; they were seen as important in the growing resistance to the autocratic system. In justifying the elimination of science from the high school curriculum Count Tolstoy had written: 'In the study of ancient languages – and sometimes in the study of mathematics – all knowledge imparted to the students is under constant and nearly errorless control, which discourages the formation of independent opinions. In all other subjects, particularly in the natural sciences, the student's interpretation of the knowledge he acquires is beyond the teacher's control. For this reason these subjects may engender personal opinions and differing views.'[8]

In 1884 direct government control of the universities was re-imposed and any remaining elements of academic freedom were abolished. The Czar's bureaucracy took all decisions on appointments, curricula and examinations. In doing so the role of the university as a centre of research was usually ignored. Subsequent decrees restricted access to education to various social classes; thus, in what became notorious as the 'Cooks letter', Tolstoy's successor at the Ministry of Education ruled that 'the sons of coachmen, servants, cooks, laundresses and small shopkeepers' should not be allowed to enrol in state schools.[9]

These developments produced a general decline in the number of students attending university which was particularly severe in the medical faculties. In 1880 medical students formed just under a half of the student body at the University of St Petersburg; by 1899 this proportion had decreased to less than a quarter.[10]

Earlier in the century the new German universities had promoted scholarship and research on the grounds that an effective teacher needs to be an active participant in his subject. Later it became clear that finding and maintaining the right balance between teaching and research is not an easy matter. A heavy emphasis on research can produce inadequate teaching, just as too much time spent on teaching and administration can impoverish research.

By the end of the 1880s a number of influential Russians had become extremely concerned about the future of science in their country. The solution that they found was one that led to a larger degree of separation between research and university education. The efforts of the Academy of Science and of some wealthy individuals led to the establishment of research institutes that had only tenuous links with the local university. This solution was also adopted in Germany and elsewhere in Central and Eastern Europe. It has persisted ever since in these countries where much more research takes place in institutions which play no part in undergraduate education than is the case in North America, for example, where the kind of two-tier university introduced by Gilman at Johns Hopkins has in the main served to keep a large proportion of research within a university context.

One of the earliest and most richly endowed of the new institutes in Russia concerned with medical research was the Institute of Experimental Medicine. This was founded in St Petersburg by a rich aristocrat, Prince Oldenburgski, who had been inspired by Pasteur's work in Paris and wished to see a Russian equivalent of the Pasteur Institute. The research it supported was intended to be directly concerned with the causes of diseases such as cholera and the plague.[11]

In 1890 Pavlov's situation was changed quite abruptly. The day after the offer of a chair at a Siberian university he was elected professor of pharmacology at the Military-Medical Academy and so could stay in St Petersburg. Less than a year later he added to this chair the post of director of the Physiology Department at Oldenburgski's Institute of Experimental Medicine. Here he was able to undertake on a generous scale the work on digestive processes that he had started a few years earlier, even though this did not have as direct a relationship to disease as the founder might have liked.

Pavlov came to spend most of his time at, and owe primary allegiance to, the institute which, as well as providing ample space and facilities, gave him considerable independence from the university and its internal politics. With his odd ways, uncompromising integrity and insistence on wearing civilian clothes Pavlov was an unconventional figure at the Military-Medical Academy and was detested by its rector. Pavlov's success in attracting students needing project supervision and in ensuring that most finished in time did not endear him to many colleagues in this era of dwindling student numbers. The rector did what he could to impede Pavlov's research by withholding normal professorial privileges; no research assistants were assigned to him and permission to travel abroad was blocked for Pavlov and for anyone who worked for him.[12]

Apart from improvement in his personal finances, the major benefit conferred by Pavlov's continued links with the academy, first as professor of pharmacology and then as professor of physiology, was the constant flow of eager student labour. In return he gave regular series of lectures, but this was about all; he was never very much involved in either the academy or university affairs in general. The Institute of Experimental Medicine insulated him from the turmoil that almost tore apart Russian society and its universities over the next three decades, and provided a setting in which the study of conditioning could begin without serious intrusions from the world outside.[13]

How Wolfsohn, Snarsky, Tolochinov, Pavlov and Babkin began to experiment upon conditional reflexes

A complicated illustration of the digestive tract in a mammal, which Lewes included in *The Physiology of Common Life*, had caught Pavlov's imagination at an early age; it is reproduced here in Figure 5.3. He became fascinated with the problem of discovering the operating principles of such an intricate system which he liked to compare to a complex chemical factory. How is it that just the right amount of the right kind of chemical is released at exactly the time it is needed as a particular foodstuff passes down the gut? The exquisite integration of processes involved in digestion must depend on an interlocking set of reflexes; what are these components and how are they integrated?

Early in his scientific training Pavlov rejected the then prevalent technique of using 'acute' preparations to study physiological problems; that is, to carry out operations that allow the study of some particular part of the body only in the short while before death occurs. Pavlov regarded such an approach, which he

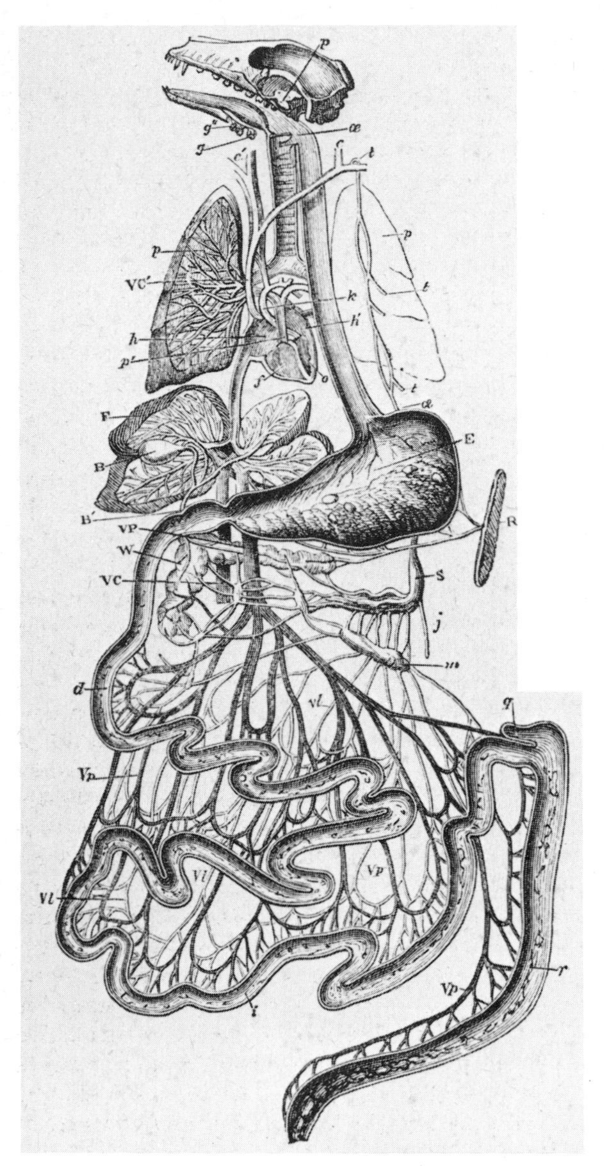

Fig. 5.3. The diagram of the digestive tract from G. H. Lewes' book on physiology which captured the young Pavlov's interest

called the 'vivisection' method, as a crude violation of an organism, one likely to be a major source of errors since specific effects of a particular operation are often masked by general shock resulting from surgery.

In contrast Pavlov normally employed what he termed the 'surgical' method whereby operations are carried out in order to reveal the normal workings of some process and so need to have a minimal effect on the animal. In such chronic preparations, the animal can usually resume its normal life; when recovery from the immediate impact of surgery has occurred, the only lasting change is that the experimenter now has access to events that formerly could not be observed.[1]

A good example of this approach is provided by one of Pavlov's earliest studies of digestion. The purpose of the operation was to establish an isolated part of the stomach wall, a 'gastric pouch', which would still contain the nervous pathways serving this part of the gastrointestinal tract. In 1879 Heidenhain had almost achieved this, but his operation involved cutting fibres of the vagus nerve. Pavlov worked out a way of establishing a pouch that left all the neural connections intact. This gave him a preparation in which, once the dog had recovered, he could study during the animal's normal feeding the release of gastric juices that were uncontaminated by food or saliva and that were isolated from any possible chemical, as opposed to neural, control.

This stomach pouch operation was particularly difficult and it is an example that also illustrates Pavlov's perseverance. It took almost six months, about thirty dogs and the rejection of expert advice that the operation was impossible, before he was successful.[2]

As this early study indicates, despite his opposition to 'vivisection' methods Pavlov was prepared at times to carry out studies resulting in the deaths of his subjects. In this he differed from Sechenov, who confined his work to frogs and refused to experiment with any warm-blooded animal. Pavlov's view was that, as long as we slaughter and hunt animals and kill thousands of our fellow men in wars, one could not object 'to the sacrifice of a few animals on the altar of the supreme aspiration of man for knowledge in the service of a high ideal, the ideal of attaining to truth'.[3] In practice it was very rare for Pavlov to sacrifice any animal. His experimental methods depended both on rapid, delicate surgery and on very high standards of animal care to ensure that after surgery his subjects lived long lives in perfect health.

An outstanding feature of the new laboratory at the Institute for Experimental Medicine was that it had excellent facilities for keeping dogs. It seems also to have contained the first surgery in the world which was specifically designed for animals and in which anaesthetic and aseptic conditions were used from the very beginning. With ample space for housing and plenty of animal caretakers the dogs were treated like favoured pets.

One surgical preparation became particularly common in the laboratory. This involved two operations. One was to establish a tube, or fistula, leading outside the body from a dog's esophagus so that, when required, 'sham feeding' could take place whereby any food taken by the dog into its mouth subsequently dropped out through the fistula before reaching the stomach. The second operation was to insert into the stomach a second tube which could then be used to collect gastric juice. This preparation was initially devised to study the effects of oral stimulation on gastric secretion without allowing food to enter the gut.[4] It was subsequently important for both financial and intellectual reasons.

For some years gastric juice became very popular around St Petersburg as a remedy for certain stomach complaints. As Pavlov was able to supply gastric juice in relatively large quantities and of a particularly pure quality by using the sham feeding preparation, the proceeds from its sale became considerable, to the extent of almost doubling the laboratory's income when this already far surpassed that of any comparable Russian laboratory.[5]

The other reason sham feeding became significant was that it turned a phenomenon whose existence had hitherto been controversial into an everyday event. It soon became well-known to everyone in the laboratory that a dog would secrete gastric juice as a response, not just to the presence of food in its mouth, but also to the sight of food; or, for that matter, to the sight of anyone who regularly came to feed him. Since the effect was held to be caused by the dog's psychological state of expecting food, it was labelled a 'psychic secretion' and considered to be quite different from the physiological secretions routinely studied in the laboratory. Until the end of the 1890s little serious attention was paid to the phenomenon.

Meanwhile research proceeded in a steady, productive fashion. By this time Pavlov had adopted the routine he was to keep for the rest of his life. From the beginning of September until the end of May he would spend all seven days of the week at work. Every day at the Institute started at nine o'clock with a precision that allowed clocks to be set by his arrival and ended equally punctually at six in the evening. A day's duties consisted of some mixture of surgery, supervision of the several current experiments, administrative jobs to do with the laboratory and some paperwork; it was interrupted only by the half-hour he allowed himself for lunch which he took alone. On returning home he would dine with his family and afterwards sleep from seven until nine o'clock. Then, with the possible exceptions at the week-end allowed in the agreement with Sara, an hour of further social intercourse with the family over cups of tea would be followed by further work in his study from eleven to one o'clock in the morning. This schedule was maintained until the summer, when he would go to the country for three months to spend his time on gardening, cycling and reading novels; physiology or any matter to do with science rarely intruded.[6]

Although a very regular routine is an essential ingredient of work with animals, no one else engaged in such research has ever imposed so precise and ordered a pattern upon their whole life. It was obsessional to a degree that would be considered deranged in any context but that of science or religion. Pavlov insisted that military punctuality be shown by all members of the laboratory, at least during their working hours. Years later an assistant was ten minutes late for the beginning of an experiment and explained that he had been delayed by the revolution that was erupting in the streets outside; Pavlov's reply was: 'What difference does a revolution make when you have work in the laboratory to do?'[7]

Pavlov may have seemed uncomfortably strange and remote to his colleagues and peers, but he was highly popular with students. He was far from being the traditionally authoritarian professor; his lectures were among the few in which students were encouraged to interrupt when they did not fully understand some point. The demonstrations that formed a crucial part of these lectures were prepared with great care. Science was presented as a collaborative enterprise in which mutual trust should extend even to the newest member. Pavlov's laboratory was renowned for its atmosphere of friendly cooperation and for the amount of attention Pavlov himself would give even to a new student spending just a few months there. Unlike many of his contemporaries, he opened his laboratory to both women and Jewish students. He could be irritable and explode with anger over a mistake; yet the most junior member knew that he could approach Pavlov at almost any time for help or advice on an experiment, or even to ask him to take it over for a while so that the student could take a break.[8]

Compared to an equivalent of today, Pavlov remained remarkably free from the various commitments that can compete strongly with active involvement in the work of a laboratory. None of his time was spent on writing grant proposals or reviewing those of others. He did not serve on committees, help edit journals or organize scientific conferences. He was a poor correspondent; in comparison both to the typical eminent scientist of later generations and, for example, to the British evolutionists of a generation earlier, Pavlov wrote very few letters to scientific colleagues. He did not even devote much time to writing reports of his research; most of the experiments were described only in the dissertation of the student who carried them out in the first place, although with considerable guidance from Pavlov where necessary. Frequently no other form of detailed report was ever published.

This reluctance to make public his research is also seen in the only two books he ever produced, which were both rewritten versions of a set of lectures. The first of these appeared during the calmly productive era of the 1890s. This was the *Work of the Principal Digestive Glands* of 1897. It described in a general way the findings that had been obtained in his laboratory over the previous nine years. It contained a passing reference to psychic secretion, but there was no hint that this was later to provide the central focus of Pavlov's research.

It was a number of years before this remarkable shift of interest and outlook was complete. Two phases can be discerned. The first involved the discovery that methods routinely employed in physiological experiments could equally well be used to study psychological phenomena; in this phase some of his students seem to have anticipated Pavlov's interest. The second phase led to the conclusion that psychological phenomena have to be described and explained in physiological terms or else they cannot be understood. It ended with a decision to devote the resources of the laboratory to this task. The transition from a dualist to a materialist position involved in this second phase made it the more dramatic and painful of the two. But somehow this change was made more rapidly by the fifty-year-old scientist who for decades had kept to a very fixed outlook on his science and on life, and had done so profitably, than by many of the much younger students and co-workers around him.[9]

The study of psychic reflexes began in 1897, the year that Pavlov's book on digestion was published, when a student named Stefan Wolfsohn carried out experiments on the first glands in the digestive tract, namely, the salivary glands. Some three years earlier a technique for inserting a fistula that made possible the measurement and extraction for analysis of saliva had been developed by one of Pavlov's assistants. By a striking coincidence this crucial first step in the study of conditioning in Russia occurred at exactly the same time as Thorndike at Columbia University in New York began to test his dogs and cats in puzzle boxes; later Wolfsohn and Thorndike both submitted their dissertations in the same year, 1898.

Initially Wolfsohn simply checked out in a systematic way a claim, made by Claude Bernard, that the set of reflexes involved in salivation is highly sensitive to differences in oral stimulation. Thus, the quantity and constitution of spittle varies widely according to what has been placed in the mouth. Dry food will induce the flow of a large quantity of particularly watery saliva, as will the insertion of some dilute acid or a handful of sand. In contrast, small stones and most wet foods produce very little salivation, while certain foods cause the glands to produce saliva containing a high concentration of mucous. In every case investigated it turned out that the salivary response obtained was beautifully adapted either towards the digestion of the particular foodstuff or towards the elimination of a potentially harmful irritant.

Wolfsohn next started to measure the salivary response obtained when a dog was simply shown various things, and discovered that it was equally sensitive to appropriate visual stimuli. Thus, when sand had been repeatedly inserted into its mouth, a dog would start to salivate at the sight of sand and the secretion would have the same characteristics as that produced by sand-in-the-mouth. This secretion was regarded as a psychic secretion like the one seen with the gastric juice preparation. It seemed clear to Wolfsohn and to everyone else in the laboratory that it was caused by the dog learning to expect that sand would shortly enter its mouth; that is, it involved mental, and non-physiological, processes.

This was very much the view of Anton Snarsky, the next student who began to work on this problem; it seems that he did so without a great deal of encouragement from Pavlov. Snarsky's main contribution was to show that apparently arbitrary signals could be as effective as the normal appearance of some substance in eliciting psychic secretion. One of his experiments can be seen as the first deliberate attempt to produce a psychic reflex. He coloured some acid black before introducing it into a dog's mouth and found that after a few repetitions profuse salivation would also occur to water that was coloured black or to the sight of any bottle containing black liquid.

For a while arbitrary stimuli, such as the black colouring used by Snarsky, were termed 'artificial' stimuli in Pavlov's laboratory, while those of the kind investigated by Wolfsohn – for example, the normal appearance of some substance – were termed 'natural' stimuli. Snarsky also showed that some apparently natural stimuli need to acquire their effectiveness in eliciting salivation. Thus, it turned out that a dog would initially show no salivary reaction at all to the smell of aniseed; but once aniseed oil had been placed in its mouth a few times, then its odour alone became an effective stimulus for the secretion of saliva. Later experiments by one of Snarksy's successors showed that a dog hitherto fed only on milk similarly displayed no reaction to the smell of meat until it had experienced meat-in-the-mouth.

Snarsky must have been an independent-minded and stubborn young man. In trying to make sense of his results he decided on a particular account of the dog's thoughts, feelings and desires during the course of its experimental trials and stuck to the account even though Pavlov favoured another. Their arguments became heated, while their differences remained puzzlingly unresolved. As Pavlov wrote many years later, it was 'an incident which had no precedent in our laboratory. We considerably diverged in our interpretation of this internal world; further attempts failed to bring us to a common conclusion, contrary to the usual laboratory practice, according to which new experiments undertaken by mutual agreement generally led to the settlement of all differences and disputes. Snarsky clung to his subjective interpretation of the phenomena, while I, astonished at the bizarre character and scientific barrenness of this approach to the problem, began to seek for another way out from this difficult situation.'[10]

By 1902 when Snarsky was at last allowed to submit his dissertation, Pavlov had begun to find a way out. Since no empirical means were available for resolving discrepancies between theories based on subjective processes, the solution was to reject all such theories. 'After persistent deliberation, after a considerable mental conflict, I decided finally in regard to so called psychical stimulation to remain in the role of a pure physiologist, that is, an objective observer and experimenter.'[11] Pavlov rarely referred to Snarsky afterwards and, when he did, it was with a degree of anger that seems accountable only if Snarsky had become a symbol for the set of beliefs that Pavlov himself had previously held and only with great difficulty relinquished.

About a year before the final break with Snarsky, Pavlov had been joined by a co-worker who became a close friend, but who was no more sympathetic than Snarsky towards the attempt to find a new way of understanding psychic reflexes. Ivan Tolochinov collaborated with Pavlov on the first experimental study of the phenomenon that came to be known as *extinction*. They found that the reactions to visual stimuli described by Wolfsohn depended on the continued application of the 'physiological' stimulus. For example, the sight of some familiar food would continue to evoke the appropriate kind of salivation only as long as its appearance was usually followed by its ingestion; if, instead, a dog was repeatedly shown the food, but not allowed to consume it, then the salivary reaction disappeared.

The fact that our mouths, or those of dogs, water at the sight, or even the thought, of something appetizing was in no sense discovered by Pavlov and his colleagues. However, that such effects persist only if certain specifiable conditions hold was indeed a new contribution from his laboratory.

Conditioning was still a side issue and most of the experiments in the laboratory were concerned with much more familiar kinds of process further down the digestive tract. It was in this context that the disturbing preliminary report of an experiment carried out in England by two physiologists, Bayliss and Starling, became known early in 1902. This shook the nervist foundations on which Pavlov's studies of digestion had been based.

Pavlov's fundamental assumption was that the integrated activity of the digestive system is based exclusively on the transmission of information via reflex arcs from sensors at one point in the tract to glands lower down. Thus, receptors in the mouth sensitive to taste, volume, texture and so on can give advance warnings of the arrival of a certain substance to various gastric glands by means of neural impulses; these impulses stimulate the glands to produce the appropriate chemical environment by the time the substance has descended to the stomach. The significant aspect of the experiment by Bayliss and Starling was to show that advance information may also be conveyed by chemical signals, or hormones. It turned out that the pancreas was at least partially under humoural control; under certain conditions the intestines release a hormone, pancreatic secretin, which triggers secretion from the pancreas.

At first Pavlov did not accept the evidence. Then in the Autumn of 1902, when Bayliss and Starling's full report was published, he asked one of his assistants to repeat their experiment. The whole staff of the laboratory gathered to watch in silence as the replication was performed. The effect of secretin soon became obvious. Pavlov disappeared into his office. He returned a half hour later, simply saying: 'Of course they are right. We cannot aspire to a monopoly of discovering new facts'.[12] From that point on the amount of laboratory activity devoted to the digestive system declined, while that to conditioning increased.

The international reputation earned by Pavlov's book of 1897 began to produce invitations to give lectures at prestigious conferences abroad. In April 1903 he attended the International Congress of Medicine in Madrid. Instead of discussing his work on digestion he chose to deliver the first public report of experiments on conditioning.[13]

In Madrid Pavlov still used the term 'psychical reflex' and had not yet made a full commitment to an entirely physiological account of conditioning. By that time he had been joined by a young researcher, B. P. Babkin, whose results were finally to persuade Pavlov that nothing was to be gained from trying to explain the phenomena of conditioning in subjective terms. As a student in 1902 Babkin first carried out a dissertation project on the pancreas. Afterwards he became a full-time assistant in the laboratory and began to follow up Tolochinov's experiments on extinction.

Babkin discovered two further effects. One was *spontaneous recovery*. A hungry dog was repeatedly shown at intervals of a minute or two a dish containing meat powder that he was prevented from eating. Once the salivary response had died away nothing happened for a few hours and then the dish was again presented. Babkin found that the salivary response now occurred once more and at almost full strength. The second effect was also a form of recovery from extinction and became known as *disinhibition*. Once he had extinguished the salivary response to the sight of meat powder – again by displaying it without allowing a dog any opportunity to eat – Babkin found that he could also restore the response by interpolating some other strong, although irrelevant, stimulus. In the initial experiments this was achieved by using another salivary reflex, that elicited by inserting a few drops of dilute acid into the mouth; a few minutes later the sight of meat produced as great a flow of saliva as ever. Later experiments showed that any kind of intense stimulation would have the same disinhibitory effect. For example, if the sudden slam of a door or the burst of an unfamiliar light occurs shortly before, or even during, the presentation of a stimulus that has just been extinguished, the latter's ability to elicit a response is temporarily restored.[14]

This work provides an interesting contrast to the concerns of the British evolutionists and early American animal psychologists of this era. In trying to understand how behaviour can be adaptive it is clearly as important to find out how and why reactions that are no longer appropriate to a certain situation disappear as to explain how they become established in the first place. An emphasis on intelligence and on the solving of problems made Western psychologists concentrate almost exclusively on the acquisition of new patterns of behaviour. It was many decades before they paid very much attention to extinction, which then turned out to present enormous theoretical problems. On the other hand, in Pavlov's laboratory concentration on phenomena related to extinction marked the pioneer experiments described here and continued throughout his lifetime. Correspondingly little experimental or theoretical effort went into the study of acquisition.

The immediate effect of Babkin's discoveries was to confirm Pavlov in his conviction that explanations in terms of commonly available psychological concepts were useless. All attempts by himself, and by others whom he questioned about these issues, to explain spontaneous recovery and disinhibition by appealing to the dog's beliefs and to changes in its awareness of the stimuli seemed ludicrously inadequate.[15] He stopped using the term 'psychic reflex' and, since the reactions he was investigating seemed to differ from more familiar reflexes only in that certain conditions had to be met for them to become established and be maintained, he started to call them 'conditional reflexes'. Later the Russian word was mistranslated into English as 'condition*ed*' and it has stayed this way ever since.[16]

By the time an account of Babkin's research was published in 1904, Pavlov knew that he was to be the first physiologist and the first Russian scientist to receive the Nobel Prize. This was for his work on digestion, but, rather than incline him to return to this field, it helped provide the confidence needed to continue with conditioning. The reaction of his fellow physiologists to the new direction he had taken ranged from alarm to complete dismay. Many advised him to 'drop that fad'. Even in its early days the Nobel Prize conferred such distinction that its recipients needed no longer be too concerned with the opinions of their scientific peers. It provided a licence to explore areas considered by general consensus to be too treacherous for lesser mortals.

The new field continued to prove very fertile as many further discoveries were made and the power of conditioning procedures became more apparent. The distinction between natural and artificial conditioned reflexes was now abandoned, since it seemed that even the most unlikely form of stimulation could become as effective a signal, or *conditioned stimulus*, as the smell or appearance of some very familiar food. For example, even though an event such as briefly cooling a small and arbitrary area of the skin had presumably never been remotely connected with feeding behaviour in the whole lifetime of a dog or, for that matter, during the evolutionary history of the species, it could easily be endowed with the property of eliciting saliva. All that was needed was a number of trials in which cooling was shortly followed by the

Fig. 5.4. Ivan Pavlov with B. Babkin, on the right, and G. V. Anrep, in the middle

arrival of meat powder. Examples like this persuaded Pavlov and many later students of conditioning that there was normally little to be gained from looking at the natural history of the animal used in their experiments.

The final point in Pavlov's steady change of outlook was reached by 1906. He moved beyond the belief that to understand conditioning a physiological, not psychological, approach was needed to the much more ambitious view that a physiological analysis of conditioning provided a means, and the only scientific means, of understanding the brain. On October 1st of that year he gave the Thomas Huxley Memorial Lecture at the Charing Cross Hospital in London.[17] After a brief survey of his general ideas on conditioning and of recent experimental data he ended with the following comments on the relationship between physiology and psychology. 'The investigation of conditioned reflexes is of even greater importance for the physiology of the highest parts of the nervous system. Hitherto this department of physiology, throughout most of its extent, has been cluttered with foreign ideas, borrowed from psychology, but now there is a possibility of its being liberated from such harmful dependence.' As an unconscious echo of Descartes' lords and masters of creation, Pavlov went on to suggest the advantages stemming from an understanding of the brain. 'The conquest which physiology has yet to make consists for the most part of the actual solution of those questions which hitherto have vexed and perplexed humanity. Mankind will possess incalculable advantages and extraordinary control over human behaviour when the scientific investigator will be able to subject his fellow men to the same external analysis as he would employ for any natural object, and when the human mind will contemplate itself not from within but from without.'[18] Science had not yet been put to the kind of use that would make a reader from later in the century sense the cold chill in such a message.

Vladimir Bechterev and Objective Psychology

The study of conditioning began during a period of dramatic political developments in Russia. When the new Czar, Nicholas II, began his reign in 1894 there was widespread hope that he would end the long era of repression and revive the reforms of the 1860s. He did nothing to encourage such hope. On the contrary the government began to react in a heavy-handed manner even to suggestions for minor changes in the political system or for minimal increases in personal freedom. As a result, opposition to the Czar's authority became deeper and more extensive than ever. In 1901 the first mass demonstrations by students took place in the streets of St Petersburg. In 1903 the first general strike was staged by industrial workers in the south. A year later the series of Russian withdrawals in the face of the Japanese armies marked the first major military defeat of a European by a non-European nation in many centuries; faith in the Czarist system ebbed further. In 1905 a revolt by the peasants broke out on a scale not seen in Russia for over a century.[1]

Early in January, 1905, a deputation of unarmed workers wishing to present a petition to the Czar was fired upon by troops. The day became known as Bloody Sunday and these killings triggered the further series of upheavals – strikes, demonstrations and mutinies within the armed forces – that has since been known as the 1905 Revolution. The Czar's ministers uncertainly mixed efforts to put down the rebellion by force with the granting of some reforms. Thus, in April a degree of religious toleration was allowed whereby a Russian subject might leave the Orthodox Church for some other variety of the Christian faith without incurring any penalty.

The universities were at the centre of this political strife. The lives of almost all academics and scientists were affected to some degree by a general belief that profound changes in Russian society both were needed and were about to happen. At the 1904 congress of the country's leading medical society, for example, it was resolved that the fight against diseases could only be generally successful 'under conditions guaranteeing the broad spread of information about their causes and prevention and, for this, full freedom of person, speech, press and assembly are necessary prerequisites'.[2] This Pirogov Society, whose meetings Pavlov attended, had developed its own plans for combating an outbreak of cholera in Southern Russia, but these had been blocked by the government.[3]

In August, 1905, the universities were given back the autonomy that had been taken from them in 1884. They became the scenes of mass meetings that could now be held without interference from the police. From such meetings in the University of St Petersburg emerged the first workers' council, or soviet, of which one prominent member was Leon Trotsky. This began to act as an alternative government for a large part of the city. In October a general strike forced the Czar to make a further crucial concession: a government manifesto announced that a representative assembly, the Duma, would be established and given some legislative powers, and the manifesto promised major changes in the laws on civil rights.

The educational and other reforms enacted during the 1905 Revolution were later gradually eroded. The Czar's manifesto divided opposition to the government and from early in 1906 the Czar steadily regained his power. Within three years measures such as the ban on any kind of student society, the exclusion of women from universities and restricted quotas for Jewish students had been brought back. By 1911 it became common for wholesale expulsions of students from the universities to be carried out and for uniformed police or detectives to attend most lectures. But before the heavy and inept hand of repression descended there was for a while an air of excitement and confidence in the future like that of the sixties when Sechenov had published his *Reflexes of the Brain*.

Pavlov was no political radical like Sechenov. The sole extent of Pavlov's participation in the events leading up to the 1905 Revolution appears to have been the addition of his signature along with those of forty-two other scientists to a memorandum of 1900 enumerating the defects of a secondary education biased strongly towards the classics.[4] Yet, as a scientist searching for a consistent framework for the new study of conditioning, Pavlov eventually adopted a materialist outlook identical to the one advocated by Sechenov forty years earlier. By that time Sechenov was dead.

As a young man Pavlov had read Sechenov's *Reflexes of the Brain* and as an old man he admitted that he may have been unconsciously, but nonetheless strongly, influenced by the book. However, there is no indication that Pavlov made any effort to study Sechenov's work or even to meet him.[5] One suspects that until his interest in conditioning was fully developed Pavlov's attitude towards Sechenov may well have been a mixture of respect for the very early work on inhibition and regret that this had not developed into a steady and substantial body of experimental data, due to Sechenov's unfortunate interest in philosophical issues and his tendency to involve himself in political affairs.

A contemporary of Pavlov, Vladimir Bechterev, was in many respects more like Sechenov. Bechterev shared Sechenov's interest in a broad range of scientific subjects and similarly participated in life outside the laboratory. Pavlov disapproved of Bechterev, and one of the reasons that Pavlov found the shift to a physiological analysis of psychological phenomena so painful may well have been because this meant the adoption of a point of view already identified with Bechterev. Pavlov viewed Bechterev's work as chaotic and careless to an extent that debased science, so that it must have been difficult to make what was in effect a public admission within the medical world of St Petersburg that he, Pavlov, now agreed with Bechterev on this matter.

Bechterev was almost eight years younger than Pavlov, but had made much quicker progress up the educational ladder. He entered the Military-Medical Academy at the age of sixteen and obtained a first degree in medicine in 1878, one year earlier than Pavlov, who had been older when he first entered university and had obtained a natural science degree before beginning medical studies. In contrast to Pavlov's student involvement in experimental physiology, Bechterev developed an interest in psychiatry. He spent three years training in a hospital specializing in mental and nervous diseases, which ended when he submitted a doctoral dissertation on body temperature in certain forms of mental illness.

From 1881 Bechterev studied abroad, visiting the laboratories of a range of famous physiologists and psychologists. These included du Bois-Reymond in Berlin, Charcot in Paris, where a new approach to the study and treatment of neuroses was arousing wide interest, and Wundt, whose laboratory of experimental psychology in Leipzig was still very new. Bechterev spent most of his time in Leipzig, but in the laboratory of Flechsig, a leading neuro-anatomist. Within four years Bechterev's papers on topics in both psychiatry and neuro-anatomy had made him famous enough to be invited to return to Russia and become professor of psychic diseases at the University of Kazan. This was a remarkable tribute in an age of university recession.

At Kazan Bechterev was enormously energetic. He set up a department for the study of mental disease, was involved in the founding of a new psychiatric hospital and began the first Russian laboratory of experimental psychology. He also managed to complete a scholarly book on the anatomy of the nervous system which became a standard handbook in Russia.

While still in Kazan Bechterev began to develop ideas about a new approach to psychological phenomena. One aspect of this appeared in his clinical work. He came to the conclusion that much more was to be gained from studying in an objective manner the changes that occurred in a neurotic patient's life than from attempts to analyse his subjective experience. The second aspect was related to his studies of brain function. Some of his experiments involved training dogs to perform tricks such as begging, making dancing movements or offering a forepaw and then determining in which areas of the brain a surgical lesion abolished such skills.[6]

In 1893 Bechterev returned to St Petersburg as professor of psychiatry at the Military-Medical Academy. He continued to be as busy an organizer as ever; he set up new clinics and laboratories, extended the work on neurological diseases at the hospital attached to the academy, founded a 'Society for Normal and Pathological Psychology' and started a journal that published papers in psychiatry, neurology and experimental psychology. His involvement was not confined to founding and editing the journal: he was the author of fifteen out of the forty-five papers in its first volume, and continued to maintain this proportion over the next few years.[7] In many ways his activities mirrored those of his nearest American equivalent, G. Stanley Hall.

His research associates and the many students who chose to carry out experiments in his laboratories saw little of him. Consultation about an experiment that he had first suggested took place on the run, if it occurred at all. Although able to work for eighteen hours each day, his commitments would always overflow. He needed at most five hours of sleep, so that consultations with colleagues or patients could be made for after midnight and many of his papers were written as he sat in bed with his wife sleeping beside him. And yet everything was still done in a great rush.[8] A new student was given a bewildering degree of freedom, in complete contrast to Pavlov's laboratory where it was insisted that a student first repeat an experiment already carried out at least once in the laboratory – which served both to train the student and to check on the replicability of earlier data – before starting a more original piece of work whose aims and details were usually quite closely specified in discussion with Pavlov.[9]

Pavlov and Bechterev were colleagues who came to share the view that the study of conditioned reflexes – 'association reflexes' was the term strongly preferred by Bechterev – was central to a scientific study of the mind and both were unrelentingly energetic in their devotion to science. In all other respects they differed greatly.

Bechterev was an inspired innovator, a genius at initiating all kinds of projects, but not as good at carrying things through to completion. He clearly had little taste for the repetitive tedium and thoroughness demanded for most experiments. When Pavlov warned his students against spending too much time reading and advised that it was usually fatal for a

Fig. 5.5. Vladimir Bechterev

scientist to start to write books, particularly textbooks, he almost certainly had Bechterev in mind. Bechterev read several languages, covering a wide range of topics in neurophysiology, psychology or psychiatry, and he wrote an enormous number of papers and an impressive number of large books. He has been described as always being surrounded by paper; 'in a carriage, in a railway car or sitting at conferences – everywhere and always – he had galley proofs and new manuscripts with him'.[10]

Bechterev's accomplishments extended beyond science. He was the principal founder of the first major educational institution to be established in twentieth-century Russia. This was the Psycho-Neurological Institute, which opened in 1907 and whose student enrolment quickly became the second largest in St Petersburg outside the university. Inspired by the mood of the 1905 Revolution it remained independent of the state system of higher education and offered a novel mixture of subjects; its degree course combined the study of history and philosophy with that of science and medicine.[11]

The major contribution by Bechterev to animal psychology began with a paper, called *Objective Psychology*, which appeared in 1904.[12] This expressed his dissatisfaction with current psychology and summarized his views on the direction that the subject should take. He argued that introspection was a totally inadequate method for understanding the mental processes even of normal, educated adult human beings and had no contribution whatsoever to make to the wider range of problems with which psycholgoy should be concerned, such as mental testing, psychiatric disorders and animal behaviour.

Following Sechenov, Bechterev viewed the reflex as a key concept for the new objective psychology and, also like Sechenov, he worried about the fact that in all but simple reactions there is often little correspondence between the intensity of a stimulus and the intensity of the reaction it may produce. However, in the forty years since Sechenov had first discussed this issue, technology had devised means of detecting all kinds of previously imperceptible reactions. By measuring slight changes in heart, pulse or breathing rates, in glandular secretions or in skin resistance it should become possible, Bechterev suggested, to measure human reactions to any kind of stimulus. This should allow the objective study of processes such as thinking that had previously seemed to be accessible only to introspection. He summed up his general rejection of the prevailing dualist approach to psychology in the following words: 'It is evident that objective psychology has no need of metaphysical terms such as the "soul", "intelligence", "will", "imagination" . . . objective psychology excludes them completely as useless material'.[13]

The first discussion of objective psychology appeared at the same time as a report by Tolochinov on the conditioning research he had carried out in Pavlov's laboratory. Bechterev already knew a little about this work, but its significance for his own general views had yet to sink in. But within three years conditioning experiments with dogs were being carried out in the laboratory as well. Some involved the recording of salivary reactions, but in general Bechterev was much more interested in the use of skeletal movements rather than secretion from some gland, since he felt that the former were much more important for human psychology. He had the good fortune to be joined by a man named V. P. Protopopov who was evidently an ingenious experimenter and someone who could work in a productive and independent way under the conditions of Bechterev's laboratory. In 1908 Protopopov worked out a convenient method for studying dogs which could also be applied in experiments with human subjects.

The situation used by Protopopov was one in which a dog stood with one paw resting on a metal plate; a voltage could be applied to the plate of just sufficient intensity to make the dog's leg flex. The standard procedure was to switch on a light, a sound, or both together, at the same time as shock was delivered to the dog's paw. After repeating this for a number of trials one of the stimuli was presented alone and, all being well, was found to elicit flexion of the leg in the absence of any shock. No surgery was needed, nor special expertise. With near identical equipment one could study the conditioning of a leg movement or the jerk of a finger in a human subject.[14]

For Pavlov the study of conditioning provided a major insight into the brain's basic functions. Bechterev's interest was somewhat different. He was much more committed than Pavlov to the view that the brain's various activities can be precisely localized, and consequently a major question for him was often *where* in the brain does such-and-such occur. The early research in Kazan on brain lesions that would abolish trained movements was an attempt to locate the new neural connections that must underlie some acquired skill such as begging or offering the paw. Conditioning appeared to offer a promising new technique for answering such old questions; set up a well-defined conditioned reflex and then determine where in the brain lesions will cause it to disappear.

Conditioning also offered a means of keeping the promises made for objective psychology. A critic might scoff at the idea of studying perception other than by using human observers to report their perceptual experience; Bechterev would show that the perceptual world of a dog could be perfectly well explored in an objective manner by means of conditioning. After first establishing some specific event – for example, a tone of fixed intensity and precise frequency – one could vary some parameter of the stimulus by small degrees – change just its intensity, for example – and look for a corresponding change in the intensity of the conditioned response. Such a change would indicate that the animal had perceived the difference in the stimulus.

A description of Protopopov's procedure and a discussion of the importance of conditioning was included in a book which Bechterev published in 1910. *Objective Psychology* expanded on the arguments he had put forward six years earlier in his paper of the same title, and added new material on conditioning and animal behaviour. It was a clear, systematic and stimulating book which showed the author's familiarity and understanding of an impressive range of contemporary research. It should have added greatly to Bechterev's reputation. Unforunately by the time it appeared the inevitable confrontation with Pavlov had occurred.

Bechterev had made two particular claims on the basis of his students' work which Pavlov did not believe. One was in the area of perception, and the other was to do with brain localization. Some experiments in Bechterev's laboratory showed that, after exposure to a procedure involving lights varying in colour, a dog's conditioned response to one light could be distinguished from the response to another. Bechterev took this result to mean that dogs can distinguish between different colours.[15] In Pavlov's opinion the possible importance of intensity differences, which would give rise to variations in perceived brightness, had not been adequately controlled. Experiments in his own laboratory gave no sign of any canine perception of colour. No one decisive study resolved the issue, but general opinion then, as ever since, held that Pavlov was correct.

The other major controversy began earlier and came to a dramatic climax in 1909. Some of Bechterev's early research, begun when he was still in Kazan, had been on areas of the cerebral cortex that can be stimulated to produce some salivary secretion, and hence were termed 'salivary centres'. Prompted by the new work in Pavlov's laboratory, Bechterev's interest returned to these centres and in 1906 one of his students, named Belitsky, claimed that removal of the cortex in this area abolished conditioned salivary reflexes.

This claim was challenged by results obtained by a number of Pavlov's co-workers. They found that after apparently similar cortical ablations reflexes that had been conditioned prior to surgery remained and new ones could still be established. Pavlov seized on this discrepancy as an opportunity to show up the low quality of work in Bechterev's laboratory. He discussed the issue in public at a meeting of the Russian Society of Physicians in 1907 where, to the embarrassment of his students, he ended by appealing to the authority of his own reputation, proudly quoting from Sechenov's autobiography a brief comment that Pavlov was believed to be the most skilful surgeon in Europe.

The controversy became increasingly bitter. Further papers reporting experimental results from Bechterev's laboratory that supported Belitsky's original claim were challenged by contrary results obtained by Pavlov's students. After two further meetings of the society that were marked by prolonged and heated argument, Bechterev and his associates remained silent at subsequent meetings whenever an experiment from Pavlov's laboratory was reported or discussed.

Finally, Bechterev entrusted a further student, named Spirtov, with the task of repeating one of the apparently crucial experiments carried out by Pavlov's workers. Spirtov obligingly showed that the claims made by Pavlov's group were wrong and those made by Bechterev were right: in two dogs with appropriate cortical ablations conditioned reflexes established before surgery were lost, and Spirtov was unable to establish any new reflexes. Bechterev decided on a public demonstration and took the two dogs to the next meeting of the society. Spirtov tested them in full view of the audience which could see for itself that a visual stimulus that had previously served as a conditioned stimulus now no longer elicited a drop of saliva.

At the end of this demonstration Pavlov came up to the front and, despite protests from Bechterev and Spirtov, insisted on testing the dogs himself then and there. He simply poured a little dilute acid from a test tube into the dogs' mouths a number of times and showed that subsequently the sight or sound of the acid splashing in the tube was sufficient to elicit salivation. This impressive demonstration of expertise left the crowded room in no doubt as to who was the superior scientist; it was much more effective than the appeal to Sechenov two years earlier. Pavlov was now not just Russia's only Nobel prize winner; he had publicly triumphed over the only other major scientists with a comparable reputation in Russian neuroscience.[16]

Following this encounter and the publication of *Objective Psychology* Bechterev's involvement in animal research declined rapidly. Conditioning studies using human subjects continued in his laboratory, but a new interest in child development and a renewed interest in psychiatric issues became his main preoccupations.

Pavlov's later work

The conventional experimental approach in nineteenth-century physiology was based on the intensive study of a few individual subjects. To answer a question about the function of some organ one set out to obtain a suitable 'preparation'. A considerable number of animals might be used in the course of developing appropriate surgical techniques, for example, but once an effective method had been developed the organ could be studied in a single animal. If there were any doubt surrounding the reliability of some aspect of the results, then the experiment might be repeated in another animal or two. As we have seen, it was routine in Pavlov's laboratory for the first task given to a new student to be that of attempting to reproduce in a new dog the results obtained by one of his predecessors.

When Pavlov shifted his research from conventional physiology to conditioning he retained his commitment to the study of individual animals. No other approach could have been so productive during the early years following the shift. Within an amazingly brief period Pavlov's experiments had uncovered a whole range of important phenomena; hardly a salient characteristic of conditioning was left for subsequent workers to discover. But, after entering a new field and locating most of the highly robust and general effects, what comes next? One possibility is to devise a systematic theory that covers the known phenomena and then to test its implications. Another is to continue in an exploratory vein with imprecise and largely implicit theories as a guide, but with the problem that the phenomena now become more elusive and variable. Under these conditions it can be dangerous to draw general conclusions from results obtained from just one or two animals.

The dispute with Bechterev appears to have diverted Pavlov from such issues. It was acceptable to use very few animals in the research on the effect of brain lesions, which was undertaken mainly for the purpose of refuting Bechterev's claims. Similarly, a limited number of subjects was satisfactory in the perceptual studies carried out during this era, which also reflect Bechterev's influence.

From about 1910 onwards the problem of variability kept on intruding. It occurred, for example, in some phenomena Pavlov termed 'induction' and 'irradiation' effects; they could be clearly obtained in some dogs, less so in others.[1] There was also the problem of sleep: from the early experiments onwards it was found that many dogs became quite drowsy and unresponsive in the experimental situation, which was, after all, designed to be as lacking as possible in all stimulation except what was occasionally provided by the experimenter; some animals proved to be impossible to use as experimental subjects for this reason.[2]

In 1911 a particularly interesting experiment was carried out by one of Pavlov's students named Erofeyeva; the situation she used turned out to be one in which dogs varied considerably in the way they reacted. Erofeyeva used a mild electric shock applied to the dog's skin as a stimulus that preceded the delivery of food. She found that the defensive movements and signs of distress her dogs initially made in reaction to the shock disappeared, and instead the shock came to elicit calm salivation. This effect was subsequently termed *counter-conditioning*; it seemed to show that conditioning methods could be powerful enough to neutralize an aversive event and even turn it into an attractive one. It seems, however, that later attempts to replicate Erofeyeva's specific results were not always successful.[3]

Two or three decades later it became common for researchers faced with such unexplained variability to use experimental designs employing groups of animals which could be compared with appropriate control groups in order to isolate the causes of variability. Pavlov employed such an approach very rarely and then only towards the end of his life.[4] Instead he tended to fall back upon an appeal to possible difference in his subjects' temperaments and inherited dispositions to explain variability. In later years distinguishing different 'types' of dogs in a systematic fashion became one of his major enthusiasms.

The counter-conditioning study by Erofeyeva is interesting for another reason. Seeing the potential of her techniques outside the laboratory she became an early proponent of natural childbirth methods.[5] This marks the first attempt to develop applications of the study of conditioning. Pavlov himself did not discuss this particular by-product, but about this time he began to develop an interest in neurosis. In certain studies some dogs would remain in a distressed state for some time after an experimental session. They could perhaps be loosely described as having been made neurotic by the experiment. Treatment with bromide was very often followed by the dog's return to a normal state. Pavlov became a great believer in this particular therapy.[6]

The evidence supporting such beliefs was slight and unsystematic. The occasional abnormal state of a dog could have occurred for a number of reasons besides the one preferred by Pavlov, that it was due to the exposure of a certain type of nervous system to a particular experimental procedure. Similarly, recovery following bromide treatment did not mean a great deal in the absence of systematic comparison with recovery rates when no bromide was given. This general kind of weakness arising from the study of individual subjects can often be overcome by repeating experimental manipulations a number of times in the same animal, but this is hard to do with something like 'neurosis'.

Pavlov's interest in problems of clinical psychology was an excursion into territory in which Bechterev was the major Russian figure. For the first time Pavlov also began to suggest in public some possible implications of conditioning for matters of very general human interest. On learning of Erofeyeva's experiments on a visit to St Petersburg the English physiologist, Charles Sherrington, lightly commented that now he understood the psychology of martyrs.[7] Pavlov made such remarks with a good deal more seriousness. He started to discuss the 'reflex of purpose' and that of 'freedom'. Ready physiological solutions were produced for complex psychological problems; for example, he suggested that 'the tragedy of the suicide lies in the fact that he has an inhibition, as we physiologists would call it, of the reflex of purpose – most often a momentary and only rarely a continued inhibition'.[8]

Bechterev was scornful of Pavlov's ventures into human psychology and at the same time feared for the effect they might have on his own attempt to found a psychology based on the concept of the reflex; he felt that 'the adversaries of the objective method in its application to the investigation of human personality are given a weapon' by the simplistic nature of Pavlov's claims.[9] In connection with some speculation by Pavlov on the inheritance of a servile attitude, Bechterev wrote as follows. 'History shows that a slave's child, when educated in a family of free citizens, becomes just as freedom loving as his fellow citizens and does not betray any sign of inherited slavery. All these facts are universally known to an extent which would make it unnecessary to mention them here, if this problem had not been recently touched upon by the authoritative physiologist, Professor I. Pavlov, who, disregarding the long and instructive history of this important problem, solves it negatively. Professor Pavlov bases his conclusion solely on his observations of one lively dog, with abundant and spontaneous salivation . . . The data which he cites in favour of the existence of an innate "freedom reflex" and a "slavery reflex" in dogs are absolutely inadequate; we have still less reason for extending these conclusions to man who according to Pavlov's statement, also possesses an innate "freedom reflex".'[10]

A new public controversy between the two men over issues more philosophical than in the earlier disputes was prevented by the occurrence of the 1917 Revolution. Bechterev was highly sympathetic towards the new government and for some years represented university students on the local Workers and Peasants' Soviet. His general political outlook, as indicated, for example, by the quotation above, should have further endeared him to the communist party. In contrast, Pavlov was often quite outspoken in criticizing aspects of the new regime. However, by now Pavlov had an unrivalled international reputation, whereas Bechterev's was in decline.

The scorn Bechterev felt for Pavlov's views on human behaviour had been expressed in a large, rambling and barely readable book which had appeared just before the Revolution and was titled *General Principles of Human Reflexology*. One important reason for the subsequent lack of interest in Bechterev among English-speaking psychologists has no doubt been because this was his only book to be translated into English. Contemporaries have noted that he appeared to age rapidly over his last ten years or so of life. After the Revolution his Psycho-neurological Institute was renamed the V. M. Bechterev Institute for Brain Research and its funding was continued by the communist government. But after he died in 1927 at the age of seventy he was quickly forgotten.[11]

Pavlov lived on. Perhaps it was his strong belief in physical exercise, the extreme regularity of his life, or his abstinence from potentially debilitating activities, but, whatever the reason, he continued to direct his now several laboratories and to broaden his interests in as energetic a fashion as ever. The Great War, the Revolution itself and the extreme rigours caused by the economic upheaval and civil war that followed had never quite halted his research, even though there were periods when there was scarcely enough food for himself and his family, let alone the dogs, and even though most of his previous assistants and would-be students were tending the wounded and dying on one front or another.

In 1921 Lenin decreed that Pavlov, now 72 years old, was to receive exceptional treatment from the local soviet. The compatibility between Pavlov's science and Lenin's philosophy, together with the desire to demonstrate to the rest of the world that the work of a prominent scientist could flourish within a communist society, made Pavlov's own political outlook of little import. He was provided with new facilities, new physiological institutes were opened and now there were more co-workers than at any previous time. So generous was his funding that by 1930 his critical attitude towards the Soviet government had very much softened.

His extensive resources allowed Pavlov to pursue all of his varied interests. Basic studies of conditioning were continued by many of his students, but the topics that Pavlov chose to talk about in the many addresses he was now invited to give were usually mental illness, sleep or personality. Even in this difficult period immediately following the Revolution he had begun to follow up his interest in human mental disorders. In the summer of 1918, instead of his usual holiday, he worked in a psychiatric hospital where his contact with patients confirmed him in his view that their disabilities could be explained in terms of the derangement of inhibitory processes in the brain.[12] Occasional examples of 'experimental neurosis' in dogs continued to occur, as, for example, in some replications of Erofeyeva's counter-conditioning study and also in some situations where a very fine perceptual discrimination was required. Particularly striking effects of this kind developed in an experiment in which food followed the presentation of a circle, but not that of an ellipse with almost equal ratios. In September of 1924 a severe flood in Leningrad reached the area where Pavlov's animals were housed and there was great difficulty in rescuing them. Subsequently many dogs reacted in a very irregular fashion in their experiments; this was attributed to the effect of the traumatic experience on 'weak' nervous systems.[13]

There were also some entirely new departures. As elsewhere, Lamarckian inheritance had been widely accepted by Russian biologists during the nineteenth and well into the twentieth century. Pavlov was no exception, as one might expect for someone so greatly influenced by Herbert Spencer. But the issue became a live one again in the 1920s and Pavlov, as a committed experimentalist, decided to test his belief. He tried to detect the possible effects of establishing conditioned reflexes in one generation of mice on their descendants; the choice of animal reflected the speed with which Lamarckian inheritance was commonly believed to operate, as had Thorndike's choice of chicks to look at the same problem over twenty years earlier. At first some positive signs were obtained, but eventually Pavlov decided that his data did not indicate any genetic effect of an animal's conditioning history.[14] This displayed an ability to question a long-held belief that would be creditable in someone of half Pavlov's age and authority. Another departure that also involved the use of a new kind of experimental animal was to study the problem-solving abilities of chimpanzees in order to test claims being made for these animals by some researchers in American and German laboratories.

Despite its huge quantity the work carried out in Pavlov's laboratory after the Revolution had a fairly limited impact outside the Soviet Union. The evidence he gave in support of his various claims was never very substantial or conclusive. Moreover there was no easy way of finding out in more detail about the research since it continued to be reported only in brief,

Fig. 5.6. Pavlov lecturing

non-technical papers. Pavlov once explained that he had intended to prepare a systematic presentation of his work, but then the revolution had occurred. In 1924 he gave a series of lectures in Leningrad that formed the basis for a book, *Conditioned Reflexes*, which provides the only overall account of his ideas and which, after its translation into English in 1927, has provided the primary means by which his work has become known outside Russia. However, even this book failed to contain the detailed exposition which would have allowed serious assessment of his research. Pavlov never seems to have felt that the work had reached a ripe stage for taking stock. As he commented in *Conditioned Reflexes*, 'new problems are perpetually arising, and at the same time an equally large number of questions are still left unsettled. We often feel compelled to turn our attention from problems which directly confront us to some unexpected new phenomenon which introduces fresh problems or which necessitates a revision of old points of view.'[15]

Pavlov believed himself to be a pioneer extending the domain of physiology beyond the limits set for it by the Berlin school. Yet very few physiologists outside Russia followed his lead in using conditioning as a method for studying the brain. Pavlov saw this failure as reflecting philosophical cowardice, as clinging to the idea that the mind should be left alone and as fear of the thorough-going materialism that Pavlov himself had adopted. From the perspective of post-revolutionary Russia it seemed that scientists still living in capitalist societies, which were rigidly divided into a ruling class that took decisions and a proletariat that provided physical labour, might well find it very difficult to escape from making an equivalent distinction between mind and body.

Whatever the truth of such a view there was another more prosaic reason why neurophysiology in general was little affected by Pavlov's work. The brain processes that Pavlov invoked to explain conditioning, neurosis or differences in personality made little sense to most physiologists from the 1920s onwards. For someone dedicated to understanding the brain Pavlov remained singularly uninterested in its anatomy or in the fundamental changes in neuroscience that began around the time he started the study of conditioning; namely, general acceptance that the nervous system is composed of individual nerve cells separated by synapses and that neural action consists of the transmission along the axones of nerve cells of brief states of depolarization of the cell membrane, or 'spikes', which produce interactions at synaptic junctions. It was difficult to interpret Pavlov's references to 'waves' of excitation or inhibition, to 'weak' nervous systems or to the production of neurosis by the 'clash of the two antagonistic nervous processes' in terms of contemporary views of the nervous system. This problem was of major concern to one of Pavlov's most ardent admirers outside the Soviet Union, the Polish scientist, Jerzy Konorski.[16]

Another problem which also perturbed Konorski was later very prominent in debates over conditioning among Western psychologists. Many aspects of an animal's behaviour are profoundly affected by the past consequences of its behaviour; this is particularly true for skeletal activity or 'motor movements', but may not be true at all for the kind of glandular secretion that Pavlov studied. As discussed in earlier chapters, the Spencer–Bain principle was introduced to explain such learning and the development of motor habits as a result of past successes; following Morgan and Thorndike, animal psychology in North America concentrated upon this kind of issue. In complete ignorance of this tradition Konorski and his friend, Stefan Miller, who were medical students together in Warsaw of the late 1920s, were thrilled by Pavlov's work, but did not believe that his theory could be extended to what is now known as instrumental conditioning. They carried out a number of experiments which involved the use of a dog's leg movement as a response as well as the Pavlovian measure of salivation. The results suggested the conclusion that there are two distinct forms of conditioning, thus anticipating by a few years the outcome of debates among American psychologists as to whether Pavlov's conditioned reflexes could provide the basis of motor habits.[17]

Pavlov rejected the criticisms made by Miller and Konorski, but was sufficiently interested, impressed and generous to invite them to work in one of his laboratories. Some of his own students studied this problem and, following their work, Pavlov suggested a way in which his theory could explain why, when the delivery of a reward depends on the occurrence of a certain response, the response often occurs more frequently. The suggestion was sketchy and unconvincing.[18] But what is relevant here is the way this issue of instrumental conditioning illustrates the very narrow empirical base for many of Pavlov's general claims. The almost exclusive concentration on the measurement of saliva for over thirty years was an excellent research strategy in many respects. However, to convince neurophysiologists and psychologists of the general importance of his ideas Pavlov needed in the long run to test them out across a broad spectrum of behaviour.

This latest interest in the analysis of instrumental conditioning was not at all typical. There is little discussion in Pavlov's published work of such fundamental conceptual questions as the conditions which lead to the development of conditioned reflexes or the form which they take. The phenomenon of *overshadowing* provides a useful example to illustrate this point.

Most of the theoretical statements in Pavlov's books on conditioning imply that temporal contiguity and repetition are sufficient conditions for the formation of a conditioned reflex; that is, if a neutral signal repeatedly occurs just before an unconditioned stimulus such as food, the signal will inevitably come to elicit a conditioned response. Overshadowing appears to provide an important counterexample to this assumption. In at least two independent studies carried out by Pavlov's students it was found that a signal of a kind that could become an effective conditioned stimulus when presented on its own remained ineffective if it was always presented in compound with some other stimulus.[19] Thus, to take one of these experiments, a set of lights in front of a dog was regularly switched on just before food was delivered and soon came to evoke salivation; with other subjects the same visual stimulus was always combined during training with a muffled sound to one side of the dogs and in a subsequent test, when the lights were presented on their own for the first time surprisingly they had no effect, whereas these subjects salivated to the auditory stimulus when this was presented alone. The presence of the auditory stimulus and its effectiveness as a signal for food had in some way overshadowed the lights.

If temporal contiguity between a conditioned and an unconditioned stimulus is all that is needed for conditioning to take place, why should the presence of another signal matter? The brief remarks Pavlov made about overshadowing show that he appreciated its potential importance, yet he did not develop a theory explicit enough to make its significance clear. Overshadowing is an interesting example because when over forty years later, it was at last studied intensively it became one of a small set of phenomena that were central to the development of new theories of conditioning.[20]

It would be silly to fault Pavlov for not providing theories of the kind now current; and that is not the intention. The point is that Pavlov did not develop anything like an explicit and coherent system, despite the impression he conveyed at times and despite the way his work was often regarded later, both by Western psychologists and even more by scientists within the Soviet Union.[21] It was not the kind of thing that Pavlov liked or was good at. His skills were partly the technical ones needed in his kind of experimental work, but above all that of sensing what is an important question and knowing how to devise a suitable experiment to answer it. It is an ability that is easy to underestimate; often, after the event, it seems all too obvious that some particular study was an appropriate one to carry out, but few people in psychology have ever showed the consistency in producing fruitful empirical work that Pavlov displayed for so many years. This is what continued to give him delight. He once commented to Babkin: 'Of course we strive to reach the highest goals in science . . . but do you not agree that what impels us to work in the laboratory is the satisfaction we get from our work'.[22]

By the mid-1920s experiments were proceeding on such a scale that Pavlov organized a weekly conference in order to keep in touch with the work of his students and associates. These became famous as his 'Wednesdays'. Age, power and fame now separated him from his colleagues. He still commanded unbounded respect, but now considerable deference too. Although a brash young outsider like Konorski could argue his case with Pavlov, from the published records of 'Wednesdays' held in the early 1930s it appears that his own students and co-workers now considered it unwise to disagree with Pavlov over important matters.[23] There was no longer a fine for using the word 'consciousness' in his presence, but the atmosphere of friendly co-operation that had distinguished his laboratory in pre-revolutionary days had also gone. The vast hierarchical system he now headed was like a model of the autocratic government of Czarist days.

Active and involved in research to the very end, Pavlov died in February, 1936 at the age of 86.

Concluding discussion

In simple terms the phenomenon now known as Pavlovian, or classical, conditioning can be described as a change in an animal's behaviour that results from a temporal relationship between two events; such a behavioural change is most marked when the first event is at first of little interest to the animal, but it shortly precedes another event that is of considerable significance. As noted in earlier chapters, this phenomenon had long been known in a general kind of way. It had also occasionally been regarded as an important one, as in David Hartley's suggestion on its relationship both to general functions of the nervous system and to speculative analyses of the association of ideas in the human mind. Likewise, nearly a century after Hartley's ideas were published, Herbert Spencer gave conditioning a prominent place in his first sketch of mental evolution. Yet no one looked at the phenomenon closely until Pavlov and that is the first reason why he occupies an important place in the present history.

The above description of classical conditioning is deliberately neutral. It is not the way it has usually been described in textbooks of psychology. With some few exceptions conditioning has been viewed for over seventy years from the perspective provided by Pavlov, that of the 'conditioned reflex'. This chapter has recounted in some detail how the first experiments on conditioning were carried out as a peripheral result of the large scale study of digestive processes carried out in Pavlov's laboratory during the 1890s. It has discussed the various factors, ranging from the discovery of secretin to the revolutionary climate of those years, that played a part in the slow double shift of the focus of research in Pavlov's laboratory from digestion to conditioning and of his philosophical outlook towards a materialist position on psychology. In view of the widespread extension of experimental methods to various aspects of the life sciences which took place around the turn of the century, it is not surprising that the first experiments on conditioning also date from this period. But this still leaves such questions as why they took place in Russia, why the results were interpreted in terms of reflex action and not in terms of the animal's understanding of its environment, and why extinction and inhibition were regarded as central problems.

One of the reasons why this kind of experiment started in Russia is that Pavlov's laboratory could provide more appropriate facilities than almost anywhere else in the world. Pavlov's experiments were not the kind that anyone with sufficient ingenuity and spare time can carry out with some string and wax. Like most subsequent studies of conditioning they took a good deal of time, were labour intensive and required an expensively high standard of animal housing and maintenance. Furthermore, the topic was sufficiently unconventional and the theoretical framework so distasteful to many physiologists that the research might well have been dismissed or ignored if carried out by anyone commanding less prestige than Pavlov. In this context it is interesting that very few people at the time, and only a handful of scholarly books since, paid any attention to the report in 1902 by a little known American scientist named Twitmyer on the conditioning of a knee-jerk reaction in human subjects.[1]

Another factor was the turbulent state of Russia and the identification of even medical science with opposition to prevailing laws and social institutions. This was a country with huge and increasingly developed resources, with scientists, writers and musicians the equal of any in the world, and yet with the kind of political system that had disappeared long ago from other European nations. In the land of the Czars the basic scientific outlook of questioning authority and of testing belief was itself a revolutionary attitude. The previous chapter pointed out the contrast between Sechenov's place in Russian society and that of his teachers, particularly the Berlin physiologists, in Germany. Sechenov helped to prepare an intellectual climate which made the extension of physiological method into psychology more acceptable to scientists in Russia than to those elsewhere. It was Sechenov too who ensured that early Russian research on conditioning paid a great deal of attention to inhibitory effects.

This returns us to the nagging problem that Sechenov's programme for psychology was begun by Pavlov, who of all the leading Russian physiologists at the end of the nineteenth century had probably the least in common with Sechenov. Many of Sechenov's students became productive scientists and held leading positions in the Russian medical world by the 1890s. Yet none of them studied the brain or made a contribution to psychology.

The solution to this problem is to be found in the intensity of Pavlov's belief in science. Although more cautious and conservative in general outlook than Sechenov and many of the latter's students, Pavlov's faith in the enlightening effect upon society of the spread of scientific knowledge was as strong as theirs. Moreover, Pavlov's dedication to a scientific point of view was so complete that, if experimental results challenged even the most cherished and long-held belief, then he would not hesitate to change the belief.

The religious quality of Pavlov's dedication to science is seen both in the integrity with which he applied the scientific method and also in the otherwordly, utterly regular pattern of his daily life. Many young Russians of that era, often students who had been expelled from a university after some disturbance or another, were prepared to sacrifice their lives in an attempt to assassinate a Czarist minister. The commitment to science shown by Pavlov took a different form, for which the only obvious inspiration was not some Christian martyr, but his uncle, the abbot. In a monastery the austerity and rigid adherence to a fixed sequence of daily rituals function to strengthen belief; in a scientific laboratory equally regular routines can be employed to test beliefs and establish new ones. Pavlov's late commitment to the view that the mind can be explained only in physiological terms reflected the unusual extent to which he was willing to change his outlook to one wholly consistent with his experimental findings. At a time of life when most people rest content with the philosophy they have arrived at, and when many scientists may first begin to show some tolerance for more subjective points of view or other metaphysics, Pavlov moved in the opposite direction.

Pavlov's research was not impelled by a search for discoveries that could be directly beneficial to mankind. Even though his career was sustained by the institutional framework of professional medicine which provided the resources for his work, he was not very interested in questions concerning the cure or prevention of illness. Throughout his studies of the heart and the digestive system of the 1880s and 1890s and during the first decade of research on conditioning his only concern was with the question of how the system normally functions.

His studies might well have carried on in this manner and continued to focus upon the central phenomena of conditioning for the last twenty-five years of his life. A major factor that led Pavlov to explore the wider implications of his research was his conflict with Vladimir Bechterev. Even before hearing of the early work on conditioning in Pavlov's laboratory, Bechterev had begun to develop ideas about a new, objective kind of psychology that would work in close partnership with the neurosciences and would apply itself to problems in the real world, notably those related to mental illness. After severely bruising Bechterev's reputation, Pavlov started to expand his own interests beyond central questions on the nature of conditioning into topics which Bechterev had previously studied, such as neurosis, as well as new ones, such as the basis of individual differences.

The old man whom Shaw's black girl met in the forest was proud of the light he had shed 'on the great problems of human conduct'. Pavlov's claims about sleep, personality and experimental neurosis ensured that his conditioned reflex attracted a great deal of attention. In recent years these claims have not been widely respected; most have seemed either too simplistic or inadequately supported by convincing evidence. Pavlov's lasting achievement for psychology has been the enormous amount of basic information on the nature of conditioning in animals that was gathered in his laboratory for nearly forty years, after Wolfsohn first discovered that the kind of spittle a dog produced at the sight of a handful of sand was identical to the spittle released when the sand was placed in its mouth.

> Psychology, as the behaviorist views it, is a purely objective, experimental branch of natural science which needs introspection as little as do the sciences of chemistry and physics. It is granted that the behavior of animals can be investigated without appeal to consciousness. Heretofore the viewpoint has been that such data have value only in so far as they can be interpreted by analogy in terms of consciousness. The position taken here is that the behavior of man and the behavior of animals must be considered on the same plane; as being equally essential to a general understanding of behavior. It can dispense with consciousness in a psychological sense.
>
> John Watson, *Behavior: An Introduction to Comparative Psychology* (1914)

6
Comparative psychology and the beginning of behaviourism

American psychology expanded rapidly around the turn of the century, becoming a prominent aspect of the general growth of higher education and scientific research. The study of animal psychology at first occupied a relatively minor position; experiments involving animals were carried out in just a few of the new psychology laboratories, many of which were, however, among the most influential centres.

For the first decade or so of the twentieth century animal psychologists in America pursued their research interests in a self-contained way, being mainly concerned with specific issues and little with general implications. The early experimenters were young people either in the process of completing their doctoral dissertations or holding relatively junior positions within a department. Thorndike had stopped working with animals soon after completing his thesis and there was no major figure in American psychology with an active interest in the subject.

The experiments were on a small scale. There was little financial support even for buying and feeding animals and no ready-made technology. A good deal of technical ingenuity was required and being a skilful handyman was a considerable asset. The situation was quite different from the one that Pavlov and his associates by now enjoyed in Russia.

The empirical questions addressed by the American researchers were well-defined and experimentally tractable. Some of the studies were directly inspired by Thorndike's paper of 1898; the question of whether animals can learn by imitation was particularly popular. The major new development was of precise methods for studying the senses of an animal and these involved various forms of discrimination training. For example, just as Bechterev and Pavlov argued in St Petersburg over whether dogs have colour vision, American investigators asked the same question of rats, raccoons and monkeys.

Studies of the sensory abilities of animals were carried out in Germany as well as in America and Russia. A debate over the appropriate kind of terminology to use in such work, and in studies of invertebrate behaviour, first arose in Germany and then in France. In 1899 Beer, Bethe and von Uexkuel advocated the use of an entirely objective terminology in describing and reporting the results of such research. Words like 'see' or 'explore' were to be banned because of their subjective overtones. Instead descriptions should be limited to a technical vocabulary which could be defined in terms of physical concepts only.[1]

American researchers by and large agreed that the possible advantages from a new kind of objectivist terminology for behavioural experiments were not worth the cost of clumsy circumlocutions and of an ugly and forbidding jargon. However, there were large differences over the conceptual issue that the suggestion about terminology reflected. Is the behaviour of an animal to be used as the one way of discovering what its subjective experience is like, as Morgan's version of comparative psychology argued? Or is it to be studied for its own sake? And to repeat again the question that Romanes had been so concerned with over twenty years earlier, how do we tell whether a given species 'possesses consciousness'? As the years went by this kind of question remained as unsettled as ever, yet strangely this seemed to make no difference to progress in settling specific empirical issues.

This realization had far-reaching consequences outside animal psychology. Whereas in Germany work on the sensory abilities of animals was categorized as physiology, in the United States the same kind of research was carried out within psychology laboratories and generally published in standard psychological journals, side-by-side with papers on reaction times, child development, schizophrenia or the nature of human consciousness. By 1910 two of the leading figures involved in animal work were becoming widely known within professional psychology. Both Robert Yerkes (1876–1956) and John Watson (1878–1958) were busily involved in a variety of professional activities, giving Watson in particular a prominence that was only partly based on his specific research achievements. Nevertheless it meant both that there was increasing pressure on them to explain the point of including animal work within departments of philosophy and psychology, and that their opinions on such matters would be listened to. The two men were friends who, although separated geographically, closely collaborated on a number of projects. Their general views on psychology were at first quite similar, but later began to diverge.

Yerkes continued the evolutionary tradition described in earlier chapters. His research increasingly concentrated on the problem of comparing the learning or problem-solving abilities of different species. The aim was to detect different kinds of psychological complexity, ways in which animals behave that reflect more than just the acquisition and performance of habits; and ultimately, by progressing towards the 'higher' end of the scale, to understand the ways in which the human mind shares some mental processes with other animals and the ways it does not. The style of research was entirely different from that of Romanes, but the general conception was very similar.

Watson developed in a different direction. As one of the first to decide that animal research could make much better progress by putting aside questions concerning the subjects' inner mental lives and concentrating on their behaviour, Watson later proposed that the same would hold true for human psychology. He presented this point of view in 1913 in a paper that provided the opening manifesto for the behaviourist movement.

Considerable interest was shown in Watson's general ideas on psychology. One reason was widespread recognition that American psychology was in a very unsatisfactory state. Its rapid growth had taken place in the absence of agreement on the intellectual directions it should take. Most psychologists believed that their subject was, or should be, a science, but continuing debate over questions to do with the mind–body problem made psychology seem still more like a branch of philosophy. William James, to whom many looked to provide a lead, and James Mark Baldwin, one of the younger would-be leaders, were as content to be regarded as philosophers as psychologists.

The major topics in this chapter are Watson's own career up to 1913 and the cross-currents of research and ideas within animal psychology from which behaviourism emerged. The two universities which employed Watson, Chicago and Johns Hopkins, were as lively as any place in the world. One of the many burning issues at the turn of the century occurred in the context of a topic which has rarely impinged upon psychology, the behaviour of very simple organisms. A controversy in this field foreshadowed in an interesting way many of the debates that subsequently surrounded behaviourism. Watson came to know both of the major American protagonists in this controversy very well; one was Jacques Loeb (1859–1924), who was at Chicago when Watson arrived as a graduate student, and the other was Herbert Spencer Jennings (1868–1947), who was later one of Watson's colleagues at Johns Hopkins. This chapter starts with an account of their disagreement.

Jacques Loeb, Herbert Jennings and lower organisms

The new forms of life revealed by the increasingly powerful microscopes of nineteenth-century biology were of great interest for a variety of reasons. Above all there was the connection between these micro-organisms and disease. But also an understanding of these creatures might bear upon more abstract issues. Many of the scientists of the 1880s who, like George Romanes, were concerned with the evolution of mind were cautious about extending the possession of psychical properties too far. However, many others were not. A book appeared in 1889 which was called *The Psychic Life of Micro-organisms* and proposed that single-celled animals, the protozoa, can perceive objects, discriminate between them, and display purposive action. The author was a French psychologist, Alfred Binet, who later became better known for his invention of the intelligence test.

Binet's ideas were similar to those advocated by the German evolutionist, Ernst Haeckel, whereby any form of life was conceived as having both a physical and a psychical aspect. Indeed, Haeckel's view that 'all nature is ensoulled' was one that extended psychical properties to plants. It was a belief that Romanes himself adopted later in life and shared with many other evolutionists. As late as 1908 Francis Darwin, a son of Charles Darwin and an eminent botanist in his own right, gave a presidential address on consciousness in plants to the British Association. But by this time most biologists had come to reject any but purely mechanistic explanations for the reactions of both plants and simple animals. They were persuaded to do so largely as a result of the work of Jacques Loeb.

Loeb was born in 1859 in the Rhine Valley. Both his father – a merchant with intellectual interests who liked France better than Prussia – and his mother died while Loeb was still in his teens. They left him with enough wealth to gain an education and consider a university career. He studied medicine in Berlin, Munich and then Strasbourg. His first experimental work was concerned with the effects of certain brain lesions on the behaviour of dogs. This was supervised by Friedrich Goltz, who was a major opponent of theories of brain function that stressed the specific localization of various psychological processes. In 1885 Loeb obtained an assistantship in Berlin that allowed him to continue this work, but a year later he abandoned it, mainly because his distaste for inflicting brain damage on dogs had grown too strong. He now took up another assistantship in Wuerzburg, retaining his general interest in the relationship between physiological and psychological processes and in experimental research, but felt no commitment to any particular problem or outlook.

A colleague in Berlin had been very much interested in horses, and in the effects of exercise and other factors upon their fitness. One of the other factors was light; a question of social, as well as academic, interest in that era was whether light promotes health and efficiency and, if it does, by what means. Loeb began to take part in experiments that examined changes in the reactions of simple animals to light. He also became intrigued by the study of movement in plants. In contrast to work on this topic by Charles and Francis Darwin, the plant physiologists at Wuerzburg believed that the tropistic motions of plants, their tendencies to move in certain directions with respect to light, gravity or moisture, could be entirely explained in terms of familiar physical and chemical forces acting locally within the symmetrical structure of a plant. Loeb extended these ideas in experiments on the reactions of fly larvae, cockroaches and caterpillars to light and gravity. He proposed his general theory of animal tropisms in 1889, the year Binet's book was published.

In later years the theory of tropisms became complicated, especially where it had to deal with reactions to changes in stimulation in cases where steady stimulation produced no effect. Nevertheless the general idea behind Loeb's early work was straightforward enough. It can be illustrated by the example of a simple heliotropism, whereby an animal moves either to or from a light. Initially, when the light first appears, the animal is likely to be at an angle to it and thus the light shines unequally on the two sides of any bilaterally symmetrical animal. The photosensitive areas on the side receiving more light release more energy than those on the other side. This difference produces an uneven reaction of whatever locomotor apparatus the animal possesses, causing it to rotate like a tank with one track moving faster than the other. When the animal is lined up with the light source, stimulation to the two sides is equated and, given a tendency to move, the animal ends up moving either directly towards or directly away from the source of light.

These notions and Loeb's earlier work on brain lesions had already aroused considerable interest among American biologists when, in 1891, the scarcity of jobs in Germany and the barriers to employment arising from his Jewish origins caused Loeb to emigrate to the United States. Within a year he had obtained an appointment at the brand new University of Chicago. The study of tropisms became a very popular topic and quite reasonably concentrated upon much simpler forms of life than the creatures with which Loeb had begun. A considerable number of experiments were carried out during the 1890s in several institutions, notably at Chicago and also at Harvard, where there was a lively group of young experimentalists.

The aim of most of these experiments was to record movements to systematic changes in various forms of stimulation, including chemicals, electricity and temperature, as well as light and gravity. The experiments were cheap and easy to run; results were more or less guaranteed. For many American biologists of the time they provided an excellent introduction to experimental methods and design. At the same time, if Loeb were correct, the research would lead to the discovery of general principles for explaining behaviour along purely mechanistic lines.

Although some of the work on tropisms in plants had involved fairly detailed explanations in terms of physical chemistry, Loeb's promise that a moth's attraction by a candle could be similarly analysed without any appeal to insectival curiosity, remained rhetoric. Even in the work with simple organisms there was rarely any attempt to get at the underlying physics or chemistry of a particular reaction. It was enough to conceive of doing so. In fact, by the time he reached Chicago Loeb himself had moved away from the reductionist viewpoint with which he had become strongly identified. In its place he began to develop

Fig. 6.1. Jacques Loeb in 1907

much more technological view of experimental work, one that saw research as a way of learning how to *control* natural processes rather than as a means of providing a detailed analysis of their mechanism.

The question, prompted by Loeb's claims for the importance of tropisms, as to how exactly they were going to provide a general basis for understanding more complex behaviour does not seem to have been one that caused loss of sleep among the busy experimenters. Loeb never provided much of an answer to this question, even in a book of 1899, *Comparative Physiology of the Brain and Comparative Psychology*, which repeated the earlier claims.

There was a further reason for an interest in simple organisms in the 1890s. The importance Haeckel had placed on recapitulation theory and on the protozoa, single-celled creatures, as a bridge between plants and animals, inspired research on the development of embryos and on the reaction of cells in general to various forms of stimulation. A leading figure in this work was Max Verworn, a zoologist who had come to Jena to work with Haeckel in 1887. Some of his experiments were on tropisms – or 'taxes' as he termed them – and a particularly striking discovery was that the paramecium, a creature large enough to be just visible to the naked eye as an elongated particle, showed a curious reaction to electricity. If a current is passed through a drop of water containing paramecia by inserting two electrodes in it, more and more of the animals swim towards the cathode as the current is gradually increased, until at a certain level motion ceases with all the paramecia pointing towards the cathode. With further increases in the intensity of the current they begin to swim in the opposite direction.

In 1896 Verworn was visited by an American, Herbert Spencer Jennings, who had just obtained a fellowship for travel in Europe. Jennings became fascinated by the behaviour of paramecia and worked in Verworn's laboratory, examining their reactions to forms of stimulation more natural than electricity. His observations of paramecium behaviour made then, and after his return to the United States, convinced him that the theory of tropisms was a totally misguided way of approaching the behaviour of lower organisms, let alone that of more complex animals. In 1900 an attack on his work by one of Loeb's students attracted considerable attention to the conflict between Jennings and Loeb.[1]

The differences between the two men in social and intellectual background could hardly have been greater. Most of Jennings' childhood was spent in a minuscule town in northern Illinois, where his father was the local doctor and a founder of the town's small literary society. The father's personal qualities and medical skills ensured that he continued to be treated with respect even after reading the work of Herbert Spencer had caused him to lose faith in religion and publicly to disavow his previous beliefs. His new enthusiasm led him to name one son 'Herbert Spencer' and another 'Darwin'. Another of the father's enthusiasms took the family to the deserts of southern California for five years until they were completely penniless and had to return to Illinois. It was clearly a stimulating and unconventional household for a boy to grow up in, one that put great store in intellectual integrity.

To his father's immense satisfaction Jennings proved to be a clever boy who did very well at school. But there was no money to send him to college. For four years after leaving high school Jennings took on a variety of teaching jobs, mainly in rural, one-room schools, until he was able to enter the University of Michigan in 1890. By this time he had developed a love of biology. Eventually he was able to pay for his way through college by working during semesters as an assistant in biology courses and in vacations for the Michigan Fish Commission and for surveys of the natural history of the Great Lakes.

Apart from confirming his enthusiasm and extending his knowledge of biology, the major impact from his undergraduate education came from a course in philosophy given by John Dewey. This presented a detailed critique of the work of Herbert Spencer. Jennings could now look detachedly on the beliefs he had acquired almost as long ago as the name that went with it. He wrote that Dewey 'set one free from my heretofore compelled adherence to such doctrines, a change which though the process was painful, as all upheavals of established principles must be, was very welcome. I was left again in the condition of suspense of judgement; the great questions were entirely reopened.'[2]

From Michigan Jennings went to Harvard for graduate studies. He knew the kind of research he wanted to do and already had a considerable amount of material. His dissertation on the early embryology of a freshwater invertebrate, the rotifer *Asplancha*, earned him a doctorate within the short time of two years. He had also learned a great deal about marine invertebrates during a summer spent in Newport, Rhode Island. Although the empirical work for his thesis was of an observational kind, his contact with two of the younger faculty members of the Harvard biology department, who were studying tropisms, had convinced him of the value of the experimental approach before his travel fellowship took him to Germany and to Verworn's laboratory in Jena.

By the late 1890s the job market in America was again very poor for the many well qualified biologists seeking some form of permanent employment. On his return from the year with Verworn, Jennings managed to obtain a succession of temporary teaching and research appointments over a number of years. Many of these allowed him to continue the work on paramecia, which was partly stimulated by the attacks by Loeb's students who regarded his ideas as a disguised form of vitalism. In 1906 he published a book, *The Behavior of Lower Organisms*, which reviewed twenty years of research on protozoa and simple multi-celled animals and expressed his own ideas on behaviour. It has deservedly been considered a classic work ever since.

A major difference between Loeb and Jennings lay in the extent to which they had observed behaviour. In comparison to Loeb's limited involvement in research of this kind, Jennings had spent many years studying a wide variety of primitive animals. This had led him to the general conclusion that the principles underlying behaviour are very complicated even at this level. With Loeb clearly one of the authors he had in mind, Jennings wrote of tropisms and reflexes that 'there are some accounts of behavior in which only these definite reaction forms are described and only those conditions are dealt with in which these appear in the typical way. Such accounts have given rise to a widespread impression that behavior in the lower animals differs from that of higher forms in that it is of a fixed, stereotyped character, occurring invariably in the same way under the same external conditions. This impression is in a high degree erroneous.'[3]

Jennings' experience had also made him very sensitive to the diversity of behavioural mechanisms. Thus, even when different creatures behave in a very similar sort of way to a certain kind of stimulation, he had found that this does not by any means guarantee that a single general principle is to be found. He was particularly struck by the example of reactions to gravity; the same basic movements, moving either up or down, were the effects of a variety of different causes in different species. He was scathing about claims to the contrary. 'We have been assured by various writers that the reaction to gravity must be explained in the same way in all cases, but this is evidently said rather in the capacity of a seer or prophet than in the capacity of a man of science whose conclusions are inductions from observations and experiment.'[4]

Tropism theory was based on two assumptions which Jennings considered false. One was that an organism is passive, remaining quiescent until some form of stimulation impinges upon it; he pointed out that even as simple a creature as an amoeba can display a great deal of varied activity in a totally invariant environment. The other assumption was that the reaction to a stimulus can be understood without taking into account the state of the organism.

The claim that living organisms are inherently active is one that has been discussed here earlier, notably in contrasting the views of Johannes Mueller and those of his students of the Berlin school. Jennings' second main point, that a theory of behaviour has to explain the state-dependent nature of reactions to external events, was intriguingly novel. For example, in terms of their routine experimental procedures Pavlov and his students knew full well that a dog would not produce spittle to the sight of meat powder or some conditioned stimulus unless previously deprived of food. Similarly, Thorndike knew that his cats and dogs would not learn to escape from his puzzle boxes unless in a 'state of utter hunger'. Yet neither in Pavlov's nor in Thorndike's account of behaviour is there more than passing attention to the point Jennings considered to be so central, namely that changes in an animal's condition can produce wide variation in reaction.

According to Jennings, changes in an animal's state could take a number of different forms. The simplest is where a stimulus is just repeated a number of times; what is most frequently observed is merely that the initial response becomes progressively weaker – the phenomenon is now termed *habituation* – although sometimes an entirely new response may begin to appear as the first weakens. In either case it means that prior stimulation has caused some relatively long-lasting internal change so that, when the stimulus is repeated, the reaction is now altered.

Another case discussed by Jennings is where a reaction to one form of stimulation depends on what other stimulation is also present. Thus, the reactions of a paramecium to a touch from the slender tip of a glass rod depend on the temperature and chemical composition of the liquid around it and to a major degree on whether it is swimming freely or in contact with some object.

Finally, in certain cases the variability in the behaviour of lower organisms appears to be due to changes in internal states comparable to those which in more complex animals might be labelled 'hunger' or 'satiation'. Even in a single-celled creature like the paramecium lack of food or of sodium can produce some changes in its behaviour. Jennings quoted more dramatic examples from the coelenterates, the group of multi-celled organisms with simple nervous systems that includes hydra, sea anemones and jelly-fish. When the tentacle of a hydra comes into contact with food-like objects it can discharge nematocysts, spiked particles which can pierce the skin of an insect larva, but will do so only if the hydra has been without food for some time. The food reaction of a sea anemone, which serves to bring objects grasped by its tentacles into the mouth, is triggered by a much wider variety of solid bodies if the animal has been without food for some time; whereas, if recently fed, the animal reacts in a languid fashion even to pieces of crab meat.

Jennings argued that, like a general theory based on tropisms, one based on reflexes also has difficulty in accommodating what would now be called 'motivational' factors. He did not use this particular term; 'motivation' was still a new word, not yet in wide circulation. Instead Jennings talked about the 'regulatory' function of behaviour in a way that reflects the ideas of Claude Bernard. Bernard had suggested that an organism be considered as a set of systems whose function is to maintain a constant internal environment; from this point of view behaviour is seen as an extension of bodily mechanisms that work to ensure that the balances of temperature, water, energy, oxygen, minerals and other substances are all kept within tight bounds. Swimming away from some chemical will usually be a much more effective means by which an amoeba can keep its body chemistry unchanged than any internal process.

The stress on complexity, on diversity of mechanism, on the spontaneity of much behaviour and on its regulatory function divided Jennings from Loeb. However, there were many beliefs that they shared. For one thing they both held that a great deal could be learned about the behaviour of complex organisms from the study of the very simple. Also, like Loeb, Jennings argued strongly against the view that the possession of a nervous system transforms an animal. A central theme in *The Behavior of Lower Organisms* is that a detailed comparison between single-celled animals and those with primitive nervous systems reveals only similarities.

Moreoever, although this was less obvious, they agreed about the importance of experimental work and of describing and analysing behaviour in wholly objective terms. Jennings paid tribute to Loeb in this respect. In replying to accusations that his book marked a retreat from the objective approach, Jennings said that when writing it he had considered that 'the battle against psychic explanations had already been won' and that 'everyone must recognise the tremendous service done by Loeb in championing through thick and thin the necessity for the use of objective, experimental factors in the analysis of behavior'.[5]

The descriptive language Jennings employed made it easy to mistake his approach for a less thorough-going objective one than Loeb's. There had been a proliferation of technical terms during the brief period since research of this kind had begun, and Jennings did not feel that this had been helpful. 'To the present writer, after a long-continued attempt to use some of the systems of nomenclature devised, descriptions of the facts of behavior in the simplest language possible seems a great gain for clear thinking and unambiguous expression . . . Less attention to nomenclature and definitions and more to the study of organisms as units, in relation to the environment, is at the present time the great need in the study of behavior of lower organisms.'[6] Jennings was not perturbed by the connotations that descriptions in the 'simplest language' usually bring with them. Thus, the behaviour of a starfish placed on its back is referred to in terms of the creature 'trying' to right itself; Jennings implied by this only that the situation evokes a series of different reactions which terminates when the starfish turns itself the right way up.

There was a further, more subtle reason for using the kind of term that would usually be reserved for much larger animals. Jennings' descriptions were intended to convey the complexity of the activities of paramecia, hydra and the like. Towards the end of the book the question is raised of whether consciousness exists in such lowly creatures, and it is promptly dismissed for the reason that no statement concerning consciousness is open to verification or refutation. Nevertheless Jennings goes on to consider why we tend to attribute consciousness to some animals. He explained that he had become 'thoroughly convinced . . . that if Amoeba were a large animal, so as to come within the everyday experience of human beings, its behavior would at once call forth the attribution to it of states of pleasure and pain, of hunger, desire and the like, on precisely the same basis as we attribute these things to the dog . . . If it were as large as a whale, it is quite conceivable that the attribution to it of the elemental states of consciousness might save the unsophisticated human being from the destruction that would result from the lack of such attribution'.[7]

In places Jennings strains to convince the reader of the mammal-like qualities of his microscopic creatures. This stands out particularly in his treatment of their ability to learn. There was still little evidence at all of learning even in coelenterates, and yet there is a good deal of discussion of this topic. The treatment is almost direct from Herbert Spencer, although clearer. Like Spencer, Jennings includes both the idea of conditioned reflex – Jennings had no knowledge of Pavlov's work on this topic – and that of instrumental conditioning, caused by the differential consequences of a series of movements, and he fails to distinguish between them.

The Behavior of Lower Organisms was published in the summer of 1906 which Jennings spent in California, just before taking up the permanent post he had at last obtained at Johns Hopkins University. During that summer he carried out a study of learning in starfish; he was fully aware of the weak empirical basis for the claims his book made on this topic. It was his last piece of behavioural research.

Some of Loeb's students criticized the book on the general grounds that it is better to have a well-defined, if limited, theory that stimulates research than an ill-defined, albeit more comprehensive, account that leaves it completely unclear how to proceed any further. The point is interesting in that the most fundamental difference between Loeb and Jennings lay not in their beliefs about behaviour, but in their attitude to science. Loeb held that rapid progress, as indicated by success in controlling natural processes, depends on isolating principles of wide generality; Jennings believed that living organisms are extraordinarily complex and that any kind of real understanding would take a long time. He admitted that 'to demonstrate the complexity and difficulty of a field of work is not an achievement to be compared in value with the demonstration that this field is simple and easily explicable on a few known principles. I am under no illusion in regard to this. The clear-cut, narrow tropism theory would be of infinitely greater value for predicting and controlling the behavior of animals than anything I have offered, if only it were true.'[8] Loeb himself, concentrating more and more on embryological research, predicted that 'if the comparative physiologists follow Jennings there will never be a comparative physiology'.[9] And in a way he was right. Interest in the behaviour of micro-organisms quickly died away.

After he arrived in Baltimore Jennings began work on genetics. Mendel's work on inheritance in the pea had been rediscovered in 1900. This had prompted a flurry of studies which set out to see whether the same principles and simple ratios could be observed in a variety of other animals and plants. By the time Jennings began to work within this field it had become clear that Mendel's laws were of considerable generality. This outcome encouraged the view that progress in biology comes from concentrating upon an appropriately representative organism. In his genetics research Jennings continued to work with the paramecium he now knew so well. With hindsight this can be seen as an unfortunate decision. Two years later Thomas Morgan began to work with fruit flies and this turned out to be a much better choice.

Jennings joined the Johns Hopkins biology department just two years after James Mark Baldwin had been appointed to revive psychology there. Some of Jennings' ideas on evolution were based on the 'principle of organic selection' that Baldwin, together with Lloyd Morgan, had proposed in the 1890s.[10] Baldwin welcomed the new arrival. With their common interest in genetics and animal behaviour it made great sense, since Jennings was now fully involved in genetics research and Baldwin no longer ran experiments, to try to attract to the psychology department someone with an active research interest in behaviour; preferably someone who was trained as a psychologist, rather than like Jennings as a biologist, and who worked with subjects a little closer to man than the paramecium; perhaps a psychologist who studied the new kind of rat that was becoming so popular.

The laboratory rat and John Watson's early career

The study of animal behaviour was in a curious situation at the turn of the century. It still attracted wide interest, particularly in North America, for the light it might shed on mental evolution and yet, with the notable exception of Thorndike's research, systematic, experimental studies of behaviour were limited to relatively mindless creatures, such as the simple organisms discussed in the previous section and various species of insect.

One reason for the lack of knowledge about mammals was the practical one that to sustain behavioural research with animals like cats, dogs or monkeys in an effective way requires the kind of space, labour and money that only Pavlov commanded at that time. Consequently the arrival of the laboratory rat in North America was very timely, since it was found to be easy to house and study even when resources were limited. Its various advantages as a subject for behavioural studies eventually made it by far the most widely used animal in the psychology laboratory.

The wild rat comes in two main varieties, both of them relative newcomers to Europe and the Americas. The black, ship or Alexandrine rat, *rattus rattus*, began its major invasion of Europe towards the end of the twelfth century at the time of the crusades. It was responsible for the spread of the Black Death and was involved in most subsequent major European plagues. Early in the eighteenth century it was followed from Asia by its cousin, the brown, Hanoverian or Norwegian rat, *rattus Norwegicus*. Hordes of brown rats crossed the Volga around 1727 and migrated westwards. Except in warm climates or in environments requiring especially skilled climbing, the brown rat rapidly displaced its cousin.

Human domestication of the brown rat began in the 1800s when rat baiting was developed as a sport. Albino individuals were occasionally observed and sometimes they were isolated for show or breeding purposes. It is quite probable that the less aggressive animals tended to be selected. In any case it was found that the rat is not necessarily the vicious, loathsome creature of legend and that, if properly gentled in infancy, that is, given regular and gentle human handling, even an adult male can be as tame as a cat or dog. By 1860 the rat had begun its career as a laboratory animal in France, where it was occasionally used in studies of breeding. This provides the first example of a species being domesticated for entirely scientific purposes.[1]

The arrival of the laboratory rat in North America and its immediate appearance in psychology laboratories was partly due to a young scientist from Switzerland named Adolf Meyer. Meyer had obtained a medical degree and written a dissertation on the reptile brain at the University of Zurich. A year spent in England, where he became familiar with the work of Hughlings Jackson and of Huxley, had inspired him with respect for what he was later to term 'Anglo-Saxon science'. Hearing of the new universities in the United States and of the exodus of biologists from Clark to Chicago, he emigrated in 1892 at the age of twenty-six to contact Henry Donaldson in Chicago. It seems that the first laboratory rats to reach North America may have been sent to him shortly after this. Within a few months of arriving in Chicago Meyer found employment as a pathologist at a state hospital sixty miles away and held an honorary teaching appointment at the university. By the time he left he had persuaded Donaldson of the advantages of the rat in studies of the nervous system. Donaldson later started a rat colony with animals given to him by Meyer.[2]

In 1895 Meyer was appointed to the Worcester State Hospital in Massachusetts and again immediately became associated with the local university, Clark. Stanley Hall invited him to give regular courses of lectures in clinical and abnormal psychology. Soon after Meyer's arrival laboratory rats were being used by a biologist at Clark, who in 1897 advised a psychology instructor there, Linus Kline, that laboratory rats 'are small, cheap, easily fed and cared for; and best of all, when placed in revolving cages, they spend most of their time, when not eating or sleeping, in running'.[3] Kline began to use rats in his laboratory course in comparative psychology and later supervised the work of a graduate student, Willard Small, who decided to study associative processes in this species.

Kline had been convinced of the importance of animal research in psychology from his reading of Lloyd Morgan and Wundt. He had gone to Clark with the sole purpose of studying animals. In this he had been somewhat deflected by criticism from Hall, who, though continuing as president of Clark, maintained close contact with the activities of the psychological laboratory, and it is Small who has been remembered as carrying out the first psychology experiments with rats. These studies at Clark took place independently of those of Thorndike and must have started at about the time Thorndike was completing his descriptions of the behaviour of cats and dogs in puzzle boxes. Unlike Thorndike, Kline and Small were concerned to study learning in situations that seemed to them as natural as possible for their subjects. In the first experiments, published in 1900, food was placed in a small box and the rats had to burrow through sawdust or gnaw through the fastening of a door to get into the box.[4] For the second set, published in the following year, the labyrinth-like burrows of wild rats suggested the use of a maze. Small constructed a replica of the maze at Hampton Court Palace and let his rats live in it. His maze is illustrated in Figure 6.2. At various times he would place food in the centre and observe the rats' progress in reaching it from a point outside.[5]

These studies were mainly observational ones and Small's tentative suggestions about their significance were more in the style of Romanes than of Thorndike, with a great deal of inference about the presumed subjective experience of the subjects. Small's papers were overshadowed by the much more provocative and analytic paper by Thorndike. However, Kline and Small had shown that rats could be conveniently used for experiments on learning in animals; Small's comments on their sense of smell and the ability of blind rats to find their way through the maze prompted the question of what sense, or senses, the rat was using in learning the maze. This became a central issue in the study of the rat for the next few years.

Animal psychology did not last very long at Clark University. These studies of the rat and some later experiments on monkeys were the major products of its laboratory of comparative psychology. Possibly because of Hall's influence and his involvement in other aspects of psychology, such as developmental studies, educational issues and a growing interest in abnormal behaviour, work on animals ceased there soon after the turn of the century. This left Chicago as the only place where rats were to be found in a psychological laboratory. Although Donaldson had been the first American scientist to work with rats, his primary interest had been in their physiology.[6] The first systematic study of the behaviour of rats at Chicago did not begin until after Kline and Small had published an account of their work at Clark. The Chicago work was carried out by John Watson for his doctoral thesis.

In terms of background and personality Watson differed considerably from the majority of young American scientists of his time. For one thing he came from south of the Mason–Dixon line; he was born and spent his childhood in the piedmont country of South Carolina. Where most of his subsequent teachers and colleagues were the sons of ministers or doctors, and generally were from comfortable and educated, even if – like that of Jennings – sometimes impoverished, households, Watson's childhood was quite different. The first twelve years of his life were spent on the small farm his family owned some six miles outside the town of Greenville. They then moved into the town when his father began to work in a sawmill.

There was a great deal of discord within the family. Watson's mother was a devout and very active member of the nearby Baptist Chapel, who greatly admired one of the local hell-fire preachers, John Broadus, and named her second son after him. Despite this christening the attempts to instil her religious faith into her children, which had completely succeeded with her first son, had no more lasting effect upon John Watson than upon his father.

A wild adolescence and a brush or two with the police while still at school are perhaps not unusual with such a background. The real mystery about Watson is that, although there does not seem to have been a single aspect of his childhood that encouraged intellectual pursuits, he did sufficiently well at high school to go to the local Baptist college, Furman University, and there developed a considerable interest in philosophy.

Furman was very small, with a total faculty membership of just eight professors, and it was on the outermost periphery of the American academic world. However, one of the professors had just completed a sabbatical year at what was now the centre of this world, the University of Chicago, and he taught Watson philosophy and psychology. The up-to-date reading for his courses included the latest American textbooks on psychology by Baldwin and by Ladd, as well as translations of Wundt.

After graduating from Furman in 1899 Watson taught in a one-room country school a hundred miles from Greenville, where the children were impressed by the rats he had tamed and taught to perform various tricks. The following July his mother died. By the end of the month he had applied to the University of Chicago to do graduate work and had written to John Dewey, the professor of philosophy there, to explain his reasons for doing so. By September all was arranged, and at the age of twenty-two Watson left Greenville.

In 1900 the group of distinguished men teaching in Chicago included Loeb, Donaldson, Dewey and George Herbert Mead, a second philosopher who subsequently was highly influential in the development of social science in America. The psychological

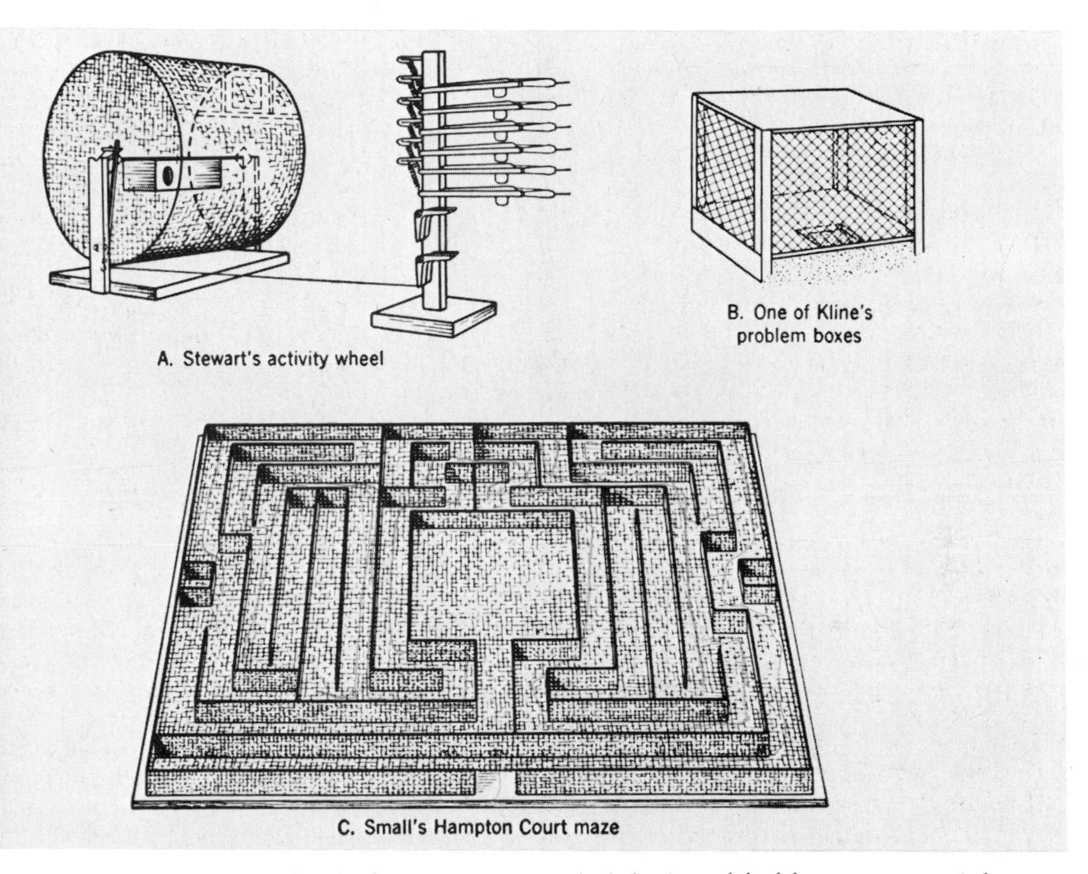

Fig. 6.2. Apparatus used in the first experiments on the behaviour of the laboratory rat, carried out at Clark University around the turn of the century

laboratory was then still part of the philosophy department. It was organized by a younger and less well-known man, James Angell, another of the early American psychologists whose interest in the subject had been stimulated by William James and who had followed studies at Harvard with a year in Germany.

Watson had come to study philosophy and had been very much attracted by Dewey's reputation. But he was quickly disappointed by Dewey, later claiming that he could never understand what Dewey was talking about. Watson continued to have a great deal of contact with Mead, for whom he had high respect, but his dedication to philosophy ebbed away and was replaced by an interest in some neurophysiological research which Loeb was doing and in the work of the psychological laboratory.

Watson had very little money and had to support himself by means of various part-time jobs. These included employment as an assistant janitor, a post whose duties included dusting Angell's desk and looking after Donaldson's rats. The situation improved a little when he was awarded a university fellowship at the end of his first year. He was now ready to start research and was inclined to work on a problem suggested by Loeb. But by now Angell had become an influential figure in his life and Angell dissuaded him from this plan on the grounds that Loeb was not a 'safe' supervisor for a new student like Watson. Instead Angell and Donaldson together suggested a research topic and jointly supervised his work.

The aim of the project was to test a current theory that learning could only occur once the process of myelinization of nerve fibres in the brain was well advanced. The study was a developmental one: Watson tested the learning abilities of rats at various ages and looked for a correlation with progressive changes in various parts of their brain, which were studied using histological techniques. This study was an early example of research in physiological psychology or psychobiology, in that both behavioural and physiological variables were measured within the same set of experiments. In all his subsequent work Watson used only behavioural measures, but the major achievements of the most eminent of the psychologists associated with Watson early in their careers, Karl Lashley and Curt Richter, were in psychobiology.

To test the learning abilities of his rats Watson adopted and modified the situations used by Kline and Small. Like them he measured the time taken by rats to get into a small box containing food, either by burrowing through sawdust to an opening or, his modifications, by pressing on a lever to open a door. This last situation is shown in Figure 6.3. Some of Watson's experiments involved mazes, which were simpler than the Hampton Court model. The results of the various experiments refuted the hypothesis he had set out to test; Watson found that very young rats, which had barely been weaned and in whose brains little myelinization had taken place, showed rapid learning.

Small had commented that many of his rats were 'timid and flighty' and difficult to handle. In contrast Watson noted that his subjects became exceedingly tame, showing no sign of distress when placed in some novel test apparatus. The difference clearly reflects the lavish attention Watson devoted to his rats from the time that they were weaned. Unusually intensive gentling from an early age can produce adults whose behaviour is so strikingly different from those that have received cursory handling as to suggest that they are from a different strain.[7]

Watson enjoyed working with animals. He liked the opportunity to use his manual skills in setting up equipment and felt that his research was getting at real issues. In comparison the few experiments using human subjects in which he became involved seemed highly artificial. He worked intensively, seven days a week with scarcely a break, in the way that Thorndike had worked, until after fifteen months or so he suffered a minor breakdown. The bout of depression was cured by a month's vacation in the country staying with friends; his childhood fear of the dark had returned and he had to sleep with the light on. Vowing to avoid pushing himself so hard, he returned to finish the final phase of his thesis work. Angell had taught Watson how to carry out experiments and now he taught him how to write about them.

In 1903 Watson was awarded his doctorate and agreed to remain in the department as an assistant, with the responsibility of setting up an animal laboratory. Donaldson lent him $350 to have his thesis published, with the title *Animal Education*, an intriguing variation on the title that Thorndike had used.

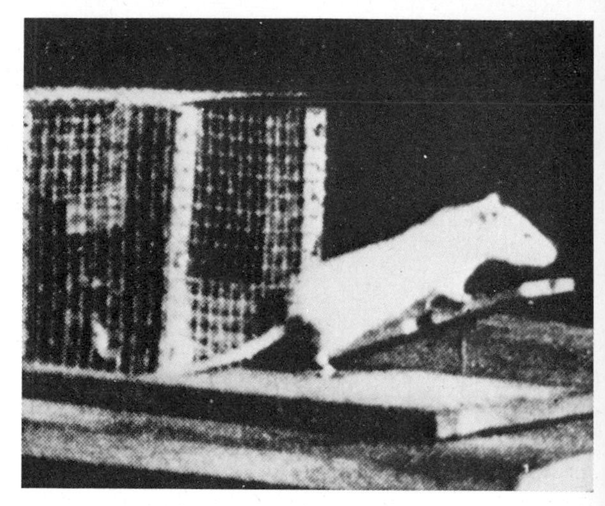

Fig. 6.3. One of John Watson's rats operating a lever to open the door to a puzzle box containing food

This monograph, like the research it reported, was a solid workman-like job, without flamboyance or pretension. Watson agreed with Thorndike's 1898 suggestion that the associative basis of trial-and-error learning in animals was little related to the associations of the human mind, but similar to the human learning of motor skills, such as Thorndike's examples of swimming and juggling.

The development of a new line of research often requires a considerable amount of tedious investigation to find the most suitable techniques. In the case of studying a new kind of animal this may involve details of diet and housing, feeding schedules, effective rewards, convenient testing routines and appropriate equipment. Much of the work that went on at Chicago under Watson's direction was of this sort. A particular issue which engaged Watson's attention was what senses are used by a rat in learning a task. Do rats learn to find their way through a maze using only their sense of smell? Can they form associations on the basis of just visual or auditory stimuli? These questions arose in the course of his thesis work, but then there was no time to investigate them. Later he was joined by a graduate student named Harvey Carr and together they performed a series of experiments on this topic which formed the major body of research on rats that Watson carried out at Chicago.[8]

Some preliminary experiments showed that the behaviour of rats in a maze was little affected by whether it was well-illuminated or not. Other results suggested that neither auditory nor smell cues were essential to them finding their way around. Watson became convinced that the matter could be best settled by using animals in which one or another sense had been eliminated by surgery. In the Spring of 1905 he went to Johns Hopkins University to learn the appropriate surgical techniques. The visit provided an opportunity to meet Mark Baldwin; the two men got on well together and a number of talks that Watson gave at Hopkins were well received.

After returning to Chicago Watson used his newly-acquired skills to remove the eyes from one group of rats, destroy the middle ears of a second group, lesion the olfactory bulbs of a third group and cut off the whiskers of a final group of rats. He was unable to detect any impairment in their ability to learn a maze. The one positive result of these experiments puzzled Watson: he found that the intact rats appeared curiously sensitive to the absolute orientation of the maze, in that rotation of the maze through 180° from its normal position in the experimental room disrupted the animals' behaviour.

On the basis of these studies Watson claimed that the behaviour of rats in a maze is controlled only by their kinesthetic sense, that is, by the internal feedback sent to the brain by receptors in joints and muscles. The experimental approach was crude and the results did not in fact demonstrate this claim. In an appendix to the report he admitted the validity of criticisms pointing out that rats may be able to use more than one modality to negotiate a maze, so that, for example, a blind rat might use its sense of smell, whereas a rat without olfactory bulbs might use its sight. As a consequence he finally tested just one unfortunate rat, in which the eyes were removed, the olfactory bulbs lesioned and the whiskers cut off, and reported that its ability to learn to get to food in a maze was unimpaired. Nonetheless research many years later showed that Watson's critics were correct and that rats can use alternative senses to find their way in a maze, the preferred sense depending on the characteristics of the maze.

The appearance in 1907 of the paper reporting this research involved Watson in his first public controversy. There was an outcry over what was seen as the needless cruelty of these experiments, unjustified by relevance to any medical problem. Watson made no claim that such research made any contribution to human psychology and specifically rejected the idea that a person placed in such a situation would learn in the same way. He felt that it was plain that kinesthetic stimuli would be of little importance to a human subject, who under comparable conditions would no doubt employ visual imagery to solve the problem of getting through a maze.

Watson came to regard the behaviour of a rat in a maze as a chain of discrete responses, controlled by kinesthetic feedback, which becomes increasingly integrated as training continues. Learning was seen as the development of a complex motor habit which, when once initiated, progresses in an invariable, automatic fashion, in the sense that it is unaffected by external stimuli. This view was confirmed for him in a subsequent study with Carr, which he was to remember with delight years later when it became known as the 'kerplunk' experiment. After rats had been extensively trained in a large maze, the latter was sawn through and re-assembled so that some of the arms were much shorter: the rats now ran squarely into the ends of these arms.[9]

By this time Watson's interests had begun to move away from analysis of the sensory control of maze performance to the general problem of studying sensory processes in animals. In 1907 he and Robert Yerkes were awarded a small grant to prepare a report on methods of studying visual perception in animals. Three years earlier Watson had first written to Yerkes at Harvard where he directed the only other animal psychology laboratory in the world. Although it was a very long time before they actually met, they became close friends and for twelve years exchanged frequent letters in which they informed each other about their experiments, shared new technical information and explained their general views on psychology.[10] In connection with their joint grant Watson set up a small primate colony at Chicago to allow him to study colour vision in monkeys.

Watson also started to follow up another new interest. In the Spring of 1907 he made the first of a number of visits to the Dry Tortugas, islands lying sixty-six miles west of Key West in Florida, to study the behaviour of noddy and sooty terns. The report of this first visit appeared in a publication of the Carnegie Institute and cannot have had a wide readership even at the time.[11] Yet in many ways this paper is far more impressive than those on rats which were published in the standard professional journals. The bulk of the paper consists of painstaking naturalistic observation of the behaviour of these sea birds, noting the similarities and differences between the two closely related species. Punctuating the details of their mating, nesting, brooding and feeding patterns are the questions these observations raised. Given the thousands of birds within the crowded nesting colonies, how do the partners recognize each other? How does each pair find its nest? In what way does the arrival of the eggs, and later the hatching of the young, produce marked changes in the behaviour of the parents? Is the egg recognized? What controls the alternation of food-seeking between the two partners?

As in his research with rats, Watson was most interested by questions about the sensory abilities of the terns and about the effective stimuli controlling different aspects of their behaviour. With the questions came ideas on simple experiments that might provide answers. He dyed the feathers of some birds, changed the markings of the eggs, placed eggs in the nests of birds that had not yet laid, and looked at the various resultant changes in behaviour. In contrast to the plasticity of the behaviour of his rats and monkeys, the reactions of the terns appeared remarkably invariable. He also noted the absence of any play in the young birds as compared to mammals.

In contrast, in their sense of orientation the terns were much more impressive than rats. Working on the problem of how each bird found its own nest, Watson discovered that greatly altering the appearance of the nest had no effect. If the nest were removed a few feet away the returning bird would alight on the spot where the nest had been. If the nest were raised vertically a few feet, the bird would alight on it without hesitation, and then show a startle reaction when it peered over the edge. In some way the birds knew exactly where the nest was, but oddly nothing of what it, or its immediate surroundings, looked like.

Much more dramatic evidence for the tern's navigational ability came from a fragmentary study of homing. Watson had acquaintances release birds, which had been captured on the Tortugas, from Key West, from Cape Hatteras, 850 miles away, and from Cuba, 108 miles away. Many of the birds returned to the Tortugas in an impressively short time from each of these starting points. Watson was mystified by this ability and convinced that all current theories of 'distance orientation' were hopelessly inadequate.

In order to test the performance of terns in problem boxes and simple mazes, he decided that it was necessary to rear some young birds by hand. In doing so, he replicated the observations Spalding and Lloyd Morgan had made on the increasing selectivity of pecking movements and on the rejection of faeces. He noted that 'Lloyd Morgan, Spalding and others are unquestionably right when they affirm that young birds, if taken early enough and reared by hand, exhibit little signs of fear'. The makeshift conditions, small number of animals and brief amount of time meant that the results of the learning experiments were limited. They indicated that the terns could demonstrate trial-and-error learning – they mastered the tasks very quickly – and suggested that vision was the most important sense involved. In general the research during this first visit to the Tortugas was of an exploratory nature, but the results were impressive for just three months' labour and laid the groundwork for future visits.

In 1908 Baldwin was in a position to appoint a professor of psychology at Johns Hopkins to be responsible for establishing a laboratory of comparative psychology and John Watson was one very obvious choice. Watson had not proposed any influential theory or made some great discovery, and his research interests may have seemed narrow. But his energetic work at Chicago had become well-known, and he had displayed broader interests and scholarship in the reviews that he had contributed to the journals that Baldwin edited.[12] Watson appeared to be a coming man. He was highly regarded by his colleagues at Chicago: Carr later remembered Watson at that time for his 'tremendous energy and enthusiasm in both work and play, . . . his irrepressible spirits, his intellectual candour and honesty and his scorn of verbal camouflage and intellectual pussy-footing'.[13] Both Angell and Donaldson wrote glowing references, the latter adding the telling comment that Watson 'possessed that gift of Heaven of getting things done'.[14]

Watson felt loyalty towards Angell and reluctance to leave a laboratory that had taken years of hard work to establish. Yet he still held the lowly rank of instructor, with a salary that was small for a family man. Just after obtaining his doctorate he had married Mary Ickes, a student who, although not from the South, was from a similarly rural background, and they now had two children. The University of Chicago, like Clark, initially offered high salaries, but quickly became less generous. In Baltimore clipper ships no longer graced the inner harbour, a major fire had recently destroyed a large area of the city and the place was a backwater in comparison to Chicago. But the university was still highly regarded. Eight years after leaving South Carolina Watson attained the prestigious and well-paid position of professor of experimental and comparative psychology at Johns Hopkins.

Robert Yerkes' comparative psychology

After 1908 anyone wishing to gain an overall picture of what animal psychologists had discovered could most appropriately do so by reading Margaret Washburn's *The Animal Mind*. Washburn was one of the few women in this period to rise to a prominent position within American psychology. She had studied at Cornell University, one of the first major universities to give women full status as graduate students, and afterwards she had obtained a number of teaching appointments at various colleges. Her interest in animal psychology developed when, on returning to Cornell for a few years, she was asked to give a course of lectures on the subject. It was some time later before she carried out some limited experimental work with animals herself.[1]

Possibly because of this lack of direct involvement in some particular area of research, Washburn's book provided a fair, sensible and also very readable review of recent developments. The research described in the book was restricted to experimental studies and purely observational evidence was deliberately excluded; there was little concern with the place of behaviour in some general evolutionary framework, but otherwise she very much followed in the tradition of Lloyd Morgan.

Washburn agreed with Morgan that for the psychologist the point of studying an animal's behaviour was to find out about its conscious states and the point of this was to provide a context for the analysis of human subjective experience. Much of the research discussed by Washburn was concerned with the sensory abilities of particular species. Where Watson was simply concerned to determine whether, for example, a slight change in a light produced a detectable reaction in some animal, Washburn wished to go further and work out what kind of sensation the animal must *experience*. 'What, for instance, is the meaning of the fact that the range beyond the violet end of the spectrum, invisible to us, produces effects upon certain animals? Are they seen, or do the sensations accompanying them rather resemble those produced by an irritating chemical? What kind of sensation quality may we suppose exists in the consciousness of an animal whose responses to light are mediated by the skin, not by the eyes?'[2]

The other kinds of research described in *The Animal Mind* indicated the major influence of Thorndike's thesis. Some studies merely extended to new species Thorndike's methods for studying trial-and-error learning. Others focussed on his analysis of associative learning and his claims that animals do not learn by imitation or by passive instruction, that is, by being 'shown' or 'put through' the task.

Thorndike's own work with monkeys had left these issues unsettled and a series of young researchers subsequently tried to resolve them. Unfortunately each study of imitation left the author firm in his own opinion, but still allowed plenty of room for criticism from those holding the opposite view. Thus, in 1902 Kinnaman reported some tests on two rhesus monkeys at Clark University. Although no sign of imitation appeared for most of the tasks he used, he was convinced that one monkey had learned from observing the other how to pull out a plug and how to operate a lever.[3] Sceptical readers were more impressed by the failure to detect any positive effect in all the other tests. Watson, to take a notable example, tried out similar tests on four monkeys in Chicago and detected no sign of 'inferential imitation'.[4]

Washburn's *Animal Mind* served as a useful overview of a decade's work in comparative psychology following Thorndike's thesis of 1898. It also provided an indication of the new kind of experimental approach that was to follow. This was seen in a study by Lawrence Cole which was enthusiastically summarized by Washburn. Cole's experiments were all directed towards specific issues that Thorndike had raised, except for a final one which was the first attempt to look at whether an animal can respond appropriately on the basis of a signal that is no longer present. Such 'delayed reactions' later became a subject of great theoretical interest and one of the more specific controversies concerning Watson's behaviourism centred on it.

The subjects for Cole's experiments were six young raccoons.[5] At first they were tested in what had become the standard kind of puzzle box so that, as Cole put it, he 'might learn by comparisons the place of the raccoon in the scale of mammalian intelligence'. The raccoon turned out to be an excellent subject for such research: the animals would often work at a door fastening even when disinterested in the food outside, they could master a new task almost as rapidly as Kinnaman's monkeys and were able to learn to open a door when this required seven different operations.

Cole was much more systematic than Thorndike. Where Thorndike's innovative attempts to train his animals to discriminate between pairs of related stimuli were quite sloppy, Cole completed a well-controlled set of tests which established that the raccoon could learn to discriminate a black from a white card, a square from a circle and a high note on a harmonica from a low one. On the other hand further tests carried out with equal care revealed no sign of imitation learning; Cole suggested that the tendency to imitate might be very rare and limited to those species that in their normal environment search in groups for stores of food.

Although the tests for imitation agreed with Thorndike's conclusion, when Cole tested his raccoons for passive learning the outcome was radically different from that reported in Thorndike's 1898 paper. Raccoons which had been put through a new task learned it in half the time taken by those which

Fig. 6.4. One of Lawrence Cole's raccoons from the experiments in Oklahoma

had not. Furthermore, when two of the animals were given human guidance in raising a lever with the nose and a third was taught to use its paws to raise the same lever, when they were left on their own they persisted with the particular mode of operation they had been taught. Finally, one negative finding that was given a great deal of stress by Thorndike was found not to hold for raccoons; although Cole initially placed his subjects in a puzzle box by picking them up by the nape of the neck and lifting them in through the door, they soon began to enter the box of their own accord when given the opportunity.

Thorndike's failure to detect any form of passive learning was a major factor leading him to deny that cats and dogs are able to form associations between ideas and to suggest that all animal intelligence consists only of the connection of motor impulses to certain situations. Cole's very clear positive results on these tests naturally led him to the contrary view, that raccoons are capable of at least two kinds of learning, of associating ideas as well as forming stimulus–response connections.

Cole also drew this conclusion from the final phase of his tests where he trained the animals to discriminate between two sequences of stimuli, of which the last item was the same in both sequences; thus, the sequence *white-blue-red* indicated that, after it had been presented, food reward could be found on top of a high box nearby, whereas the sequence *red-red-red* meant that there was no reward on that trial. Since at the moment when the raccoons had to either respond or withhold from climbing the box *red* was always present, it seemed to Cole that his animals must be retaining visual images of the colours earlier in the sequence in order to solve the task.

This use of a procedure that had not been at least tried out by Thorndike was one unusual feature of Cole's study. Another was that it took place in Oklahoma. Almost all the other experiments on animal behaviour performed in America during the first decade or so of the century were carried out at either Clark, Chicago, Johns Hopkins or Harvard. And of those four places it was Harvard that produced the most research, both in terms of volume and variety.

The driving force behind the animal laboratory at Harvard was Robert Yerkes. Either directly or indirectly he was involved in a large proportion of both the research that followed on from Thorndike and the new developments in discrimination training which were summarized in Washburn's book of 1908. He remained a central figure when research on comparative psychology in America subsequently began to take a new direction.

Yerkes was also the person most responsible for the general growth of animal psychology in the United States. He worked steadily on various projects of his own, supervised the work of a number of able students and seems to have maintained a steady correspondence with anyone in America involved in behavioural work. He was also the major figure on the professional side of his science; from 1904 to 1910 most behavioural papers were published in a journal which had only covered topics in neurology and physiology until Yerkes became one of the three editors, and from 1911 onwards he became the managing editor for a new journal devoted entirely to behavioural work, the *Journal of Animal Behavior*, which he and Watson founded.

Yerkes and Watson had a great deal in common besides their interest in animal behaviour. Yerkes was also a country boy who had managed to get a college education without much help or encouragement from his parents. He was born in 1876, making him two years older than Watson, and his family owned a small farm outside Philadelphia. An uncle had encouraged his interest in medicine, but, when Yerkes was twenty-one, another relative unexpectedly offered him a loan of a thousand dollars and this gave him the opportunity of doing graduate work in biology before settling down to his medical studies. He entered the biology department at Harvard in 1897 where the study of tropisms in lower organisms was one of the livelier research topics. Within a short time he was doing work of this kind himself and enjoying it so much that the idea of a medical career was abandoned.[6]

Although Thorndike had left Harvard shortly before he arrived, Yerkes became increasingly interested in Thorndike's experiments and arranged to work for him as an assistant one summer. Due to this interest in animal psychology Yerkes transferred to the psychology laboratory in 1899. In 1902 he obtained his doctorate and, like Watson at Chicago two years later, was invited to remain as an instructor with the responsibility of establishing a laboratory for animal psychology.

Although in background, interests and career Yerkes and Watson were so much alike, they differed considerably in temperament. Yerkes was a more cautious and less ebullient man than Watson. He attributed his shyness and lack of physical strength to the scarlet fever of his childhood which had killed his sister and from which as a seven-year-old he had barely survived.

Fig. 6.5. Robert Yerkes *ca* 1908

Much of Yerkes' early work was concerned with whether reptiles, amphibia and invertebrates such as crabs were capable of learning in trial-and-error situations, and in general he found that they could learn such tasks. Other experiments looked at the sensory abilities of a variety of species; at the question of whether frogs can hear, for example. Then he combined these two interests in his first intensive study of a single kind of animal; this was an unusual strain of mouse which, because of its perpetual activity and strange movements, was called the 'dancing mouse'. In the course of studying this animal Yerkes perfected procedures for studying discrimination learning, using the apparatus illustrated in the left-hand panel of Figure 6.6. His method was novel in that, since the mice moved around within the chamber without needing any incentive such as food, punishment could be used for deterring a mouse from making an incorrect reponse rather than reward for the correct response.[7]

An important study Yerkes carried out with a student named Dodson on the discrimination of brightness differences illustrates the use of this equipment.[8] A mouse is given a choice between the two compartments shown in Figure 6.6; the door to one is marked with a light coloured card, which might

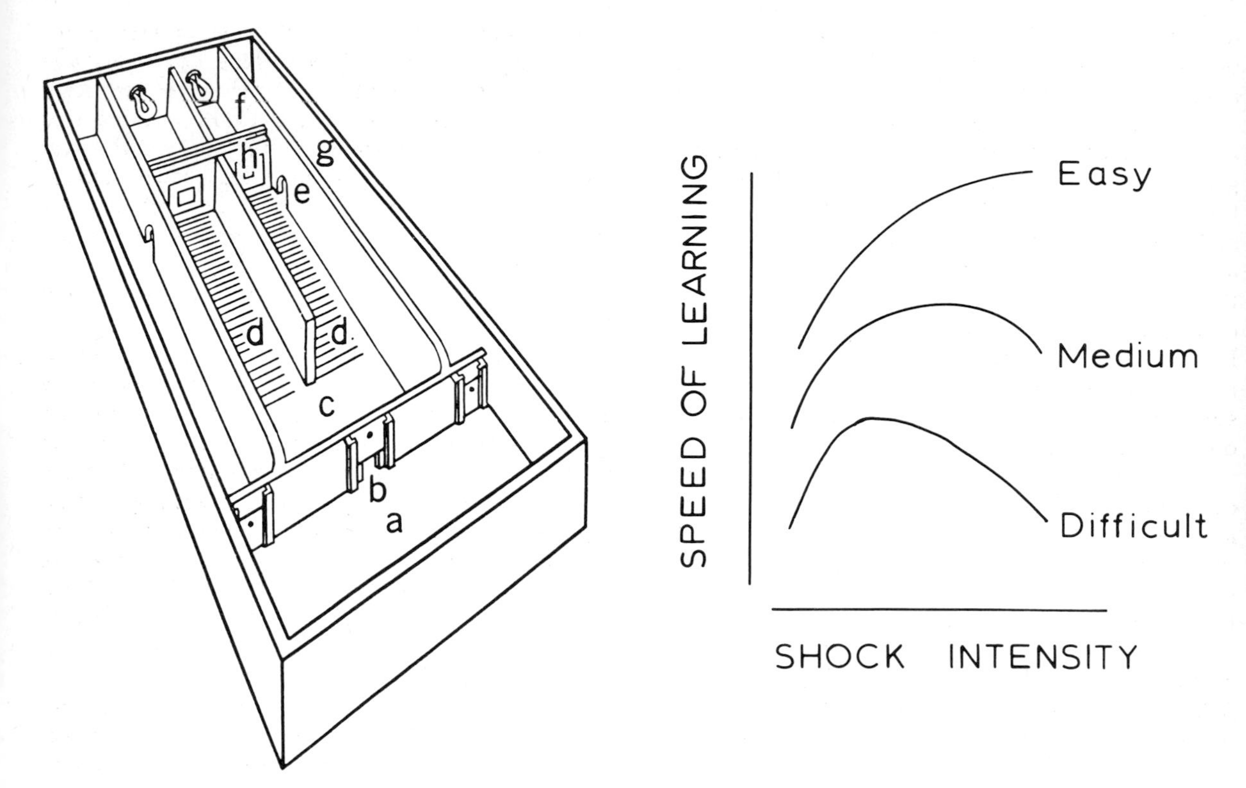

Fig. 6.6. Discrimination box used by Yerkes to study the dancing mouse, with the Yerkes–Dodson Law illustrated by the graphs to the right

be the positive stimulus for this particular animal, while the other contains a dark card. Whichever compartment the mouse chooses he can scurry out the other end and around to the starting point again. However, if he enters the incorrect compartment a brief electric shock is given.

In Yerkes and Dodson's study different groups of mice were trained by this method to learn discriminations of various levels of difficulty; as one might expect, a discrimination between black and white cards was learned more rapidly than one between two greys of slightly different intensity. The more interesting aspect of the experiment was that for each level of difficulty shocks of different intensities were used in order to find out what intensity was optimal for rapid learning of that particular discrimination. It turned out that for the easiest discrimination the highest level of shock produced the fastest acquisition, whereas for the most difficult discrimination a lower level was best. The general principle that, as a task increases in difficulty, the optimal motivation level becomes lower has since been known as the Yerkes–Dodson Law. It is illustrated on the right-hand side of Figure 6.6.

This research with Dodson was just one of the many dissertation projects that Yerkes supervised at Harvard. Many of the others were concerned with imitation learning, using at various times rats, cats and monkeys.

The collaborative project with Watson, on methods for studying vision in animals, meant that the two of them became increasingly absorbed in purely technical problems over such matters as lenses, projection bulbs and appropriate training procedures. It is this preoccupation that accounts for the distorted way in which Pavlov's work first became widely known among American psychologists. In 1909 Yerkes was a joint author of the first general introduction to Pavlov's work published in English.[9] It gave a clear account of the methods used for conditioning the salivary response, but failed to indicate that for Pavlov the point of this method was to study the general functions of the brain and the nature of learning. Instead, the review concentrated almost entirely on those studies which were most closely related to Yerkes' current interest; the only experiments from Pavlov's laboratory that the review described in any detail were ones that used conditioned salivation to test a dog's ability to make auditory or visual discriminations.

Yerkes and his co-author displayed a positive, yet restrained, attitude towards Pavlov's approach, suggesting that it was not markedly more effective than the methods already being developed in America to study perception in animals, that it was difficult to use and that it was probably unsuitable for animals other than dogs. They missed the point which Bechterev quickly perceived, that the use of the salivary reflex in Pavlov's laboratory was incidental and that some other kind of response might well be much more suitable when applying Pavlov's basic procedure in a different situation.

As far as other American psychologists were concerned, Yerkes and Watson presented the unfamiliar picture of two young scientists from different institutions collaborating in complete harmony and with considerable success in the development of a new field of psychology. However, although for some years it did not become apparent to the outside world, by around 1910 their views as to the direction this new field should take were beginning to diverge. Animal psychologists in America began to divide themselves into a 'traditional' wing, most strongly represented by Washburn, and a 'radical' wing increasingly identified with Watson. The main issue was whether behaviour provides a basis for making inferences about an animal's subjective experience, as the traditionalists claimed, or whether it deserved study in its own right since, as the radicals claimed, the contents of an animal's consciousness are beyond the scope of scientific enquiry. As noted earlier, these differences among American researchers echoed disputes that had occurred a few years earlier in Germany and then France.

In Yerkes' scientific papers the behaviour of his experimental animals is reported without much more attempt to speculate on their subjective experience than would be found in a paper by Watson. The extent to which Yerkes at this time differed from Watson and endorsed the aims of comparative psychology as defined by Morgan and then by Washburn is apparent in a textbook he wrote in 1911.[10] A central theme is the primacy of introspection as a psychological method and the analysis of human subjective experience as the ultimate goal of psychology. It contains numerous passages that Watson could never have written; for example, one allowing in Haeckel-like fashion for the possibility that plants have minds.[11] A discussion of the minds of animals contains the following passage. 'There is no question, in the mind of the person who really knows animals, that the higher vertebrates possess a great variety of sense qualities and feelings . . . of emotions, sentiments, associations, memory images, ideas, and even certain forms of judgement . . . and the more liberal among psychologists are at present inclined to believe that at least some animals, among them the dog and the horse, the raccoon and the cat, experience conscious complexes which are much like ours.'[12]

From the time his textbook appeared the research Yerkes carried out was much more specifically directed towards finding hard evidence for 'liberal' beliefs in the superior abilities of certain species. He appears to have realized gradually that, although claims about the conscious experience of animals can never be proved or refuted, questions about the possibly complex basis of behaviour are nonetheless open to various kinds of experimental enquiry. This led to a form of comparative psychology which, like Morgan's, compared the mental abilities of different species, including man, but where questions concerning the nature of animal consciousness receded into the background.

Yerkes had probably never agreed with Thorndike's denial to animals of any kind of learning except habits formed by the 'stamping in' of stimulus–response connections; 'sensation–impulse' theory as it was often called at this time. But until around 1911 he had always played according to Thorndike's rules, whereby inferential imitation was made the key issue. Over the years a number of Yerkes' students had attempted to obtain positive evidence for imitation learning with the main aim of demonstrating the inadequacy of Thorndike's theory. Cole's procedure, whereby raccoons had to delay their reaction until the end of a series of stimuli, was imperfectly designed, but it did alert Yerkes to the possibility of other, hopefully more productive, ways than imitation procedures for studying alternative kinds of learning to habit formation. The final impetus for Yerkes' approach came from one of his students, named Gilbert Hamilton, who, in the year Yerkes' textbook appeared, published a study that suggested to Yerkes yet another way of systematically studying processes more complex than the strengthening of stimulus–response connections. Furthermore, Hamilton's procedure had an attractive virtue of allowing easy and meaningful comparisons across a wide range of species.[13]

Hamilton was a psychiatrist who had become enthusiastic about new developments in psychopathology, particularly the work of Freud and Jung in Europe and that of Adolf Meyer in America. He believed that the central theme underlying their work on mental illness was recognition that a person's behaviour was the end-product of a complex set of unconscious processes that he termed 'reactive tendencies'; in the long run the only sure way of analysing these reactive tendencies was by the use of experimental methods and consequently of comparisons between animals and man.[14] Thus, when he later became interested in the causes of homosexuality, he carried out pioneering experiments on the sexual behaviour of monkeys.[15]

The achievements of animal psychology in 1911 were of little use to Hamilton. He could see the point of evaluating the sensory equipment of various species, but in studies of learning the emphasis on quantitative aspects, such as the speed or number of errors with which an animal mastered a maze or puzzle box, seemed to him superficial and misplaced. A method was needed for finding out 'what, if any, are the qualitative differences of reactive tendency that account for the fact that some mammals learn slowly, and with many errors, to meet situations which their fellows of superior age or race learn to meet quickly and with but few errors'.[16]

Hamilton had no formal connection with any university. He was employed by a millionaire with a large estate in Santa Barbara, California, but his duties as the family therapist usually left him time to carry out research. These unusual circumstances produced the first example of a method for studying learning in different kinds of animals, where performance does not critically depend on an individual's perceptual and motor abilities or on his level of activity. Hamilton's experiment was a pioneering attempt which, no doubt partly because he was outside the university system, did not receive the attention it deserved either at the time or in later years. It is worth describing at a level of detail which makes clear quite what he was trying to do.

The general form of apparatus used by Hamilton was a chamber containing an entrance door, which could not be used as an exit, and four exit doors, which could be locked or unlocked by means of a set of strings operated remotely by the experimenter. On each trial only one door was unlocked, so that when a subject pushed against it he could leave the chamber and find the reward available outside. The singular aspect of the situation was that there was no consistent signal to indicate which door was unlocked on a particular trial or which three were locked. The only rule was that whichever door had been unlocked on one trial would be locked on the next; otherwise the choice of doors was random, except that over the one hundred trials each subject received each exit door was left unlocked exactly twenty-five times.

The specific version of this chamber used for a given test varied according to the subject. These included ten human beings, five monkeys, sixteen dogs, five cats and a horse. The human subjects were not a representative sample of the race; Hamilton tested two men who worked on the estate, his own child of 26 months and some children between the ages of ten and fifteen. He regretted the absence of any ages intermediate between that of his son and the elder children, but explained that his 'wholly undeserved local reputation as a vivisectionist seemed to create a stubborn unwillingness on the part of parents to supply young children for experimental work'.

Although the study was not intended to bear directly on problems of psychopathology, this was the ultimate aim of the work. Two of the human subjects, a boy of eleven and one of the men, were of particular interest to Hamilton, since he had previously judged them to be mentally defective. His descriptions of his subjects are revealing about their attitude towards the experiment and his attitude towards them. 'Man 1', the representative adult human being, is described as follows:

'Age, 34 years. Native (Spanish–Indian) Californian. Ranch laborer in the experimenter's employ. A man of limited education, but of average intelligence for his class. He went through the trials in the stolid, unemotional manner that characterizes his work in the fields. The "boss" wanted him to walk into and out of an enclosure 100 times, and he did so without asking questions or shirking his task.'

Man 1's fellow was labelled by Hamilton as 'Defective Man A' and given the following description:

'Age, 45 years. Native (Spanish–Indian) Californian. Ranch laborer in the experimenter's employ. Limited school education, but had read history and uncritical works on socialism. He was a nervous, suspicious, "muddled" person, with a grievance against society in general, and a surprising fund of self-acquired misinterpretations relating to his social environment. He expressed a belief that my experiment was dangerous meddling with the human mind, and that it had some occult power of "making people crazy". His curiosity and his desire to argue matters rendered him available, but he seemed to be in constant dread of the apparatus, and always labored under a suspicion that it was not the simple structure that it pretended to be.'

The general measure of performance used by Hamilton was to count how many attempts to open a door a subject made over his one hundred trials. The procedure meant that an individual who understood the situation perfectly from the beginning and could always remember which door had been unlocked on the previous trial – and thus had to be locked on the present trial – would score about 200. A subject who understood the situation, but always forgot the outcome of the previous trial, could score 250 by using a sensible strategy. A subject with little understanding of the situation, a poor memory and a tendency towards inappropriate 'reactive tendencies' would produce a very high score. This happened with the horse, an 'eight-year old gelding of Western breed', whose total of 461 attempts to open a door exceeded that of all other subjects despite the stableman's belief in his 'smartness'.

Hamilton was gratified by the way these scores fell into what seemed an eminently reasonable pattern. The human subjects were best, followed by the monkeys, the dogs and then the cats. Within each group the older subjects made fewer errors than the younger: Man 1 was better than the children, just as the dogs were better than the puppies and the cats better than the kittens. Moreover Defective Man A scored 217 to Man 1's surprising 200, while the retarded boy performed worse than any of the other children except Hamilton's own baby. Hamilton analysed in some detail the different patterns of responding exhibited by the subjects and a considerable part of his report was devoted to discussing the qualitative differences he observed. However there is no point here to any further discussion of his findings, since the study was one of those that is more interesting for the possibilities it raised than for its actual results.

Yerkes was no doubt as aware as anyone of the weaker aspects of Hamilton's study. For example, especially in view of the results Yerkes himself had obtained with Dodson when varying the level of shock in their study with dancing mice, the differences obtained by Hamilton might have been due to variations in motivational level; for example, his son may have made more errors than any of the monkeys simply because the toys he found beyond the exit provided a less effective incentive for performing well than the food provided for the monkeys. Hamilton recognized this problem, but did not discuss an equally serious weakness of his procedure, which is that it confuses what are likely to be very different kinds of mental process; thus, as implied above, a subject can perform badly either because he forgets the last trial or because his understanding of the situation is poor.

Hamilton's test was deliberately made insoluble in the sense that there was no way in which a subject could know which particular door was the open one on a given trial before pushing at the three possible ones in turn. In this respect it retained the features of a trial-and-error task and did so in order to make it easier to study the kinds of reactive tendencies subjects used. Yerkes subsequently used a procedure he termed the 'multiple choice method' that was superficially similar to Hamilton's four-door situation, but with the crucial difference that the correct choice on any trial could in principle be predicted by a subject.

The kind of arrangement he used is illustrated in Figure 6.7. In the version shown here a subject is presented with a choice of entering anything up to nine compartments. The number of compartments that were unlocked varied from trial to trial, and could be as small as three or as large as the complete set of nine. If the subject entered the correct compartment he would find the reward, but if he entered one of the other unlocked compartments he would be confined there for a minute or so. Yerkes selected the correct choice on a given trial by some systematic rule. The simplest one was that for any set of compartments the one on the extreme right of this set was always correct. Thus, if the compartments labelled 9, 8 and 7 in Figure 6.7 are open, 7 is correct; if 7, 6, 5 or 4, then 4 is correct; if 5, 4, 3, 2 and 1, then 1 is correct; and so on.

It was already known that animals can learn very rapidly a discrimination based on position; thus, if in a two-choice situation the left-hand compartment always contains food and the right-hand never does, an animal comes to enter the reward compartment within a very small number of trials. Such a performance may indicate that the animal has learned to approach a particular place or that it has learned the relationship implied by the description 'right-hand' compartment. Yerkes' procedure ensured that except for the two compartments on the far right no one compartment contained reward more frequently than any other. Consequently a subject can only perform well on this task by learning the rule imposed by the experimenter, since developing a tendency to approach a particular compartment, a *position habit* as it became known, could raise performance only a little above chance level.

Yerkes first tried out the procedure using a simplified form of the situation just described with some patients in a Boston mental hospital. The results were encouraging and he went ahead using the apparatus shown in Figure 6.7 to test in separate experiments some crows and then a pair of pigs.[17]

He used a series of four problems with the pigs.

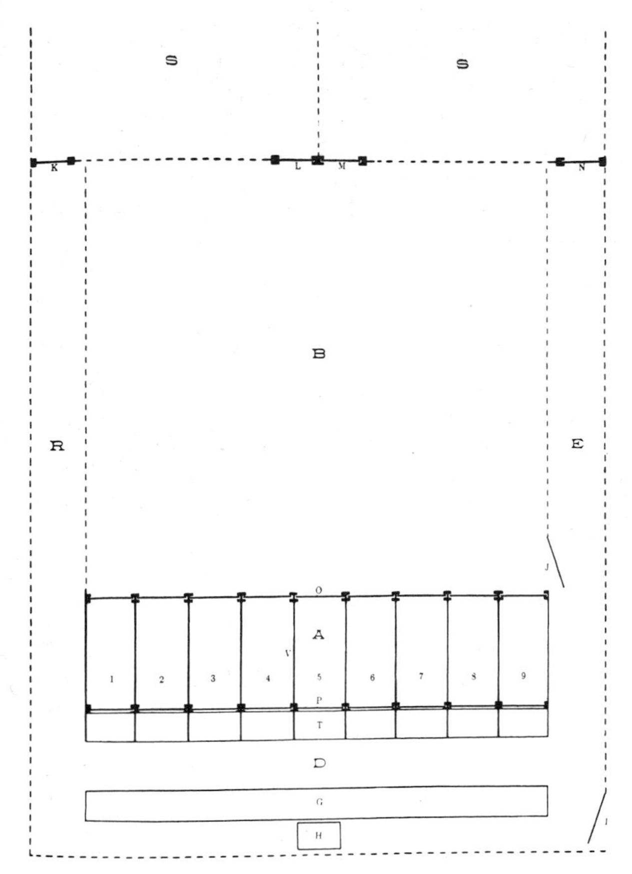

Fig. 6.7. Ground plan of multiple choice apparatus used by Yerkes to study the behaviour of pigs

The first employed the rule described above, whereby reward was available in the compartment on the extreme right of the available set. This was easily solved; a criterion of ten correct trials in succession was reached within forty-five trials by both pigs. In the second problem the correct compartment was always the second from the left of the present set and in the third problem it alternated from extreme left to extreme right on successive trials. To Yerkes' great surprise the two pigs also performed well on these two tasks, although about five hundred trials were required before the same criterion was reached. However, there was no indication that the animals could solve the fourth, and final, problem in which the correct compartment was always the middle one of an odd numbered set.

Although uncited, two older figures seem to lurk behind this research. The general idea of arranging animals along some dimension of intellectual ability, like Cole's earlier reference to placing raccoons on a 'mammalian scale of intelligence', is highly reminiscent of Herbert Spencer's linear concept of evolution. And the procedure itself seems admirably designed to find out whether a particular animal is capable of adjusting its behaviour according to what Lloyd Morgan had called an 'abstract relationship'.

Yerkes himself talked of studying the extent of 'ideation' in animals. It seemed that he now had an interesting method for doing so and one which, like Hamilton's, could easily be used with almost any kind of species. The study of pigs was reported in 1915. By this time another procedure had been developed by a graduate student at Chicago named Walter Hunter. This appeared to provide a way of studying another aspect of ideation in animals and was also easy to use on a wide range of species. Hunter's procedure was termed the delayed reaction test and represented an alternative, and better controlled, method for studying the kind of issue that had interested Cole.

Watson's former student, Harvey Carr, had remained at Chicago to run the animal laboratory after Watson left. He had wanted to find out whether rats could perform well in a discrimination situation in which the opportunity to make a choice only became available a short time after the discriminative stimuli had disappeared. Early results had not been promising. When Hunter arrived in 1911, freshly graduated from Texas and with an interest in animal psychology inspired by Washburn's book, Carr persuaded him to take up the problem.[18] At first Hunter had no more success than the two earlier students. But then, impressed by Cole's glowing report, he decided that he would use raccoons and became as pleased with these animals as Cole had been.

The situation used by Hunter consisted of a release box made out of wire netting or glass, from which an animal could see three exit chambers.[1] When it was released, choice of the correct exit would allow the animal to leave the experimental chamber and find food outside; if the animal chose one of the two wrong exit chambers, it would turn a corner to find a locked door. In addition to his four raccoons Hunter used seventeen rats and two dogs in appropriate versions of this basic arrangement. He also tested five children, ranging in age from two to eight years, in a situation which required them to press one of three buttons in order to sound a buzzer and obtain the candy reward; otherwise this was equivalent to the situation used with the animals.

The first of the two main parts of Hunter's procedure involved a simple discrimination whereby a subject was trained to go to whichever exit, or button was lit by an electric bulb. Once this was mastered th

Fig. 6.8. Pig entering choice area of the multiple choice apparatus whose ground plan is given in Figure 6.7

second stage began. The aim here was to train the subject to choose the exit that had been lit, after the light had already disappeared. When the light was switched off the subject was kept in the release box for a delay which was gradually extended as training continued until performance started to deteriorate.

Hunter found that the rats could tolerate only very brief delays, ten seconds at the most. The raccoons were better, the best of the four being able to perform well with a twenty-five second delay. Of the dogs one was very poor, but the other could work well with delays of up to five minutes. As for the children, each of them outclassed all of the animals. The oldest, an eight-year old girl, *M*, reached delays of twenty-five minutes; Hunter believed that she could have gone on to much longer intervals had time allowed.

The behaviour displayed by subjects during the delay period was systematically recorded and these results turned out to be at least as interesting as the length of delay an animal could tolerate. Hunter found that the rats and dogs relied entirely on orientation cues. When released the rats would head directly to whatever exit they were pointing towards; it seemed that the only way for them to respond correctly after a delay of even one or two seconds was by maintaining orientation of both head and body towards the exit that had contained the light. The reason why one dog could do so well at very long delays was because he was able to keep his head pointing in the right direction all the time. The most interesting aspect of the children's performance, and frequently that of the raccoons too, was the absence of any orientation; whether they chose correctly or not after a long delay appeared to be unrelated to any activity during the delay period. Distracting noises, which produced marked immediate reactions, left their choice performance unchanged. The raccoons would sometimes head off in one direction and then alter course to choose the correct exit, something that was never observed in rat or dog. The children were unaffected by whether they left the room; *M*, for example, could spend a long interval chatting outside and still make a correct choice when she returned.

Various control procedures established that subjects were not using some subtle external signal, such as unconscious cueing by the experimenter or a slight increase in the temperature of the exit chamber that had just been lit. With respect to the raccoons and children, Hunter commented: 'We have exhausted our ingenuity as to objective possibilities of explanation, and as a consequence are forced to conclude in favor of an intra-organic factor'. He suggested that his was the first study to demonstrate conclusively that animals can under some circumstances base their behaviour on ideas. 'The type of function here involved is *ideational* in character. By applying the term "ideas" to those cues, I mean that they are similar to the memory idea of human experience so far as the *function* and *mechanism* are concerned. They are the residual effects of sensory stimuli which are retained and may be subsequently re-excited.' Because this conclusion applied to raccoons and children, and not to rats and dogs, the delayed reaction procedure seemed to provide a powerful tool for distinguishing levels of mental processes in different species, as well as a way of rebutting Thorndike's claim that all animal intelligence is limited to the connection of impulses to the sensations of stimuli that are currently present.

In less than twenty years since Thorndike had begun his doctoral work a handful of people had created an active new field of experimental science. For the first decade or so the major issues had been those raised by Thorndike, together with the study of the senses. Within this period a considerable amount of precise knowledge about the learning and perceptual capacities of various species had been gained for the first time. And this was not just a matter of making more specific what was already known in a general sort of way. For example, it was a genuine discovery that flew in the face of accepted belief to establish that among mammals man's perception of colour is shared by few other species apart from fellow primates and that rats, Cole's raccoons and even bulls live in a black, grey and white world. Equally there was hitherto complete ignorance over whether a creature such as a turtle was capable of learning anything and an entirely mistaken impression of the effectiveness of imitation or passive learning. Following this initial period new developments, such as the work by Hunter and by Yerkes, meant that animal psychology remained a lively research tradition.

Such intellectual health was in complete contrast to the precarious institutional position of animal psychology. Within the few departments where such work was pursued it was in general regarded with suspicion, if not hostility. It was not at all clear to most American psychologists that such activities merited a place within their subject. At Harvard Yerkes was under considerable pressure to switch to human psychology. He held the low rank of instructor and in 1908 he failed to obtain a promotion to assistant professor which went instead to a dilettante philosopher. Yerkes considered this to be a personal humiliation. It was followed by a move to prevent the awarding of doctorates to graduate students whose only research was with animals. Yerkes rejected incentives to specialize in educational matters, but his use of patients at the Boston Psychopathic Hospital as the first subjects on which to try out his multiple discrimination procedure and his textbook on human psychology mark his reaction to accusations concerning the narrow irrelevancy of animal psychology.[20]

Until well after the First World War animal research did little more than maintain a toehold within academic psychology in America. However, as described in the final chapter, by the end of the 1920s the rat laboratory acquired a central place in almost every major psychology department. To understand how this shift occurred and to appreciate the significance of one contributory factor, Watson's behaviourism, a general picture is required of American psychology early in the twentieth century.

American psychology at the beginning of the century

It would have been a difficult task in 1908 to produce an equivalent to Washburn's book that summarized recent and distinctively American contributions to human psychology in a way that would be comparably illuminating to a layman. Books were written; in fact, a puzzling feature of the early protagonists of a scientific psychology is the effort they devoted to translating German tomes on psychology or writing their own textbooks, at the expense of getting on with their science. The obituaries written years later by their successors are full of information on these books, on the laboratories these pioneers founded, the journals they edited and the students they trained; yet substantial contributions to the understanding of the human mind receive little attention. Symptomatically it became common among the many university laboratories of psychology established in the 1890s to publish collections of 'minor studies'; major studies were rare.

The lack of direction that characterized psychological research in America at the turn of the century stemmed from a central dilemma. The problem was to find a way of reconciling three conflicting aims. One was to keep psychology focussed on the study of the mind as traditionally conceived, and thus to concentrate on the nature of subjective experience. The second aim was that it should be scientific, in the sense that the validity of its factual or theoretical claims should be subject to assessment by the kind of empirical methods employed in other sciences, and preferably by experimental methods. The third aspiration was that psychology should be interesting, either by relating directly to fundamental questions of everyday life or by making significant practical contributions.

This third consideration steadily increased in importance. Doctorates in psychology were awarded more rapidly than academic posts were being created and hence there was a pressing need to show that training in psychology provided knowledge and skills that the outside world should value. In the early part of the century the two most obvious fields for psychology to make its contribution to society were education and mental health. As we have seen Hamilton earned his living as a psychiatrist, while many of Yerkes' other students, such as Cole, ended up more typically in jobs related to the training of teachers. In order to provide their graduate students with useful training and preserve the intellectual coherence of the subject, university psychologists needed a body of theory, knowledge or methods that would be applicable in one or other of these fields.

For over twenty years several major contenders for this role in turn attracted great interest. As described in this section, early on there was the developmental psychology of G. Stanley Hall and also that of James Mark Baldwin, while these were followed by the importing from Europe of new theories of psychopathology, notably that of Sigmund Freud; these in turn were followed, as described in the final chapter, by a period of intense enthusiasm over mental testing. In comparison, a research degree in animal psychology seemed to offer little of practical value. Watson and Yerkes believed it might be good preparation for a career in education, but until well into the 1920s few agreed with them and fewer still shared Hamilton's faith in the benefits animal research would confer upon psychotherapy.

To a small, but highly influential, group of psychologists represented by Edward Titchener concern with potential applications of psychology was a serious mistake at this stage in the subject's development; to proclaim the general usefulness of a psychological degree was to run the economic risk of selling unripe fruit. The fundamentals had to be settled first and these very much included the question of consciousness. This was the most popular topic in the journals of the pre-war era so that the subject could easily be seen as a sub-branch of philosophy rather than as the independent natural science many wished it to be.

When Watson was challenged to defend the inclusion of animal studies within psychology, he replied by attacking the notion that the primary concern of the subject should be with the nature of conscious experience. Independent of the strength of his arguments on such matters, in the long run what led to the increasing attractiveness of behaviourism was its promise of a combination of usefulness and scientific respectability that previous alternatives had failed to provide. It would make contributions to education that would be at least as beneficial as Hall's genetic psychology or the development of mental tests, it would provide methods for treating the mentally ill that were at least as effective as psychoanalysis and it would make direct contact with the realities of everyday life in a way that academic disputes over the nature of mind singularly failed to do. All of this was to be achieved while maintaining, as no previous approach had been able, the belief that at its core psychology should be a laboratory-based experimental science.

At the end of the nineteenth century the only area in which American psychology was expected to supply some expertise was that of education. A great many people involved in the business of reforming, first, primary schools and then high schools, as well as those with more private concerns about the best way to bring up children in a changing world, looked to psychology for delivery of the goods first promised by Hall almost twenty years earlier. One result was the emergence of what was called 'genetic psychology' and this took two competing forms.

The early 1890s were stormy years for Hall. In 1890 his wife and daughter died in a fire in their home. In 1892 the troubles at Clark University, caused by lack of funds and by the way he, as president, had reacted, left the size of the faculty reduced to a third of what it had been and the number of students reduced even more drastically. In 1893 strife over journal policy had led to the founding of the *Psychological Review* by Baldwin and James Cattell in direct rivalry to Hall's own *American Journal of Psychology*. And in 1894 he reached the age of fifty and continually experienced feelings of malaise, which he described as 'the early psychic symptoms of old age'. Yet within a short time he bounced back with renewed vitality and, although continuing as president of Clark, vigorously developed his own version of genetic psychology.

Hall's solution to the dilemma surrounding psychology's future was to brush aside the claims of empirical validation. Hall had earlier been as responsible as anyone for the growth of experimental psychology in America. His new genetic approach, however, now rejected his former view that psychology should be based on the laboratory and instead embraced an evolutionary perspective for studying the evolution of mind in general and the psychological development of the individual, a perspective that encouraged the collection of observations rather than experimental results. Hence the term 'genetic' – from 'phylo*geny*' and 'onto*geny*' – which held an entirely different meaning from the one it acquired twenty years or so later. The interest in child development stemmed partly from the consideration that, from a biological point of view, a particularly remarkable feature of *homo sapiens* is the uniquely long time he needs to reach maturity. Among other things some recent European research on play in animals and children had attracted attention to the general question of the evolutionary function of an extended childhood.[1]

Hall became deeply involved in the child study

movement and organized the collection of questionnaires on a whole range of topics. Unrestrained by any feeling of obligation to be systematic, critical or consistent in interpreting the results produced by this method, he was now able to pour out paper after paper in which loosely substantiated claims were combined with evolutionary, moral and religious overtones. Some of the ideas were sensible, even prosaic and of lasting value; for example, he was responsible for the introduction of routine eye and ear tests in American schools and for a novel concern for hygiene. However not many of his suggestions were as straightforward as this.

At first Hall was mainly interested in infancy. Thus, in one paper of 1897 he discussed childhood fears which, he decided, emerged in a natural sequence that reflected the pre-history of the human race and included fears of loss of physical support and orientation, water, animals, darkness and disease. By the turn of the century his main concerns became the nature of adolescence, the 'paradise of the race', and the related practical issue of high school reform. In 1904 his ideas on these topics appeared in what was his major work, a two volume book entitled *Adolescence: its Psychology and its Relations to Physiology, Anthropology, Sociology, Sex, Crime, Religion and Education.*

One influential reviewer with more interest in the topic and sympathy for Hall's aim than many of his colleagues was Edward Thorndike. He wrote of Hall's book: 'Torrents of rhetorical enthusiasm over youth, love, genetic psychology and other matters will irritate the scientific student and probably will befuddle the general reader. One has to gyrate about from whales to vital statistics, to the lives of saints, to Jacksonian epilepsy, to the Hopi dancers, until one prays for a range of knowledge equal to President Hall's to empower him to see the unity and organization of the book or any chapter in it.'[2] Hall's *Adolescence* convinced those who had still been uncertain that this was not the direction psychology should take.

Hall's genetic psychology was erected on the evolutionary framework provided by Spencer and Haeckel. Its central pillars were the Lamarckian principle of the inheritance of acquired characteristics and the recapitulation principle by which individual development is held to mirror in some manner the history of the race. Within biology these two principles were already being seriously questioned even in the early days of Hall's genetic psychology. The person with the best grasp of what was wrong with this aspect of Hall's theories and with a better understanding of evolutionary theory than anyone in America was Baldwin.

Fig. 6.9. James Mark Baldwin

Despite their many differences Baldwin did agree with Hall in his dissatisfaction with experimental psychology and his belief that psychology should become genetic. Baldwin's understanding of evolution, one that like that of his friend and contemporary Lloyd Morgan, rejected Lamarckian principles and made a clear distinction between cultural and biological inheritance, produced a different outlook on child development from that of Hall. In particular it put little weight on human instincts and a great deal of emphasis on understanding the way that a child interacts with and learns from its environment, particularly its social environment. Thus, while Hall's ideas on evolution led him to emphasize 'biological' determinism, maturation and 'natural' norms for many aspects of childhood, for Baldwin the central problem was learning. 'How does the individual organism manage to adjust itself better and better to its environment? How is it that we, or the amoeba, can *learn to do anything*? This latter problem is the most urgent, difficult and neglected question of the new genetic psychology.'[3]

Baldwin's interest and ideas were prompted by the birth of his first child, Helen, whom he studied and tested in a fairly systematic way. From these observations, and the study of his second child, plus his impressive knowledge of both biological and psychological research, he developed a theory of cognitive development that is a direct forerunner of Piaget's work and a theory of social learning that also anticipates some recent theories of personality and social development.[4]

Baldwin believed very strongly in theory. He objected to 'that most vicious and Philistine attempt, in some quarters, to put psychology in the strait-jacket of barren observation, to draw the life-blood of all science – speculative advance into the secret of things – this ultra-positivistic cry has come here as everywhere else, and put a ban on theory. On the contrary, give us theories, theories, always theories!'[5] But even the less positivistic of his fellow psychologists asked that theories should at least be accompanied by suggestions as to how their validity might be checked. With all the time devoted to editing three journals, establishing new laboratories, writing many books and co-ordinating a massive *Dictionary of Philosophy and Psychology* there was little left for working out satisfactory empirical methods for studying development.

A third group sensitive to the appeals for a psychology that could be applied to educational problems regarded Baldwin, and James too, as guilty of leading the study of the mind back into philosophy. To a number of psychologists inspired by Cattell and based mainly at Columbia the future of psychology lay in the development of quantitative methods that would produce precise factual information of direct use to the classroom teacher or administrator of an educational system. The major line of development was the mental testing approach stemming from Galton's ideas. In the hands of psychologists like Thorndike this was combined with experimental work that took those aspects of classroom instruction that could be easily quantified and attempted to examine them within a laboratory situation. A much debated example was the question of massed *versus* spaced learning. In a classroom is it better to provide one hour per day of French lessons or a block of five hours just once a week? In a laboratory will a person, or rat, taking part in an experiment on learning benefit from long periods of time between successive trials?

In the first decade of the century this plain, empirical approach had not shown much progress. Indeed the Galtonian tradition seemed to have foundered. A study of 1901 revealed little correlation between the various simple measures of individual differences promoted first by Galton and then by Cattell; a person's reaction time was only tenuously related to the time he took to name colours or his ability to estimate time, and none of these indices predicted academic achievement.[6] At the same time a different type of test attempting to assess abilities of a more intellectual kind, which had been developed by Binet and Henri in France, was dismissed on the basis of a cursory evaluation.[7] Full appreciation of the French work did not come until 1910 when two of Hall's students began to develop American versions of the Binet tests, despite initial discouragement from Hall.

As mentioned earlier, the view that it was premature for psychology to turn to practical problems was held by Edward Titchener, an Englishman, who after studying for two years with Wundt, came to America in 1892 to establish a psychological laboratory at Cornell University. As the latest arrival from Leipzig he came to represent experimental psychology of the traditional German kind and this position was maintained by a steady stream of translations of Wundt and other German texts. These eventually ensured that American psychologists' view of their German roots came to be dominated by Titchener's particular perspective, which emphasized and distorted certain aspects of Wundt's work and totally ignored others.[8]

The aim of Titchener's psychology was to analyse conscious states. The main method was that of introspection as practised by trained observers who carefully noted the sensations they experienced in some experiment on perception or reaction times. He termed his approach *structuralism* to indicate his concern with the nature of consciousness and the elements of which it is composed.

Few psychologists of his generation completely shared Titchener's views. He became increasingly isolated as he withdrew from contact with his peers and became an aloof figure to his students. A bruising controversy with Baldwin in 1895 began this trend and established an alliance, wholly unlikely on intellectual grounds, between Hall and Titchener. He took no part in the activities of the American Psychological Association and in 1904 started an informal group called the 'Experimental Psychologists'. At meetings of this group papers reporting research on animal, child, abnormal or applied psychology were banned, no matter how experimental they might be. An intriguing aspect of the subsequent history of academic psychology in America is that, despite Titchener's social and intellectual isolation and the ultimate rejection of his particular kind of psychology, his influence and that of the Society of Experimental Psychologists, which developed from his group, remained enormously pervasive.

Many psychologists of the time shared Titchener's belief in keeping psychology a pure, experimental science, but the enthusiasm for evolutionary theory of their undergraduate days left a lasting effect on the way they viewed psychological phenomena. Furthermore, most of them had been introduced to psychology by William James' *Principles*. In contrast to Titchener's claim that psychology should determine the *structure* of consciousness, James discussed its *function*. 'Consciousness . . . has in all probability been evolved, like all other functions, for a use – it is to the highest degree improbable *a priori* that it should have no use.'[9] Titchener dismissed those who preferred the Jamesian view as 'functionalists', and the name stuck.

One of the main centres of functional psychology was Chicago, where the influence of local philosophers like Dewey and Mead reinforced its pragmatic and evolutionary aspects. James Angell, Watson's supervisor, was seen as a leader among the younger functionalists. Contrasting his approach with the tradition represented by Titchener he wrote that psychology should be concerned with 'the identification and description of mental *operations* rather than with the mere *stuff* of mental experience' and that mental activity should be seen 'as part of the larger stream of biological forces'.[10] Nevertheless a good part of the research supervised by Angell was, like that of Titchener's students, based on laboratory experiments with adult human subjects, dealing with a range of traditional issues in perception and memory and putting little emphasis on the development of theory or on immediate relevance to practical issues.

After finishing the *Principles* James himself turned mainly to issues in philosophy and religion, but nonetheless kept up with new developments in psychology at least as well as most of the younger generation. In particular he took a lively interest in what was happening in Europe. Although he completely ignored the later work of Wundt on social psychology and language – and thus, since Titchener ignored this too, ensured that it had little impact in America – James followed closely the important new approach to psychiatry of Wundt's student, Emil Kraepelin, and even more the work on neurosis and hypnosis therapy by Pierre Janet and others in Paris.

Attitudes towards mental illness at the turn of the century were dominated by a belief in the overwhelming influence of heredity and its causation by specific pathologies of the brain. The proportion of cases of 'nervous' and 'mental' disorders in America attributed to heredity rose from around 20 per cent in the 1840s to around 90 per cent in the 1900s. The only personnel with professional training in the state hospitals for th insane were neurologists whose main task was viewe as identifying from *post mortem* examination of patient's brain the specific lesion responsible for h disorder. The discovery in 1905 that general paralysi of the insane, GPI, was associated with an identifiab pathology of the brain and was linked to syphilis gav great encouragement to the general commitment to a organic approach.[11]

A few of the newly-designated 'psychiatrists' ha serious misgivings about the prevailing wisdon Adolf Meyer in particular was sufficiently skilled i neuro-anatomy to worry about the continuing failur to find any sign of abnormality in the brains c schizophrenics. Also he had gained unusually exten sive clinical experience since his arrival from Switze land and believed current neurological theories c insanity to be totally inadequate; there were too man cases of lasting recovery from severe depression o schizophrenia for him to believe that such illnesse were caused by irreversible deterioration of the brain Meyer and some of the young psychiatrists began t look to psychology to provide an alternative basis fo understanding mental illness to that offered b neurology. Various forms of psychotherapy wer tried, mostly in and around Boston, where Jame remained the most eminent representative o academic psychology.

The new developments in psychopatholog greatly affected James' outlook. Although not particu larly interested in educational matters, he was all i favour of psychology being of practical value. James attitude towards Titchener's insistence on 'pure psychology was one of scorn; 'The function o Titchener's "scientific" psychology (which "structur ally" considered is a pure will-of-the-wisp) is to kee laboratory instruments going, and to provide plat forms for certain professors.'[12] James began to talk o the 'subconscious mind', a term that was self contradictory as far as Titchener was concerned. I 1901 James wrote that 'the menagerie and th madhouse, the prison and the hospital have bee made to deliver up their material. The world of mind i shown as something infinitely more complex than wa suspected; and whatever beauties it may still possess it has lost at any rate the beauty of academi neatness.'[13]

One problem that had troubled many readers o James' *Principles of Psychology* in 1890 was that although it put great emphasis on the nature o consciousness, it was not entirely clear about wha 'consciousness' meant; James had been widely criti cized for inconsistency in his treatment of th mind–body problem. With a new perspective gained from the notion of subconscious mental events he now began to concentrate upon this problem. In addition he was encouraged by the mounting attacks on the idealist movement that had dominated philosophy in America and England during the latter part of the nineteenth century, reading 'the signs of a great unsettlement, as if the upheaval of more real conceptions and more fruitful methods were imminent, as if a true landscape might result, less clipped, straight-edged and artificial'.[14] James' contribution in the summer of 1904 towards a true landscape was to write at a remarkable rate a number of papers expressing his ideas on the mind and proposing a way of resolving traditional problems in philosophy which he labelled 'radical empiricism'. The first in this series, and the one most important for psychology, was called 'Does consciousness exist?'; it concluded that the answer was 'No'.[15]

James argued that the distinction between mind and matter, between 'physical stuff' and 'stuff called consciousness', that had prevailed in one form or another since the time of Descartes was a fundamental mistake. He claimed that consciousness 'is a non-entity, and has no right to a place among first principles. Those who still cling to it are clinging to a mere echo, the faint rumour left behind by the disappearing "soul" upon the air of philosophy.' According to his theory of pure experience '"outer" and "inner" are names for two groups into which we sort experiences according to the way in which they act upon their neighbours'. Thoughts and feelings have the same status in experience as external objects: 'things and thoughts are not at all fundamentally heterogeneous; they are made of one and the same stuff, which as such cannot be defined, but only experienced'.

In the final section of his paper James confronted the objection that in the last resort there remains the basic intuition exploited by Descartes that, while we may doubt all else, we know that we have thoughts. James' opinion closely anticipates the treatment of thinking proposed by behaviourist psychology some ten years later. 'My reply to this is my last word, and I greatly grieve that to many it will sound materialistic . . . I am as confident as I am of anything that, in myself, the stream of thinking (which I recognize emphatically as a phenomenon) is only a careless name for what, when scrutinized, reveals itself to consist chiefly of the stream of my breathing . . . There are other internal facts besides breathing and these increase the assets of "consciousness", so far as the latter is subject to immediate perception; but breath, which was ever the original of "spirit", breath moving outwards, between the glottis and the nostrils, is, I am persuaded, the essence out of which philosophers have constructed the entity known to them as consciousness.'

Besides philosophy James was prepared to speak and write on a number of topics like telepathy and religion which many felt that professional psychology should leave well alone. There was one topic that he agreed should be left and this was sex. James had no great inclination to criticize the conventional morality of his age. In the *Principles* he had preached the importance of will-power in controlling physical desire. 'No one need be told how dependent all human social elevation is upon the prevalence of chastity. Hardly any factor measures more than this the difference between civilization and barbarism.'[16]

With one notable exception psychologists agreed that the subject of sex should be approached with extreme caution. Although – or, perhaps, because – women in the United States enjoyed far more freedom than their sisters in Europe, the prudishness that had pervaded the Western world during the last half of the nineteenth century remained much more entrenched in America. In many European countries well-publicized attacks, usually by artists or writers, were made at the very end of the century on the perceived hypocrisy and evils of conventional morality, and these helped make possible some, albeit very limited, open discussion of topics such as birth-control, adultery or homosexuality. But there were no equivalent movements across the Atlantic.

Among psychologists Stanley Hall perceived that a subject such as adolescence can hardly be treated in realistic fashion without mentioning sex. By clothing such discussion in what James described as 'religious cant' and by allowing his style to become even more effusive when sensitive questions were approached, Hall appears to have believed that he might avoid causing offence. However most of his peers found that the style made things even worse. In his review of *Adolescence* Thorndike complained of the difficulty of understanding Hall's ideas; in private he noted that Hall's book 'is full of errors, masturbation and Jesus. He is a madman.'[17] Even before *Adolescence* appeared, Angell reflected a general uneasiness about Hall's new enthusiasm when he wrote to Titchener: 'Is there no turning Hall away from this d. . .d sexual rut? I really think it is a bad thing morally and intellectually to harp so much on the sexual string, unless one is a neurologist.'[18]

Many of the teachers and laymen who heard Hall speak or read his work were upset by the discussion of sex. On the other hand some of the people looking to psychology to provide insight into matters of personal significance regarded as pathetic the temerity shown by most academic psychologists towards anything that might make the profession lose some respectability. For them, effete arguments over structuralism *versus* functionalism, or over the existence of consciousness, might be interesting as philosophy, but were an evasion of the job that psychology should be doing. It was an opinion expressed later in vivid fashion by a biologist recalling the psychology of this era. 'After perusing during the past twenty years a small library of rose-water psychologies of the academic type and noticing how their authors ignore or merely hint at the existence of such stupendous and fundamental biological phenomena as those of hunger, sex and fear, I should not disagree with, let us say, an imaginary critic recently arrived from Mars, who should express the opinion that many of these works read as if they had been composed by beings that had been born and bred in a belfry, castrated in early infancy, and fed continually for fifty years through a tube with a stream of liquid nutriment of constant chemical composition.'[19]

It was into a generally unsettled situation that Sigmund Freud's work attracted increasing attention from American psychologists. His treatment of sex was startling, far more shocking than anything Hall had written. In 1905, for example, Freud had written *Three Contributions to the Sexual Theory*, in which he deliberately set out to explode common beliefs or pretensions that for normal individuals sexual experience does not begin until puberty. In plain, unadorned language he discussed the connection between thumb-sucking and masturbation, the resemblance between 'a satiated child sinking back from the mother's breast with reddened cheeks and blissful smile' and the 'expression of sexual gratification in later life', the sensation of pleasure enjoyed by an infant holding back its faeces for 'masturbatic excitation of the anal zone', and so on.[20]

Many American psychologists found Freud's theories hard to follow and expressed in what Angell, for example, regarded as 'needlessly repellent terminology'. Furthermore, they appeared to depend solely on rich and private interpretation by an analyst of the revelations by his few neurotic patients in a way that appeared immune to familiar kinds of evaluation. Yet, after all, Freud had been a respected neurologist with some well-received papers on the nervous system to his credit. Moreover, a recent convert to psychoanalysis, Carl Jung, who was based in a recognizably scientific psychiatric clinic in Switzerland, had begun to employ objective, quantifiable tests in the study of free associations and his results appeared to support Freud's theories.

At Clark University Meyer and Hall became very interested in Freud's work and discussed it in their lectures. The twentieth anniversary of the founding of Clark was due in 1909 and Hall decided that a major part of the celebrations should be a conference bringing together leading psychologists. It was Hall's own inspiration to invite Freud and Jung. Their acceptance ensured great interest in the Clark celebration among American psychologists, if only out of curiosity to see the notorious Freud in the flesh.

In Vienna Freud received the invitation with some amazement. He was used to having his books and papers either treated with derision or completely ignored by the academic world of Austria and Germany. Less than five years had passed since contact with Jung's circle in Zurich had marked the first expansion of psychoanalysis beyond Freud's immediate circle of mainly Jewish followers in Vienna. Participation in the Clark conference and the honorary degree Hall bestowed upon Freud constituted the first official recognition of psychoanalysis. It was also a highly important event for the movement in other ways.[21]

In late August 1909, Freud, Jung and another psychoanalyst took a ship from Bremen to sail for New York. Altogether Freud and Jung spent most of the seven weeks of their American journey in each other's company, when previously they had had only occasional, brief contact. On the voyage they analysed each other's dreams, but not to each other's satisfaction. Freud refused Jung's request for some information about his private life which would have clarified a certain dream, whereupon Jung was led to believe that Freud 'placed personal ambition above truth'; he gradually became convinced that 'Freud himself had a neurosis'.[22]

Once in America Jung became increasingly enthusiastic about the people, the countryside and the culture. Freud enjoyed the reception he was given at Clark and was delighted by Hall. 'Who could have known that over there in America, only one hour away from Boston, there was a respectable old gentleman waiting impatiently for the next number of the *Jahrbuch*, reading and understanding it all, and who would then, as he expressed it himself, "ring the bells for us"?' In general, however, Freud's impression of America was far from favourable. The lack of formality distressed him, he had difficulties with the language and most of the time he felt considerable physical discomfort; he suffered from a recurrent stomach ache, which he blamed on American cooking, and an infection of his prostate gland which made him acutely sensitive to the apparent inaccessibility of toilets wherever he went. After the conference the three psychoanalysts spent some days as the guests of an American psychologist and his family at a remote camp in the Adirondack Mountains near Lake Placid. Jung enjoyed himself hugely and entertained the party with German songs. Freud thought he was suffering from a mild attack of appendicitis. After returning at last to Vienna Freud concluded: 'America is a mistake; a gigantic mistake, it is true, but nonetheless a mistake'.

Following his American summer Jung no longer heeded Freud's disapproval and he turned to the study of the occult and the paranormal, abandoning for good the association tests and the scientific approach to schizophrenia that had gained him the invitation to Clark. America added greatly to Freud's self-confidence and he weathered the subsequent defection of Jung, and of other followers like Alfred Adler, without a wavering of belief in his own particular theories.

The visit of Freud and Jung to America in general considerably enhanced the reputation of psychoanalysis there and provided the basis for its subsequent rapid growth. But this growth occurred almost entirely outside the educational system. Most of the leaders of academic psychology at the Clark conference, who listened carefully to what Freud and Jung had to say and gained a great deal of personal respect for them, decided that this was not the way they wanted psychology to go. Psychoanalysis seemed at close quarters just as dogmatic and speculative as it had from afar; they worried about the central theme of infantile sexuality and could not see what kind of evidence could either confirm or modify Freud's views on the subject. Hall remained a loyal Freudian for a few years, but then, to Freud's dismay, decided that Adler's deflection of emphasis away from sex was more to his taste.

The Clark conference provides an intriguing view of American psychology in 1909. The guests comprised both those psychologists of whom Hall approved and those he considered too important not to invite. A group photograph from the conference is reproduced in Figure 6.10. It shows Hall in the centre of the front row, looking suitably patriarchal, with Freud to his left and, beyond Freud, the two Swiss, Jung flanked by the diminutive Meyer. Also in the front row are the two most prominent representatives of academic psychology present, Titchener and James. James is the only man from Harvard; a tradition of mutual distrust and avoidance between Harvard and Clark had been so strong for at least fifteen years that, for example, while Yerkes and his students at Harvard worked on animal behaviour, they had no contact whatsoever with those doing closely related work just an hour away in Worcester. Similarly, among the other universities that Hall did not regard kindly, Columbia was represented only by Cattell, standing behind James in the photograph, and there was no one whatsoever from Chicago. Animal psychologists were also notable for their absence from the Clark conference, unless Jennings, on the far right of the front row in the photograph, is counted; he was the only biologist there and the only representative from Johns Hopkins. Baldwin was not there either; whether or not his importance would have outweighed Hall's dislike of his old rival is a moot question, since by then Baldwin had left America for good, as is related in the next section.

John Watson's behaviourist manifesto

At the turn of the century there were psychological laboratories in most American universities, with one glaring exception: the Johns Hopkins University in Baltimore, where experimental psychology had first taken root in the New World. After Stanley Hall's departure for Clark in 1888 the philosophy department at Hopkins had been wound down and the apparatus from its psychological laboratory sold off. But by 1903 the university's financial state had improved considerably and Daniel Gilman had been succeeded by a new president who, with more sympathy towards psychology, decided to re-establish a department of philosophy and psychology. In his search for a suitable person to carry this out, he wrote for advice to Baldwin. In reply Baldwin listed the few men he considered suitable for the task and ended by making it clear that he himself would welcome an appropriate offer, for he was feeling dissatisfied with the conservatism of Princeton and with the considerable undergraduate teaching load he was obliged to carry.

The president of Johns Hopkins was delighted by the prospect of such a catch; Baldwin was one of the leading figures in psychology, known for his theoretical contributions to both developmental psychology and evolution, an experienced experimentalist with the founding of flourishing laboratories at Toronto and Princeton already to his credit and, as co-owner and editor of *Psychological Review*, a central figure in the professional structure of American psychology. Within a few months he was appointed and once again enthusiastically set about organizing a new department. Among the letters of congratulation the following from an English psychologist, James Ward,

PSYCHOLOGY CONFERENCE GROUP, CLARK UNIVERSITY, SEPTEMBER, 1909

Beginning with first row, left to right: Franz Boas, E. B. Titchener, William James, William Stern, Leo Burgerstein, G. Stanley Hall, Sigmund Freud, Carl G. Jung, Adolf Meyer, H. S. Jennings. *Second row:* C. E. Seashore, Joseph Jastrow, J. McK. Cattell, E. F. Buchner, E. Katzenellenbogen, Ernest Jones, A. A. Brill, Wm. H. Burnham, A. F. Chamberlain. *Third row:* Albert Schinz, J. A. Magni, B. T. Baldwin, F. Lyman Wells, G. M. Forbes, E. A. Kirkpatrick, Sandor Ferenczi, E. C. Sanford, J. P. Porter, Sakyo Kanda, Hikoso Kakise. *Fourth row:* G. E. Dawson, S. P. Hayes, E. B. Holt, C. S. Berry, G. M. Whipple, Frank Drew, J. W. A. Young, L. N. Wilson, K. J. Karlson, H. H. Goddard, H. I. Klopp, S. C. Fuller

Fig. 6.10. Group photograph from the psychology conference at Clark University, 1909

reflected on the difficult situation which had surrounded the appointment of Baldwin's predecessor over twenty years earlier and sounded a warning note.

'My dear Baldwin, . . . I do not know if I ever told you that I was once interviewed and cross-examined by a wealthy gentleman – his name I think was Thomas – over twenty years ago when they were starting a chair of philosophy. He told me frankly that I was not orthodox enough – and then went back and appointed Stanley Hall! "At Baltimore", he said, "we are a church-going people": and he had awful stories of the consternation that Huxley – or perhaps it was Tyndall – had produced by a special course of lectures. I hope the "church-goers" will not harry you, nor you frighten them. "Prosit Baldwin", I have just said to my wife, as she handed me a cup of coffee. In this beverage, beloved of church-goers, I drink your health.'[1]

By this stage of his career Baldwin was content to let others get on with laboratory work. Even his interest in questions concerning child development was shifting away from the kind of empirical study he had carried out with his own children towards what he called 'genetic epistemology'. This ambitious attempt to construct a theory of knowledge based on evolutionary and developmental principles became the main focus of his intellectual energies for the next few years.[2] Consequently he needed to appoint other people to handle the less philosophical aspects of the department's work.

Initially Baldwin made a number of excellent temporary appointments that ensured a good start for the department. One of the later and more permanent appointments was that of Knight Dunlap in 1906. Dunlap had obtained his doctorate in psychology at Harvard and, prior to joining Baldwin, had taught in California for four years. As a student he had become impatient with psychology's preoccupation with consciousness and introspection. He remained highly critical of Titchener's structuralism, believing that psychology should concern itself with action as much as with images and sensations and needed to be a practical subject.[3]

Dunlap's arrival at Hopkins coincided with that of Jennings who from 1906 headed the biology department. In spite of the new interest in genetics that ended his active research on the behaviour of lower organisms, Jennings maintained close contact with psychology and continued to give lectures and laboratory classes on behaviour for many years. When John Watson arrived in 1908, one of the first things he did was to attend Jennings' course.

Watson injected a great deal of energy into what was now a well-established department, but within a few months psychology at Hopkins suffered a severe blow. Baldwin suddenly departed for Mexico and soon after resigned from the university. The reason for this abrupt event was a well-kept secret until recently. It seems that in the summer of 1908 Baldwin had been caught in a police raid on a brothel in Baltimore, but by giving a false name had at first been able to prevent any scandal. Although two reporters recognized Baldwin, they kept quiet until the following winter when his reputation acquired some political significance. As acknowledgement of Baldwin's increasing involvement in educational matters, the mayor of Baltimore invited him to become a member of the school board. At this point the reporters began to leak their information. When interrogated by the president of Hopkins, Baldwin claimed that a friend had suggested visiting a negro social club and that he, Baldwin, had gone along out of curiosity, not knowing that ladies would be present.[4]

Baldwin's explanation was not considered acceptable. On the grounds that the university could not be seen to condone immorality by continuing to employ someone guilty of such an offence, he was required to resign immediately. Within the next year one or two leaders of the psychological establishment knew enough of this story to decide that Baldwin should become a forgotten name in American psychology. And, since he lived abroad from that time on, this is what quickly happened.

The event meant that, inadvertently, Baldwin made a second major contribution to Watson's career. The first had been to appoint the young man from Chicago to a full professorship. The second occurred because in the rush of his departure Baldwin turned over to Watson, as the most suitable person on the spot, both the chairmanship of the department and the considerable, and influential, journal commitments that he was now relinquishing. A year later Baldwin's name disappeared from its foremost position on the cover of the *Psychological Review* and that of John B. Watson replaced it. The only brief comment must have mystified all but the few readers in the know. It simply stated that 'Professor Baldwin has resigned his position in the Johns Hopkins University. He is advised to give his voice a prolonged rest from continuous lecturing. He will spend a year at least abroad.'[5] As for Watson, at the age of only thirty-one he now occupied a key position in American psychology.

By the time of Baldwin's departure the trustees of Johns Hopkins were much more interested in its new medical school than in the older parts of the University. Johns Hopkins had specified that part of his fortune should be spent on founding a hospital, and associated educational facilities, in one of the poorest areas of Baltimore. It was nearly thirty years after his death before there was enough money to start on this plan. The most immediate impact on the psychology department came from two of the many important innovations included in the medical school: financial support from a steel magnate made it possible to include a psychiatric clinic and, a few years later, medical students were required to take courses in psychology. By 1909 Adolf Meyer had been persuaded to take over the planning of what was to become the Phipps Psychiatric Clinic.

In the time since he had left Switzerland to meet Donaldson and find a job in Chicago Meyer had become a leading figure in American psychiatry. Part of his reputation was based on practical achievements, starting on a small scale in a hospital in Illinois and then in Worcester and finally on a grand scale in New York State. In each place he had completely reorganized ward procedures, instituting such novelties as systematic case-histories. With the help of his wife he was the first to introduce a form of psychiatric social work, whereby some attempt was made to understand the social and family background of a hospital's inmates. He left New York for Hopkins with the reputation of having turned the state's insane asylums into modern mental hospitals.[6]

Throughout this period of busy involvement in practical and organizational matters, Meyer had maintained close contact with academic psychology and he often published his views in its journals. From the perspective of mental illness Meyer attacked the idea of a psychology based on the presumed independence of mind and body, and appealed for a psychology more closely tied to biology. As noted earlier, he was at the same time highly critical of the attitude prevalent among his fellow psychiatrists, that of treating psychological disorders as simply the symptoms of some underlying organic disease which in most cases reflected some strong inherited predisposition. Instead, Meyer suggested that mental patients need to be understood in terms of the way they behave and that, in particular, a clinician should examine closely the formative years of early childhood and the environment in which the patient has lived. These views and his conviction of the importance of repressed sexuality made him take an immediate interest in Freudian theory; Meyer wrote that 'no experience or part of our life is as much disfigured by convention as the sex feelings and ambition'.[7] Baldwin's recent departure provided a reminder of the force of these conventions.

Fig. 6.11 Adolf Meyer in the newly opened Phipps Psychiatric Clinic, 1913 (the impression this photograph may convey of illustrating an early test of electro-convulsive therapy is a misleading effect of its composition)

Meyer invented the term 'psychobiology' for the combination of psychological and biological approaches he believed necessary for solving the problems of psychiatry. His major interest was in the causes of schizophrenia. In 1905 he attempted to explain this illness as a result of cumulative patterns of defective habits, and a year later he suggested that patients can be classified in terms of 'reaction types', the typical ways in which they behave in difficult or emergency situations. It appears that his idea influenced Hamilton when he decided to carry out the comparative study of reactive tendencies using the multiple choice apparatus described earlier in this chapter.

A very much younger man than those described so far arrived at Hopkins in the Autumn of 1911 to take up a fellowship offered to him by Jennings. Karl Lashley was from West Virginia, where interests that developed during his solitary childhood in a small country town had led him to study biology at the state university. From there he had gone to Pittsburgh as a graduate student and, during the first summer, had met Jennings when working at a marine biology station. Lashley was twenty-one when he arrived to start research on genetics with Jennings. From the very beginning he also spent a lot of time with both Watson and Meyer.[8]

For many among the group of psychologists and biologists at Hopkins, these years from about 1910 until America's entry into the First World War in 1917 were golden years. They were able to meet, talk and pursue their interests with relatively little competing pressure from teaching or administrative duties. Compared to most other American universities of the time Hopkins demanded little undergraduate teaching of its faculty. Also, in philosophy and psychology there were very few graduate students; the department awarded only one doctorate in the period from 1904 to 1912. Watson envied Yerkes the steady succession of able young students who came to study psychology at Harvard, attracted more by the prestige of the university than the quality of the department.

Little is known in detail about the lives of this group at this time. None of them wrote more than the sketchiest of autobiographies. Lashley refused even that. Watson burnt most of his private papers. Jennings, who kept a detailed diary for most of his life, let it lapse during this period. Their joint research suggests that the two men who spent most time together were Watson and Lashley. One collaborative project concerned the development of a baby monkey born in the small colony Watson started at Hopkins. In the summers Lashley joined Watson on the Dry Tortugas to continue the study of homing by terns. With Watson's encouragement Lashley also studied the ability of a parrot to learn to imitate sounds and began to examine the discrimination of visual patterns by rats.

Watson's editorial duties and other professional commitments kept him very busy. Nevertheless, the joint project with Yerkes on methods of studying vision in animals was finished and the report, which was mainly about apparatus and techniques, was finally published.[9] He spent almost a year and a half working with his wife on a series of collaborative experiments on the rat's sensitivity to monochromatic light. It emerged that the rat, like most other mammals, did not possess colour vision; however, Watson noted that 'from every standpoint the experiments are far from satisfactory'.[10]

Watson's continuing concern with the question of what animals can see was in contrast with the outlook of his various colleagues. Jennings, Meyer and Dunlap all wanted psychology to concern itself with action, and drop its emphasis on perception; and Meyer and Dunlap, at least, also wanted it to become useful. In many ways Watson's general views on psychology were initially much more orthodox. A major reason for his interest in the behaviour of animals was that it provided the only way of answering the kind of question about their senses which experimental psychologists had for some decades been asking of their human subjects. And as for practical implications, whether or not it turned out that the rat possessed colour vision was unlikely to advance, even in a most indirect way, Meyer's efforts to understand schizophrenia or the search for a solution to any other pressing human problem.

Within the field of animal psychology Watson was already committed by the time he arrived at Hopkins to a tough experimentalist position and to the avoidance of subjective terminology in interpreting an animal's behaviour. However, his views on the rest of psychology were far less developed and his new colleagues seem to have persuaded him to adopt a more consistent and critical attitude towards current research in human psychology. This made his professional position increasingly anomalous: here was the head of a prominent psychology department and the editor of America's leading psychology journal holding little respect for the activities of many of his fellow psychologists and at the same time pursuing a line of research which most of them believed to belong to biology, having little significance, if any, for the central issues of psychology. Privately Watson admitted to Yerkes in a letter of 1910 that he wondered whether he was a physiologist rather than psychologist.[11]

Watson was provided with an opportunity to present his thoughts on this situation when early in 1913 he was invited to give a series of lectures on animal psychology at Columbia University. His opening lecture was concerned with the relationship between animal and human psychology and for the first time he made public what he thought of the present state of American psychology. Later that year the lecture appeared in the *Psychological Review* with the title 'Psychology as the behaviorist views it'.[12]

Watson explained that he used to be embarrassed by the frequent question: 'What is the bearing of animal work upon human psychology?'. The accepted position when he had started research was that one needed to construct the conscious content of the animal whose behaviour had been studied in order to relate the results to the study of human consciousness. Eventually this was realized to be absurd; questions about an animal's consciousness were unanswerable. However, this did not matter. 'One can assume either the presence or the absence of consciousness anywhere in the phylogenetic scale without affecting the problems of behavior by one jot or one tittle; and without influencing in any way the mode of experimental attack upon them.'

What could be done about the gulf that had opened between the study of the behaviour of animals and the study of the conscious states of human beings? The choice appeared to lie between allowing the study of behaviour to develop entirely independently of psychology as currently conceived or changing psychology. Watson's message was that the only real solution was for psychology in general to adopt the 'behaviourist' approach already common in animal psychology. It was stated in a clear, direct manner in the opening paragraph of the paper.

'Psychology as the behaviorist views it is a purely objective experimental branch of natural science. Its theoretical goal is the prediction and control of behavior. Introspection forms no essential part of its methods, nor is the scientific value of its data dependent upon the readiness with which they lend themselves to interpretation in terms of consciousness. The behaviorist, in his efforts to get a unitary scheme of animal response, recognizes no dividing line between man and brute. The behavior of man, with all of its refinement and complexity, forms only part of the behaviorist's total scheme of investigation.'

Watson gave two main arguments why psychology should adopt a behaviourist view. The principal one was that the lack of progress displayed by psychology was due to its reliance on subjective data; claims based on private introspection can neither be proved nor disproved in the way that empirical claims in other sciences are open to public assessment. He discussed issues that had recently generated a great deal of controversy. One concerned 'imageless thought': do all kinds of thinking involve imagery, or do some not? Another concerned the attributes of visual sensations: are they limited to *quality*, *extension*, *duration* and *intensity*, or should *clarity* and *order* be added? 'I firmly believe that two hundred years from now, unless the introspective method is discarded, psychology will still be divided on the question as to whether auditory sensations have the quality of "extension", whether "intensity" is an attribute which can be applied to color, whether there is a difference in "texture" between image and sensations, and upon many hundreds of others of like character.'

Fig. 6.12. John Watson in 1912, when voted by students the most handsome professor at the Johns Hopkins University and just before giving his lecture series at Columbia

In attacking the kind of psychology identified with Titchener Watson was expressing dissatisfaction shared by very many of his contemporaries. The difference was that Watson advocated a much more radical alternative than anyone else. He was scornful of the efforts of the functionalists – and, by implication, of his former supervisor and colleague, Angell – to redirect psychology. As far as he could see, the difference between structural and functional psychology was primarily one of terminology, a matter of introducing 'process' or 'function' and removing 'content' and 'structure', while proceeding very much as before.

The second of Watson's criticisms was that most research in human psychology did not allow 'the educator, the physician, the jurist and the business man to utilize our data in a practical way'. He was encouraged by what he took to be the flourishing condition of those branches of psychology which had already become less dependent on introspection, such as 'experimental pedagogy' and the 'psychology of tests', and which were beginning to prove their usefulness. However, these examples were the exceptions: 'One of the earliest conditions which made me dissatisfied with psychology was the feeling that there was no realm of application for the principles which were being worked out in content terms.'

Running through Watson's paper was a firm optimism about the benefits psychology would obtain from adopting a behaviourist point of view. Within a few years, it was suggested, the advantages would become apparent both in terms of the subject's general progress and of the production of results with important practical value. Few specific suggestions were made. Watson admitted that he 'was more interested at the present moment in trying to show the necessity for maintaining uniformity in experimental procedure and in the method of stating results in both human and animal work, than in developing any ideas I may have upon the changes which are certain to come in the scope of human psychology'. He also admitted that it was not clear to date how a behaviourist psychology would handle what he termed 'more complex forms of behavior, such as imagination, judgement, reasoning, and conception'.

Watson's work over the next few years represented an attempt to confront these deficiencies, beginning with the problem of how to treat thoughts and feelings in behaviourist terms. His suggested solution first appeared in 1914 when he published a book, *Behavior: An Introduction to Comparative Psychology*, which mainly consisted of a clear and very readable review of recent work on animal behaviour and was comparable, except in its theoretical stance and more critical tone, to Washburn's book of six years earlier.

The opening chapter was the same as the paper on behaviourism of the previous year, but with two notable changes that suggest a considerable increase in Watson's confidence in his own ideas. Both changes were based on his general claim that 'there are no centrally initiated processes'. First, he asserted that thoughts and images are sensations arising from events outside the brain, these events being 'habits' identical in their properties to the kind of bodily actions that more usually go by the name, but normally difficult to observe. Watson called them 'implicit behavior'. The example he concentrated upon was thinking; he argued – without referring at all to William James – that this is really sub-vocal speech. 'If implicit behavior can be shown to consist of nothing but word movements (or expressive movements of the word-type) the behavior of the human being as a whole is as open to objective control as the behavior of the lowest organism.'[13]

It seems as though the second innovation was deliberately intended to shock. It suggested that a key problem in introspective psychology, the analysis of feelings, could at least in principle be solved by studying a certain kind of peripheral event. All feelings, claimed Watson, have their basis in sensations arising from the 'reproductive organs and the related erogenous zones'. He then worked through an example of how the pleasing effect to the male of the sight of a female animal might be explained in terms of physiological reactions. Despite Freud's visit and the attention given to his theories it was still unusual to find explicit discussion of sex in psychology books.

Only the first chapter discussed the basis of feelings in sexual sensations, but later in the book Watson returned to the idea of thinking as sub-vocal speech in a chapter interestingly called, 'Man and beast'. This title suggests what the contents confirm, namely, that Watson was proposing a revision of the Cartesian view of man's place in nature, which attacked one form of dualism in order to substitute another. The chapter starts by bemoaning the fact that even good natural scientists can show distressing lapses.

'We find among biologists generally the tendency to treat simple reflexes and habits in a perfectly objective way, but suddenly when the reactions begin to get complex we find them introducing the concept of the psychic. It is introduced as a *deus ex machina* to account for complexity in response. Even Loeb has not escaped the tendency.'[14] It then proceeds to introduce language as the dividing line between man and beast. 'The fundamental difference between man and animal from our point of view lies in the fact that the human being can form habits in the throat.'

Watson was convinced that no other creature but man is capable of any kind of language. He never justified this conviction, but from it stemmed his highly sceptical attitude towards any claim of outstanding abilities in a particular animal. In his book he cited with strong approval Pfungst's study of the talking horse, Clever Hans, and proposed that the same kind of analysis can equally be applied to subsequent examples of horses and dogs for which extravagant claims had been made. 'From our point of view it can be readily understood that the search for reasoning, imagery, etc. in animals must forever remain futile, since such processes are dependent upon language or upon a set of similarly functioning bodily habits put on after language habits.'[15]

Watson's identification of thought with sub-vocal

speech seemed implausible to most of his contemporaries and to many people since who might have been inclined to agree with his arguments against introspection and for a science of behaviour. However, it was crucial to his point of view and therefore deserves some more comment.

The kind of problem Watson worried over is well-illustrated by an example he provided, an example which, like most others, contained many personal elements. 'Someone suggests in words that you borrow one thousand dollars and go abroad for a year. You think over the situation – the present condition of your research problems, your debts, whether you can leave your family, etc. You are in a brown study for days trying to make up your mind.'[16] For Watson the aim of the psychologist is to predict the response to a given stimulus. In the present example there is a long delay between the initial stimulus, the suggestion of a trip abroad, and a later response, the decision whether or not to go.

According to traditional psychology and to everyday belief the connection between the two events is to be found in the trains of thought of which the person is aware at intervals during the intervening period. The minimal claim made by Watson is that the thoughts which, if need be, the person can report provide a poorer predictor of his eventual decision than that provided by the systematic measurement of his behaviour by an outsider. In some places a more extreme claim is made, implying that such delays between a 'stimulus' and a 'response' are literally filled by an uninterrupted sequence of muscle twitches in the throat. Watson stated that, according to his point of view, a man who 'suddenly lost his laryngeal apparatus without any serious injury to the other bodily mechanisms' should no longer be able to think.[17]

There is a very close connection between the everyday example of the person tempted to go abroad and the delayed reaction procedure employed by Hunter. The results produced by Hunter's raccoons disturbed Watson. His inclination was to dismiss them as yet another claim for thinking or images in animals that no doubt would soon be found invalid. The trouble this time was that the results were obtained in the laboratory of Harvey Carr, whom Watson had trained and knew to be a careful experimenter, and that the procedure used by Hunter was clearly very carefully controlled.

In *Behavior* Watson used the delayed response situation to illustrate his ideas on thinking. Thus, rats and dogs depend on *explicit* behaviour; they need to maintain orientation towards the place where the light appeared in order to go there when the delay is over. Watson's suggestion was that children employ the *implicit* behaviour of talking, of repeating instructions to themselves, and this allows them to find the correct place after long delays in which they move around. This left the awkward cases in which raccoons performed well without maintaining bodily orientation. 'Were Hunter a less careful investigator,' Watson acknowledged, 'we might think that there was an actual error in observation.' Apart from suggesting some unlikely possibilities of artefact in the situation, all Watson could do was to point out the difficulty it presented for him. 'There is no known mechanism of response which might account for this. It thus seems best to reserve our attempt at explanation.'[18]

Watson's book promoted ideas and attitudes that were not obvious in his earlier paper on behaviourism. Although Meyer, Jennings, Dunlap and Lashley would not have expressed them in the same way, or as effectively, they shared most of the opinions expressed in the paper of 1913. They were all against relying on introspection; all for detaching psychology from philosophy and placing it closer to biology; for changing it into an objective science of behaviour; and for making it more relevant to everyday problems. But in the book of 1914 Watson explicitly attached to his behaviourism a set of beliefs that owed very little to his Hopkins colleagues and much more to Jacques Loeb.

Meyer, Jennings, Lashley and, of course, Baldwin were all Darwinians; in contrast Watson, like Loeb, believed that nothing was gained from trying to place psychology within some evolutionary context. The others were pluralists, rejecting the idea that any one principle, whether Loeb's tropism or Watson's habit, could serve as an explanatory tool for every aspect of behaviour. Jennings' views on this have been described earlier in this chapter. As for Baldwin, his scheme of development posited the emergence of increasingly complex and varied processes. The most notable feature of Meyer's career in psychiatry was its eclecticism, his readiness to recognize important insights and encourage new theories, such as psychoanalysis, but strong resistance to the claims of any would-be universal formula. His attitude is shown, for example, in one paper where he suggested that unless 'one has a chance to use, and with a feeling of justification, a free pluralistic method of dealing with things, dogmatic restrictions kill off many a possibility of seeing things for what they are worth'.[19] Finally, Lashley's subsequent career, which in many ways was the most productive of them all, was similarly marked by its openness to new approaches and an even-handed rejection of monolithic theories based on a few simple principles.

Watson's friends at Hopkins did not share his faith in the peripheral origins of all psychological events. In their view the brain did a great deal more than connect incoming stimuli with responses; and the rejection of introspection need not entail giving up attempts to work out what kinds of process might occur within the brain. For example, Lashley later echoed Jennings' concern with the problem of motivation and the doubt that its complexities could be adequately handled, as Watson briefly suggested, in terms of stimulation arising from various bodily organs.

These differences between Watson and the others largely derived from a more fundamental difference, their contrasting assumptions about the nature of science. Watson held the same kind of positivism as Loeb, whereby the aim of a science is to *predict* and *control*; it achieves this by amassing empirical generalizations in a manner in which the construction of theory is unimportant. For the others the goal of a science was that of *understanding* some set of natural phenomena and for this endeavour theories play an essential part. The word 'control' occurred frequently in Watson's writing, but it was used very rarely by the others.

It is puzzling that Lashley and Watson should have spent so much time working together on various projects and yet have maintained such conflicting outlooks. Sometimes it seems in reading their work that each is expressing one side of a long-continued argument that has been fought on and off in many different contexts.

As an appropriate reflection of his arguments for a new kind of human psychology, Watson devoted little of his time to animals and instead began his first systematic experiments with human subjects. One might have expected that these would be related to his striking claims on the nature of thoughts and feelings; in fact he stayed for a while with familiar problems of perception. What he wanted was to make good his claim that human perception could be studied without reference to the observer's subjective experience, in the same way as one might study perception in animals. As he noted, 'it is one thing to condemn a long-established method, but quite another thing to suggest anything in its place'. It was here that the conditioning research carried out in Russia suddenly became of interest.

Earlier Watson had agreed with Yerkes that Pavlov's salivary conditioning procedures were of limited usefulness in studying animal perception. Lashley devised a way of studying conditioned salivation with human subjects, but this too seemed unsuitable for general application. A change of attitude towards conditioning came when Watson read Bechterev's *Objective Psychology* and realized that the methods used in Bechterev's laboratory might provide the approach he needed. Watson and Lashley together used various kinds of stimuli to signal the delivery of brief shocks to the fingers or toes of their human subjects; they worked with one or two children as well as with adults. Having managed to reproduce the kinds of effect that Bechterev described, they could begin to use the method for testing aspects of perception.

This research on conditioning was sufficiently advanced for Watson to make it the main topic of the presidential address to the American Psychological Association which he gave at the end of 1915.[20] His nomination as president was not an endorsement of behaviourism so much as recognition of a productive research career and, even more, the industrious performance of many professional duties over the preceding eight years. From the address it could easily be concluded that behaviourism had not advanced a great deal since its birth two years earlier and that it had not altered the scope of psychology at all.

At the time of this address there was a change in Watson's circumstances that led him into a very different kind of work. Most of the Johns Hopkins departments were moved to a new campus further from the centre of Baltimore, but the new space available for the psychology laboratory was even more limited than in the old premises. As a typical gesture of encouragement towards a new approach that might at least contain some truth and be of some use, Meyer offered Watson rooms in the Phipps Psychiatric Clinic to set up his animal laboratory. Watson's new position in the medical school complex made it easy for him to gain access to the maternity ward. Within months of his move very much more of his time was spent observing the behaviour of new-born babies than that of his rats. The studies of babies that he carried out in the eighteen months before America entered the First World War are described in a later chapter.

Concluding discussion

It may be useful to comment on three of the themes that run through the varied set of issues and studies described in this chapter, the themes of consciousness, complexity and empirical productivity.

A view very widely held among psychologists at the turn of the century was of consciousness as a non-material entity possessing both a passive aspect, as the subject of inner experience, and an active aspect, that of a causal agent. This latter aspect was most apparent in discussions of behaviour. Thus, the actions of an adult human being could be split into two categories: first, actions with a simple basis, which are 'habitual' and performed 'automatically' or 'unconsciously' and, second, those determined in a more complex manner, which involves the intervention of consciousness or mind. It was a view that had long been regarded as the common-sense one within the Western tradition and essentially reflected the old Cartesian division between instinctive behaviour and rational action; it included the important extra assumption that by careful reflection a person could understand the causes of his rational actions. Within this framework the main interest for a psychologist in studying behaviour arose from the contributions this might make to the analysis of consciousness.

During the first two decades of the century two major lines of enquiry helped to displace this kind of view from its central position in American psychology. Research in animal psychology was one of these. In studying lower organisms Loeb and Jennings agreed, despite their many differences on other issues, that one cannot draw any inferences about an organism's subjective experience from its behaviour, which should be explained only in terms of material causes; the notion of consciousness as a causal agent must be firmly excluded from the analysis of behaviour.

The early comparative psychologists who studied much larger creatures possessing backbones and nervous systems were slow to reach the same conclusion. But then they gradually discovered that they had been making considerable progress in understanding the varied topics that interested them – sensory capacities, the extent and nature of trial-and-error learning or of imitation, and so on – even though no progress was made at all on questions concerning consciousness in animals. In 1913 Watson was exaggerating a little when he claimed that all of the 'behaviour men' had come to recognize this point, but certainly the majority had already travelled a good part of the way.

The second line of empirical work to undermine any simple form of dualism was mainly European and was based on clinical, rather than experimental, evidence. This was the study of psychopathological states and of therapeutic methods, which started with studies of hysteria and hypnotism in France and culminated in the work of Freud and Jung. In the present context the primary importance of this work was that it destroyed the equation of 'complex' with 'conscious'. It was no longer possible to believe that the causes of any action that is not just a simple repetitive habit must be accessible to introspection.

The general implications of psychopathology were first clearly perceived in America by James and by psychiatrists like Meyer. Later, in the second decade of the century, the notion that one can attempt to understand bases for animal behaviour that are more complex than stimulus–response connections without getting caught up in questions concerning consciousness began to gain ground among animal psychologists. Quite appropriately, the first major study to reflect this outlook was that by Hamilton, who was trained in psychiatry and explicitly acknowledged the influence of ideas from human psychopathology. It is noteworthy that his report does not discuss the possibility of asking his human subjects *why* they reacted as they did.

The study of human psychopathology is relevant here in another way. The visit to America by Freud and Jung was in some respects like that of applicants attending a job interview or of salesmen travelling to explain their wares. Academic psychology needed to find a new departure which would be of more practical use than the analysis of consciousness by introspection. The eventual rejection of psychoanalysis made much easier the later acceptance of behaviourism. It also highlighted the already overwhelming preference for approaches that are conducive to productive *experimental* research and the pronounced suspicion towards alternative modes of empirical enquiry.

The commitment to experimentalism was particularly strong in Watson. If some outlook in psychology failed to produce a steady increase in well-founded factual knowledge then there must be something seriously wrong with it. This belief was a major factor in his continuing and fierce scepticism towards evidence of complex learning in animals and in the difference between his general outlook and that of his Hopkins colleagues. Some comments on the relationship between complexity and experimental productivity will serve as an appropriate ending to this discussion.

In a very different context John Stuart Mill once described what he viewed as the major obstacle to progress in psychology. 'All students of man and society who possess that first requisite for so difficult a study, a due sense of its difficulties, are aware that the besetting danger is not so much of embracing falsehood for truth, as of mistaking part of the truth for the whole.'[1] The sentiment was one that James, Baldwin and Meyer heartily endorsed.

To mistake a part of the truth for the whole may, nevertheless, provide a powerful impetus for research; or, perhaps more fairly, a strong belief in the power of some single principle to encompass a very large proportion of some subject often has considerable pragmatic value. Loeb's theory of tropisms and its application to the behaviour of lower organisms provides an excellent example. It stimulated a large number of studies which yielded a great deal of information about such creatures as paramecia and hydra. Later, Jennings argued that what was now known about these animals indicated that their behaviour was based on more complex and diverse mechanisms than proposed by tropism theory and that to understand these mechanisms more than just a knowledge of immediate, peripheral stimulation was needed. By the time he had convinced most researchers, active work in this field died away.

Loeb and Jennings stand as clear representatives of two distinct scientific traditions that for a while were intermixed within animal psychology. Loeb was a German experimental physiologist who distrusted evolutionary theory as he distrusted any view not grounded firmly on experimental data; Jennings was a Darwinian and, while committed to experimental methodology, could hold firm to beliefs based on other kinds of evidence. The everyday business of getting on with experiments delayed for some years the appearance of a similar division among those who studied vertebrate behaviour.

Around 1910 the two strands began to separate out. Yerkes explicitly placed his comparative psychology in the evolutionary tradition, attempting to use systematic experiments yielding quantifiable results, but within the framework developed by Romanes. Watson's views, in complete contrast to those of his friend, came to sound more and more like those of Loeb. From an attitude of suspending judgement he moved to strong agreement with Thorndike that animal intelligence is limited to the acquisition of habits. Unlike his Hopkins colleagues he refused to accept that processes more complex than stimulus–response connections could be entertained without slipping into a dualist way of thinking; to discuss 'ideation' in raccoons, as Cole and Hunter had done, was to allow the soul back into psychology, an attitude strangely similar to one that a preacher from Watson's childhood might show in condemning some minor misdemeanour because it permitted the Devil to slip back into a Christian's life.

Like Loeb, Watson denied central factors in the determination of behaviour and claimed that even the thoughts, feelings and motivational states of human beings can be adequately analysed in terms of various forms of peripheral stimulation. Although never fully expressed in any one place it is clear that the underlying rationale for these claims was the belief that they provided the only basis for carrying out good experiments producing valid, objective data on such matters. To contemplate complexity, as Mill proposed, was to follow the path taken by Baldwin, losing touch with empirical reality and regressing into philosophy. Watson hoped that the day would soon arrive when students of psychology would not be expected to know any more about philosophy than students of chemistry or physics.

7

Apes, problem-solving and purpose

> Our knowledge of the psychology of the anthropoid apes is less than our knowledge of the psychology of any other animal. But notwithstanding the scarcity of material which I have to present there is enough to show that . . . in their psychology, as in their anatomy, these animals approach most nearly to *homo sapiens*.
>
> George Romanes: *Animal Intelligence* (1882)

Early in the seventeenth century Descartes proposed that even apes do not possess minds. The two reasons that he gave for this conclusion were that they displayed nothing comparable to human language and that they had no ability to behave in an adaptive manner when confronted by a problem in some novel situation. Both of these inadequacies on the part of apes provided sure proof of their lack of capacity for thought. It is hard to know on what grounds Descartes made these claims and his critics pointed out the lack of evidence on such matters. A hundred years later la Mettrie suggested that such failings, even if real, were by no means fundamental and, if trained in an appropriate way, an ape could acquire the various skills of a gentleman, including an ability to speak.

When evolutionary theories were developed in the middle of the nineteenth century the mental capacities of the apes became a topic that provoked a great deal of debate, but there was still very little more information to fuel discussion than there had been in Descartes' time. The Darwinians themselves had remarkably little to say on this matter. The claim that the chimpanzee or perhaps the orang-utan, along with the dog, comes closest in intelligence to man was repeated here and there, but always in passing.

Alfred Wallace was the only one of the Victorian evolutionists to have first-hand experience of such creatures in their natural habitat. Early in his long travels in Malaysia and the East Indies, and three years before the idea of evolution by natural selection occurred to him during a fever, he had spent some weeks in Borneo hunting orang-utans. This was in 1855 and it was many years before he wrote about this experience. His account of these animals concentrated upon details of how he had shot each of his victims and the question of whether full-grown males ever exceed a height of four feet and two inches. The problem of mental evolution had not come to the forefront of debate and, since his travels were made possible by the prices museums and private collectors would pay for his specimens, Wallace's preoccupation with hunting and with physical dimensions is perhaps understandable. The one detailed description of the behaviour of these animals concerned a captive baby orang-utan that he managed to keep alive for a few weeks. Wallace was struck by its human qualities, especially its helplessness, which contrasted vividly with the monkey of similar age that Wallace provided as a comparison.[1]

When Thomas Huxley made public the Darwinian claim that natural selection was intended as an explanation for human evolution too, a major basis for his argument was the close similarity between human anatomy and that of the great apes. His evidence came from the specimens that travellers like Wallace had shipped back to Europe, but he did not speculate on what functions might have been possessed by the large brains that he carefully examined. Even Charles Darwin himself simply repeated a story illustrating an orang-utan's understanding of the principle of the lever and then passed on to discuss the mental abilities of more familiar animals.

By the 1880s when George Romanes made the kind of claim illustrated by the opening quotation to this chapter there was still no real evidence that an ape's mind most closely resembled that of man. Romanes made a short, but significant, reference to a tame orang-utan kept by Baron Cuvier; this had been seen to drag chairs into appropriate positions in order to climb up and operate door handles that were otherwise out of reach.[2] After all his careful sifting of the evidence on animal intelligence Romanes had little more to offer on the subject of apes than had Darwin.

The state of profound ignorance over the habits

and mental capacities of apes was maintained throughout the period of lively debate over the evolution of mind. It is curious in retrospect that so little effort seems to have been expended on finding out more about them. One reason may have been that it was a sensitive topic. Ever since Bishop Wilberforce's taunt regarding Huxley's descent from an ape, Victorian cartoonists had delighted in portraying the Darwinians in the company of some hairy, long-armed simian. To study apes made one an easy target for ridicule.

A more compelling reason was that such studies were particularly difficult. Romanes found this out when, deciding that a more substantial basis was needed for his conclusions concerning the minds of apes, he began to study a chimpanzee named Sally at London Zoo. 'It occurred to me that I might try some experiments on the intelligence of this animal. The circumstances in which she is placed, however, did not prove favourable for anything like systematic instruction. Being constantly exposed to the gaze of a number of people coming and going, and having her attention easily distracted by them, the ape was practically available for tuition only during the early hours of the morning before the Menagerie is open to the public; and, as a rule, I did not find it convenient to attend at that time.'[3]

Beginning in 1887 Romanes attempted to teach Sally to perform two separate tasks. The first involved counting: she was to hand him one, two or three straws as requested. She learned this well and later could perform very accurately up to five, but less certainly with larger numbers. By then the critical reception given to *Animal Intelligence* had persuaded Romanes to be more cautious when interpreting an animal's behaviour. He considered, but then rejected, the possibility that her performance may have been controlled by unconscious expressions and gestures on the part of her trainer. Instead, he was convinced that she possessed a true understanding of counting and even exhibited 'some idea of multiplication'. Later commentators, like Lloyd Morgan, were less convinced.[4]

Romanes also tried to teach Sally the names of colours. This was not at all successful and it seemed to him that she might be colour-blind. He stopped working with her at this point. His only other contact with primates was with the monkey whose testing had been largely in the hands of his sister, Charlotte.

There was no immediate sequel to Romanes' work and the nineteenth century had ended before any further attempts were made to test the intelligence of apes. By this time Thorndike's paper of 1898 and his subsequent brief study of monkeys had forcefully repeated the claim made centuries earlier by Descartes. To many of those who believed Thorndike to be wrong, the apes seemed most likely to disprove his general proposal that all non-human intelligence was to be explained on the simple basis of stimulus–response connections established during trial-and-error learning. More specifically there was the question of imitation. The early experiments had made it very clear that the phrase 'to monkey with', meaning to fiddle with something in an uncomprehending and often destructive way, was apt; the expression 'to ape', meaning to copy another's actions, might turn out to be equally appropriate, despite Thorndike's denial.

The case of the horse, Clever Hans and the analysis of its behaviour in 1905 by Oskar Pfungst were important in attracting the attention of German psychologists to issues in animal psychology. Pfungst's results made it seem even more likely that the results Romanes had obtained from his study of Sally were hopelessly contaminated by unconscious cueing on the part of the investigator. The German interest in animals aroused by Clever Hans was maintained by other developments that became almost as notorious. A group of three horses, the Elberfeld horses, attracted wide attention a few years later. They were said to possess mathematical abilities surpassing even those originally claimed for Clever Hans, including the calculation of cube and fourth roots of numbers. Again it became almost certain that their replies to questioners were controlled by barely perceptible movements which the questioners were unaware they had made.[5]

The realization that it is easy to be misled over an animal's talents had practical as well as theoretical implications. For example, doubts were expressed about the intelligence and tracking abilities of the dogs used by the Prussian police and in 1913 the Ministry of the Interior commissioned a study of six of the best dogs. It was found that in the absence of unintended signals from their handlers their performance was abysmal. A much more rigorous training programme for police dogs was initiated.[6]

In the face of all this negative evidence there remained the persistent belief that some species of animals must be more intelligent than others. If the intelligence of the horse and dog, so highly regarded by Romanes and other earlier writers, had perhaps been overrated, there still remained the ape. Some brief tests were carried out on the apes in the Berlin Zoo and Pfungst himself attempted to assess the ability of captive chimpanzees to learn by imitation, but without achieving a great deal.[7]

Possibly because of the wide interest in such questions Germany became the first country to provide the necessary support for systematic studies of apes and thus to make the first substantial empirical contribution towards the arguments first raised by Descartes and la Mettrie. Before describing this research it will be helpful to consider the more immediate intellectual context in which these studies were carried out.

All of the early students of ape behaviour were influenced by a strong reaction that occurred to the predominant mechanist trend of biology at the end of the nineteenth century. A common form taken by this reaction was to insist that living creatures are organized systems and not simple aggregates of their constituent elements as the conventional analytic attitude appeared to insist. In the 1890s research in embryology was an important source for such ideas and one embryologist in particular, Hans Driesch, became an articulate and influential spokesman for this kind of view.[8]

Driesch believed that any organism possesses three essential qualities that are absent from any non-living system. The first was the 'equipotentiality' of its constituent parts. In a key experiment of 1891 he found that after the fertilized egg of a sea-urchin had divided into two parts, if either of these was destroyed, then the other would nonetheless develop eventually into a well-formed, but smaller, animal. In general he believed that the subsequent development of any one part of an embryo depended as much on its relationship to the other parts as on its own constitution. The second characteristic was 'equifinality', meaning that development may reach the same end point using several different routes. Finally, Driesch believed that all living systems exhibit the property of 'self-regulation' which he defined as 'any occurrence or group of occurrences in a living organism which takes place after any disturbance of its organization or normal functional state, and which leads to a re-appearance of this organization or this state, or at least to a certain approach thereto'.[9]

Starting from experimental embryology Driesch extended his ideas to developmental biology in general, considering such phenomena as the regeneration of severed limbs or adaptation to unusual environmental changes, and then into an overall 'philosophy of the organism'. He claimed that a new biology based on principles of structure or form was needed in order to gain a deeper understanding of life; for example, such an approach would avoid the impasse both Darwinian and Lamarckian theories of evolution had reached in explaining the origins and continuity of species and the boundaries between them. Among the more psychological of the topics he discussed was the question of whether the brain displays the characteristic of equipotentiality; for example, to what extent can one area take over the functions of another when the latter is destroyed?[10]

Another extension, which is particularly important here, concerned the way that the actions of living creatures display the characteristics of 'equifinality' and 'self-regulation'. Driesch gave the example of a dog making its way to a certain place. In one case the dog might be heading there in a direct line, when a carriage crosses the line, causing him to run more quickly and make a curve in order to avoid the carriage. In another case one leg is injured so that the dog has to use three legs to get to his goal. In both cases the final end-point is reached, even though either a different route or a different set of movements from the normal ones has been employed; behaviour has in some sense been regulated so as to adjust to disturbances in an appropriate way.[11]

One of Driesch's main preoccupations was with the level at which it is appropriate to describe and analyse some biological phenomenon; a painting such as '*The Madonna of the Chair*, examined with a lens at a distance of 1 cm shows up quite differently than at 5 m away. The first time we see only blotches. Is then the study of blotches really the only task of the biologist?'[12] In psychology the problem of levels first came to prominence in the context illustrated by this example; by the turn of the century many psychologists began to question the value of trying to understand human perception by means of experiments on 'blotches'. The kind of terminology Driesch used in the context of embryology – terms like 'structure', 'organization' and phrases like 'form is not a mere sum of certain elements' – was introduced into the study of perception.

A little later the same problem was discussed and similar arguments put forward in the context of behavioural studies. We can describe someone's actions in the following way: 'moves through doorway into room; picks up and replaces cushion; looks towards floor and moves direction of gaze in sweeping movements; stops and scratches head; pats chest and then rear'; and so on. Under normal circumstances it is likely that, if asked what this person was doing, we would simply reply: 'looking for his lost wallet'. It i[s] quicker and yet in an important respect more informative. But if we are watching what an animal does, to describe its behaviour at such an abstract level mean[s] imputing a specific purpose and overall organization to its movements. Should a scientific observer avoid a[ll] purposive language and employ the more detailed and cumbersome kind of description that would be as appropriate for the movements of a clockwork toy? The early students of apes attempted to maintain a detached, scientific attitude, and became acutely aware of the problem that the more detailed and objective their analysis of what the animals did, the less informative it seemed to become.

Driesch was regarded as a maverick from early on in his work in embryology. He studied with Ernst Haeckel who, on reading an early suggestion by Driesch for a mathematical analysis of a certain problem in embryology, suggested that the author might profitably spend some time in a mental hospital.[13] By 1899 Driesch became committed to the cause that became his life's work, to demonstrate the inadequacy of materialistic science and defend a new form of vitalism. Driesch viewed materialism as the approach to nature which assumes 'that there is but one ultimate principle at its base, a principle relating to the movements of particles of matter' and, as strengthened in particular by 'dogmatic Darwinism', maintains that 'organisms are merely arrangements of particles of matter, nothing else'.[14]

He argued that, in contrast to as simple a living system as a fertilized egg that has just split into two, 'a machine . . . cannot remain itself if you remove parts of it or if you rearrange its parts at will'. Since he was unable to conceive of a machine that would possess such a property, this became his first proof of vitalism, the doctrine that living things are imbued with some non-material substance which Driesch, following Aristotle, called *entelechy*. In subsequent years he developed other proofs. His books were widely read, both in German and English, but his vitalism and the increasingly general level at which he put forward his views lost him the respect he had previously enjoyed among his fellow biologists. His appointment as a professor of philosophy and a later interest in Lamarckian inheritance and in telepathy put him beyond the scientific pale. Consequently, although many of his ideas were widely absorbed, it was rare to cite him as their source, since to do so was to invite automatic disfavour.

In summary, in the early part of this century the study of apes began against a background which contained two major elements: Thorndike's claim that apes are no more capable of thought than cats or dogs, and Driesch's arguments on the inappropriateness of a stimulus–response level of analysis for understanding behaviour. Disbelief in Thorndike's claim and sympathy for the views expressed by Driesch were particularly important in the work of Wolfgang Koehler (1887–1967), who carried out the first intensive research on apes. In his general approach, and in his choice of specific test procedures, Koehler was also guided by earlier studies carried out in England by Leonard Hobhouse (1864–1929), who was the first person to respond to Thorndike's challenge by carrying out further studies of problem-solving by animals and the first to examine the ability of an ape to learn by imitation.

Leonard Hobhouse and articulate ideas

Edward Thorndike's paper of 1898 provoked widespread disbelief. Most of his readers continued to hold that cats and dogs are capable of learning by imitation and of discovering solutions to problems by means that involve at least some primitive understanding of the situation. They saw Thorndike's claims as based on a mistaken choice of tasks that allowed his animals no opportunity for displaying their mental ability.

Among the first to react in a positive way, by carrying out studies to prove Thorndike wrong, was Leonard Hobhouse. This work refined ideas that Hobhouse had already been developing for many years. He became widely known among animal psychologists prior to the First World War, but thereafter he was almost completely forgotten.

Hobhouse was born in 1864 and spent his childhood near Liskeard in Cornwall. His father was an Anglican clergyman with a wealthy and aristocratic family background. After a boarding school education Hobhouse obtained a classical scholarship to Oxford University which he entered in 1883 to study 'Greats', the classics and philosophy. His academic career continued to be a distinguished one and, after graduating, he remained at Oxford as a college fellow, giving tutorials in philosophy.

In view of his family background and highly traditional education Hobhouse held surprisingly radical views. As an undergraduate he took part in attempts to organize an agricultural union in the countryside around Oxford. As for philosophy, Hobhouse was fiercely opposed to the Idealism then dominant in English universities. He identified with the neglected tradition in British philosophy that had seemed to have ended with John Stuart Mill. In the preface to his only book on philosophy Hobhouse expressed his attitude towards Mill and his distaste for the obscurity of much contemporary philosophy: 'Mill was guilty of shortcomings and inconsistencies, like other philosophers, but the head and fount of his offending was that, unlike many other philosophers, he wrote intelligibly enough to be found out'.[1]

Hobhouse dedicated himself to combating the climate of reaction that he believed to have settled on Britain since the mid-1880s. He saw a close connection between the misguided philosophical outlook of Idealism and the evils of current government policies. Hobhouse particularly detested the imperialism and racism of Britain's foreign policy.

Another characteristic of Hobhouse that was uncommon among his fellow philosophers was a serious interest in science. From 1888, a year after he became a tutor, he assisted with experimental work in the new physiological and biochemical laboratories at Oxford. His wish to understand current work in biology stemmed from a conviction that general problems concerning the nature of human knowledge must be set within an evolutionary framework. This also led him to the question of what animals understand about their worlds. Thus, in 1898 when Thorndike's paper appeared, Hobhouse was, with Lloyd Morgan, one of the fast dwindling number of Englishmen still concerned with mental evolution.

By then Hobhouse had left Oxford. It is not clear whether it was mainly to escape from the university's philosophers or to take a more active part in political life, but, for whatever reason, Hobhouse moved to Manchester in 1897 to become a staff leader writer on the *Manchester Guardian*. This was the foremost newspaper supporting the Liberal party and, during the period in which Hobhouse worked there, it became involved in a lone campaign against Britain's conduct in the Boer War and the jingoist attitudes this aroused; in many respects the situation resembled the one that developed in America over sixty years later in connection with the Vietnam War. Hobhouse devoted his mornings to philosophy and psychology, and in the afternoons worked at the *Guardian* offices, standing at a tall desk to write fluent articles on current affairs at a speed which amazed his fellow journalists.[2]

It was in these circumstances that Hobhouse managed to study some of the animals at the nearby Belle Vue Zoological Gardens and to complete the book that comprised his only major contribution to psychology, *Mind in Evolution*, which was published in 1901. A year later he left the *Guardian* and moved to London where he continued to work as a journalist and also became involved in various kinds of political activity. His academic interests switched from the mind to society and after a few years he obtained an appointment at London University which placed him among the first professors of sociology in the English-speaking world. He is remembered as one of the founders of British sociology, but he left no legacy to psychology in Britain.

Fig. 7.1. Leonard Hobhouse

Hobhouse's view of the animal mind differed from that of Thorndike in two fundamental respects and both were related to the essentially Cartesian assumptions that Thorndike made. Hobhouse appears to have been very familiar with Driesch's ideas and, like Driesch, he believed that an analysis of behaviour must begin by considering a living creature as an *organized* self-regulating system. 'The normal life of any organism from highest to lowest is a process of unceasing change. It involves a constant interchange of substance with the outer world, and equally constant metabolism or transformation within itself of the substance which it takes up from without and a no less constant transformation of energy . . . Throughout this unceasing process of change which differentiates it from inanimate matter, the organism preserves its own identity as clearly as the unchanging rock.' What an animal does is to be seen as part of a general system serving to preserve its identity; the behaviour of an animal is not simply a set of independent reflexes or stimulus–response units, as Thorndike proposed; what is of crucial importance is how various forms of behaviour are integrated, organized or, to use the term Hobhouse favoured, 'correlated'.

The second reason Hobhouse had for rejecting Thorndike's view reflected his evolutionary outlook

Hobhouse did not believe that there was just one kind of animal intelligence, but instead held that in the animal world there was a considerable range of mental organization, different 'methods of correlation and adjustment'. He was critical of Herbert Spencer's simple linear account of mental evolution. In now familiar terms, Hobhouse stated: 'Evolution is not serial. Its plan is not that of a straight line or even a spiral, but rather that of a tree'.[4] Thorndike must be wrong in claiming that no matter from what branch an animal might be selected its only form of adjustment would turn out to be the blind formation of habits.

The principal subjects in his initial studies at the zoo were a dog, a cat, an otter and an elephant. He used a variety of tests. Some resembled Thorndike's and used boxes, containing food, which could be opened by sliding a bolt, moving a lever or releasing a hook. In other situations a subject could only obtain a container of food by pulling on a cord attached to it in some manner or other. For example, in one test a toy bucket holding some meat was suspended from the bannisters of a stairway, in such a way that the subject, a dog in this case, could only reach it by mounting the stairs and hauling on the string.

At first Hobhouse concentrated on the problem of imitation. He did not use, as Thorndike had done, a situation in which an animal could observe a demonstration of how to respond given by a fellow member of the same species. Instead, Hobhouse more simply and more questionably looked at whether his own demonstrations of what to do appeared to benefit an animal. He quickly discovered that a major problem was that of gaining the animal's attention; his animals were rarely attentive unless there was a chance of immediate and direct reward. Furthermore, attentiveness was a characteristic that could be acquired; he observed that as one or two animals were put through a long series of tests they became less easily distracted. This suggested to him a possible flaw in Thorndike's studies of imitation. Since the untrained cat, placed in a position to observe the performance of a trained cat in the puzzle box, did not benefit at all from the opening of the door, there was no reason to expect this observer to take any sustained interest in what was happening. It seemed to Hobhouse that this was the factor leading to Thorndike's failure to gain any evidence of learning by imitation.

Even though this might well be a valid criticism of Thorndike's method, Hobhouse's own evidence on the matter was far from conclusive. At best he found that, following his demonstration of how to pull at a string or open a particular box, on occasions the animal was unusually quick to make the appropriate response on the next trial. Such evidence convinced Hobhouse that, after they had been trained to be attentive, his subjects were all capable of learning by observing his actions. Hobhouse also found that in some tests where no prior tuition was given there appeared to be a sudden transition from trials in which a series of undirected, haphazard movements occurred to a smooth, rapid performance of the response. The critical reader might well decide that this result negated the value of the occasional evidence on observational learning, since no results were given to show that such transitions were any more frequent following 'tuition' than for trials where no tuition was given. Instead, for Hobhouse the sudden emergence of a solution to a task meant that some process more complex than the formation of stimulus–response connections was involved, and this formed the basis for the second of his detailed objections to Thorndike's arguments.

Thorndike had proposed both a theory of trial-and-error learning based on the formation of connections between stimuli and responses, S-R theory, and, from his common observation of a progressive decrease in latencies, that at a more molecular level the formation of S-R connections indicates the 'wearing smooth' of pathways in the brain. By the time Thorndike published the report on his monkey experiments he had abandoned the use of a smooth learning curve as a criterion for deciding whether learning was based on S-R connections. Whether a particular instance of learned behaviour is correctly described in S-R terms is a different issue from that of what kind of neural process forms the basis for S-R learning. As Thorndike implicitly recognized, it is in principle possible for S-R learning to become complete in a single trial or, in his terms, for a single satisfying event to fully stamp in an S-R connection. Thus, the form of a learning curve may bear only very indirectly on the question of what an animal has learned. However, to Hobhouse the examples of an abrupt increase in the speed with which a response was made, which he found in Thorndike's graphs or saw in his own tests, indicated at least some primitive form of reasoning.

One major effect of Thorndike's first publication was to change the standards for studies of animal behaviour. By 1901 Hobhouse's use of a few animals, the consequent absence of control subjects, the easy-going imprecision of the procedures and the informal reporting of non-quantitative results, embedded in a book of otherwise mainly speculative content, were already dated. Thorndike's thesis research had been concisely reported in a professional journal with a comprehensive presentation of the results that allowed Hobhouse to use Thorndike's data to criticize Thorndike's theory. It is not easy to determine exactly what procedures Hobhouse used, what the outcomes of the tests were or the extent to which his conclusions followed from his data.

The reports Hobhouse provided of how the dog, cat, otter and elephant performed on the various tasks suggest that their behaviour left him a good deal more sympathetic towards the view that animals can only solve problems by trial-and-error than when his ideas has been based only on casual observation and on what he had read. If this were the only empirical work in his book, it would deserve its present obscurity. However, Hobhouse went on to test a rhesus monkey and chimpanzee and the results he obtained from these two animals provide one of the reasons why *Mind in Evolution* became so influential. The other reason was its clear and thoughtful discussion of what were to become key issues in the study of animal behaviour.

One idea that he discussed was the analogy between the behaviour of an animal and that of a self-regulating machine; for example, a steam engine whose speed is controlled by a regulator. To use a more recent terminology, Hobhouse suggested that behaviour should be viewed as part of a negative feed-back control system. Although not directly stated, his argument was that such an approach need not lead to Driesch's vitalism. A related issue, which had more immediate influence, was the question of what it means to describe an action as 'purposive'.

The actions of a spider in spinning may be described as having the purpose of completing a web. These actions are determined by the purpose only in the very indirect sense that the spider's inherited neural mechanisms, which serve to organize the behaviour, have evolved during the history of the species in such a way that only those structures which are best adapted to achieving such a purpose survived. The actions of Hobhouse's elephant, Lily, in pushing in a bolt with her trunk can also be described as serving the purpose of opening the food box in a similar sense: in this kind of case, reinforcement has acted in a selective manner within the lifetime of the individual animal to establish a neural organization which makes the response likely to occur in that particular situation.

Hobhouse argued that many human actions are purposive in a strict sense and, unlike the two previous examples, cannot be based on instinct or habit. To use his example: someone needs a book and, remembering that for once it has been left in a room upstairs, goes to fetch it from this room and not from the nearby bookshelf from which it has been fetched a hundred times before. Such an action is not prompted by events in the immediate external environment – unless *absent mindedly* the person goes to the bookshelf – but is determined by the goal of obtaining the book and knowledge of a method for achieving this goal.[5] Though this may be intuitively obvious, how can one demonstrate that an action is purposive in this strict sense in terms of behavioural criteria, given, as Hobhouse believed, that observation of behaviour is the one and only method of psychology?

One of the criteria he suggested was the complete acquisition of some response after a single successful attempt. The problem of interpreting this type of result was discussed above, and in any case Hobhouse himself did not regard it as a particularly satisfactory criterion. He was more confident about the significance of certain other phenomena: actions that resulted from some sort of perceptual learning and the occurrence of appropriate behaviour in a novel situation. 'As judged from outward action then, the certain signs of purposive action appear so far to be these two: the relation upon which it is based may be experienced without leading to the formation of a habit; and again, may be applied in circumstances differing from those in which it was originally perceived. Action may often be purposive without possessing these marks, but where we find these marks we may be sure that it is purposive.'[6]

Hobhouse provided some spatial examples to illustrate what he meant in claiming, for example, that in purposive action 'the results of experience are applied to action in a manner not determined by the experience itself'. In the example of fetching a book from upstairs, the action is not determined by the previous experience of leaving the book, although this experience provided knowledge necessary for the action. Perhaps the idea is seen more clearly in one of the tests he used. A dog is brought to an unfamiliar house and is allowed to roam around it for a while. It is then taken to Point A outside the house, where it can see its master at Point B inside the house, but cannot get to him directly. If the dog, when called, takes an appropriate route, which can be shown is unlikely on a chance basis, then its behaviour, Hobhouse would argue, utilizes its past experience of roaming through the house, but is not determined by it. It would be incorrect to view the action as a habit or the result of the formation of an S-R connnection, since the animal may never have gone from A to B before. The goal or purpose, that of reaching its master, determines the action.[7]

To the extent that a dog is able to take a direct route which is both appropriate and one that is relatively novel to him, then he displays the simplest form of purposive action, which Hobhouse called 'practical judgement'. This kind of action was characterized by some flexibility in the choice of means, which he contrasted with 'the rigidity of habit'. Rather tentatively, he concluded that some of the behaviour his animals displayed in the puzzle box situations could also be classed as practical judgement. 'In some cases there were indications that the animal was learning not merely to execute a certain movement, but to do a certain thing . . . When a cat, for example, learnt to pull out a bolt, it learnt not merely to paw, nor to paw at a particular thing, but to pull that bolt right out, a thing requiring a certain combination of repetition of minor movements. I have already referred once or twice to Jack's (the dog) efforts with the skewer. Even more remarkable was the combination of efforts by which he pulled the string up through the bannisters. We cannot in such a case apply the conception of a perceived object discharging a uniform motor reaction. There is rather a combination of movements which are not always the same except in this, that they are so adjusted as to produce a certain perceptible change in the external world.'[8]

When Hobhouse tested Jimmy, the rhesus monkey, and the chimpanzee he called Professor, he had no doubts whatsoever that they were capable of practical judgement. Furthermore, they sometimes performed in ways that seemed qualitatively more intelligent than anything he had seen the other animals do. He decided that the superiority of the two primates was best described as an ability to hold 'more articulate ideas'. 'By a more articulate idea is meant one in which comparatively distinct elements are held in a comparatively distinct relation. Thus, that a bolt must be pushed back is a crude idea; that it must be pushed back so as to clear a staple, a relatively articulate one, implying a distinction between the parts perceived (the bolt and its staples) and an appreciation of the relation between them. As ideas become more articulate, the results of experience are fully combined or modified to suit practical ends. Something like originality begins to show itself, and we have instances of what we have called "spontaneous application".'[9]

The observations that gave rise to these proposals began when Hobhouse saw the Professor use a rug from his cage to retrieve nuts which had been thrown to him by onlookers, but had fallen short of his reach. The rug would be thrown like a net over the desired object, which was then worked towards him. Hobhouse demonstrated the similar use of a small stick by pulling a piece of banana about it with it. The Professor at once tried to use this stick, but found it less effective than his rug. The next day the use of a reversed walking stick was demonstrated and this was quickly adopted as a means of retrieving a box containing sugar or a banana. The following day there occurred what was probably the most intelligent piece of animal behaviour that had been carefully observed up to that point. Hobhouse placed a piece of banana and a large stick outside of the chimpanzee's reach, and gave him a second, shorter stick. Nothing happened, and so after a while Hobhouse used this second stick to push the longer stick about. When handed the smaller stick, the Professor at once used this to obtain the large stick and with the latter at last raked in the banana.

The rhesus monkey, Jimmy, had no previous experience of using anything like a rug to retrieve objects and learned to use a stick as a tool more slowly than the chimpanzee. Nevertheless after only four trials he began to manoeuvre a stick into an appropriate position for getting a nut and later became very adept. Like the Professor, when no stick was available, the monkey would attempt to use novel objects, such as a rope or an iron bar, as potential substitutes. In contrast, although Hobhouse found it relatively easy to train his dog and the elephant to pull a box towards them with a stick which was *already* in an appropriate position, they never showed any sign of attempting to get the stick into a position where they could use it.

When other tests were carried out, some inspired by Charlotte Romanes' work with her monkey, Hobhouse was again impressed by the apparently qualitative difference between the two primates and the other animals. It appeared in the way that the monkey and the chimpanzee dealt with knots tied in ropes, or a hook securing a wire that in turn held a bolt in place. The Professor, but not Jimmy, learned to poke a stick into a tube and push out a banana inside. Jimmy moved stools or boxes into position, so that he could climb up to an otherwise unobtainable potato. In some cases there was a sudden transition from a 'sub-primate' type of fumbling to an adept performance of the appropriate action.

Why was the performance of the Professor so much more impressive than that of the other animals? Hobhouse did not have a lot to say in answer to this question. He allowed that his choice of tests may have favoured the ape, recognizing the possibility 'that "ape intelligence" lends itself more readily to the kind of experiment which human intelligence most readily devises.'[10] In particular, apart from the spatial tests in which only dogs appear to have been used, his tasks were ones that emphasized an ability to manipulate objects. 'The monkey can apply one object to another; that is why he can use a stick, a stool, a poker. It may be said, so could the dog, if he had a hand. I am not prepared to deny this hypothetical statement; but I would rejoin that if the dog had a hand, his intelligence would have developed in a different way.'[11]

In view of the controversy which surrounded tests of ape intelligence thirty years later it is worth emphasizing that Hobhouse believed that the Professor's skill was based on his past experience and did not reflect some kind of innate ability to solve certain kinds of problem which was independent of prior learning. The crucial difference between the Professor and, say, the elephant lay in *what* had been learned in the past; the chimpanzee faced a new task possessing much more detailed knowledge of the elements, and their relationships, of some previous, relevant situation, knowledge of a kind which allowed him to 'articulate' these elements into a new and immediately appropriate combination.

However, by the time this controversy arose very few people read Hobhouse any more or knew that he had been the first to use the tests that were now of such interest. And terms like 'articulate', 'correlation' and 'practical judgement' never became part of the animal psychologist's technical vocabulary. Hobhouse's language served to create a barrier that prevented an easy understanding of his ideas by later generations. An important example illustrating shifts in terminology is his use of 'assimilation'[12] to denote what later became known as 'classical' or Pavlovian conditioning, or simply 'conditioning'. The term 'assimilation', indicating the process by which a stimulus comes to assimilate some of the properties of another event or object which it frequently accompanies, was widely understood before the First World War, but little used afterwards.

Finally, there is the question of what Hobhouse meant when ascribing to an animal the ability to perceive relationships. He explained this in the following way. 'By the term perceptual relation, I intend a perceptual content in which distinct but related elements are included. In such a perception, the relations contained contribute to the character of the whole as much as the elements that are related, and in that sense the relations must be said to be perceived. It does not follow that the character of any of the relations concerned is analysed out and distinguished from the terms which comprise it. When I look at any complex object e.g. the front of a house, I am aware of a whole with many distinct parts.'[13] Hobhouse made these observations in 1901, long before such ideas and language became identified with the Gestalt school of psychology that arose in Germany a decade later.

Wolfgang Koehler's tests of chimpanzee intelligence

The stimuli used in most nineteenth-century experiments on perception were physically simple events: spots of light, pure tones of fixed frequency, monochromatic colours, pressures applied to a single spot on the skin, and so on. This research made the rarely discussed assumption that continued work along such lines would eventually lead to an understanding of the perception of complex events, more like those of significance in everyday life, without the introduction of new principles. The kind of attack on this assumption made by Driesch, in arguing that a painting cannot be understood by examining it with a magnifying glass, was expressed by a number of critics. One of them was an Austrian psychologist, Christian von Ehrenfels, who in 1890 wrote a notable paper on the subject with the title, 'On Gestalt (form) qualities'. Von Ehrenfels gave the example of a melody. No amount of careful analysis of how the individual notes are perceived can lead to the processes by which a series of notes is heard as a melody, since, for example, all the notes may be transposed and yet leave the melody intact. A property, such as 'being in the Key of A minor', cannot be found by examining the notes in isolation, but arises from their relationship with one and other. This provides an instance of a form-quality based on temporal relationships. Other types of form-quality that were described by von Ehrenfels were mainly spatial; for example, perception of a square is made up of four elementary elements, the sides, plus the form-quality of the spatial relationship between them.[1]

These ideas later made a particularly strong impression on a younger, German psychologist, Max Wertheimer, who decided that von Ehrenfels had stopped short of a very important discovery. In the summer of 1910 Wertheimer was on his way by train to a holiday in the Rhine Valley and was thinking about the relevance of form-qualities to the study of visual perception. An idea occurred to him on 'apparent movement', movement seen when two spots of light appear in quick succession. He left the train at the next station, which happened to be Frankfurt, left his bags at a hotel, bought a stroboscope at a toy-shop and started to experiment in his hotel room. Later he was offered the use of facilities at the university. Wolfgang

Koehler had just arrived there and he served as a subject in Wertheimer's experiments. A little later they were joined by Kurt Koffka, another Berlin graduate, who also took part in the research.

Wertheimer believed that the study of perception had to begin with forms or 'Gestalten'. It was a complete mistake to study 'elements', since these had meaning only as parts of a whole; no 'side' exists if there is no square in the first place. The key point of his experiments on apparent movement was that the perception of movement in this situation could in no way be understood as a conjunction of two basic elements, those of perceiving stationary points of light.[2]

Wertheimer convinced Koehler and Koffka of the truth and importance of his insight and thus began a life-long collaboration between the three men, who devoted themselves 'to a common evangelical effort to save psychology from elementism, sensationism and associationism, from *sinnlose Und-Verbindungen* ("mindless conjoining"; Wertheimer's phrase) and to bring it to the free study of phenomenal wholes'.[3] The exchange of ideas between them laid the foundation for what was to become known as Gestalt psychology. Their interaction may well have had the same quality as that among the group of men at Johns Hopkins from which, among other things, John Watson's behaviourism emerged. Although Wertheimer, Koffka and Koehler kept in contact throughout the rest of their lives, the daily collaboration of their Frankfurt days ended in 1913 when Koehler left to study chimpanzees.

This unusual opportunity arose because some years earlier the Prussian Academy of Sciences had received a bequest specifically intended for the establishment of a research station for psychological and physiological studies of apes. Tenerife, one of the Canary Islands, a possession of Spain, was selected as an appropriate site. In 1912 the first director was given a one-year appointment, with responsibilities of setting up the station, of transporting there a small number of chimpanzees from West Africa and of beginning to test the animals. In 1913 a successor was needed. Stumpf, still Professor of Psychology at Berlin, suggested that Koehler would be a suitable man for the job.[4]

The choice of Koehler was made despite his lack of experience in working with animals. The major factor appears to have been Stumpf's respect for him as one of the most promising students of psychology from Berlin. Koehler was born in Estonia in 1887 and moved with his parents to Prussia at the age of five. On leaving school he had followed the German tradition of studying various subjects at various universities. Some contact with the physicist, Max Planck, gave him an enduring interest in physics. He became particularly interested in theories of electromagnetic fields, which were leading to the development of radio communication at the time he was a student. Whereas it seems that for Thorndike the invention of the telephone contributed to the kind of theory he adopted, for Koehler the nature of certain physical processes, rather than the human artefacts based upon them, strongly and explicitly influenced the kind of explanation he sought for mental phenomena.

Despite his studies of physics, the research he carried out for his doctoral studies was in psychology. On the basis of experiments in Berlin on the perception of sound, which were supervised by Stumpf, he received his doctorate in 1909 and obtained the teaching post at the University of Frankfurt that led to his meeting with Wertheimer.

Koehler was twenty-six years old when he set off for Tenerife with his wife and their new baby. Although he had had no training in animal work, he was a skilled experimenter, knew a great deal of psychology and, above all, his three years of exchanging ideas with Wertheimer and Koffka provided a fertile basis for understanding the behaviour of the seven chimpanzees that came into his charge.

Koehler's stay on Tenerife was much longer than originally intended and it was 1920 before he returned to Germany. The First World War, which began less than a year after his arrival, meant that he was virtually isolated for much of that time. However, control of the surrounding seas by the British navy did not stop him from receiving scientific journals which kept him up-to-date with research, particularly with experimental work in America. They were seven highly productive years. In addition to the studies of problem-solving which made him famous, he carried out a number of other innovative studies, of perception, discrimination learning and memory, and wrote a well-regarded book on the nature of psychological explanation. The experiments he carried out on chimpanzee intelligence, most of which were completed within the first year of his stay, are described here and his theoretical ideas and other work are discussed in the following section.[5]

Koehler gave two main reasons for wishing to study the intellectual abilities of his apes. One was comparative: given that chimpanzees resemble man in so many other ways, how similar is the way in which they think? The second echoed the rationale for animal psychology first proposed by Sechenov: the study of

Fig. 7.2. Wolfgang Koehler

how apes solve problems allows us 'to gain knowledge of the nature of intelligent acts' in general since, in the first place, their behaviour is simpler, and, second, it is virtually impossible to set human beings tasks demanding an intelligent solution which do not resemble problems that they have already encountered and overcome.

To meet these aims Koehler needed test situations which were likely to be utterly different from anything the chimpanzees had previously experienced and yet were simple enough for the animals to understand: 'elementary problems in which, if possible, the animal's conduct can have one meaning only'. He regarded the situations developed by comparative psychologists in America as completely inappropriate for this purpose, since they were far too complex for there to be any chance of the subjects even dimly comprehending them and, in any case, they denied the animals an opportunity of seeing what the solution might be. In mazes and puzzle boxes 'the first time they get out is, therefore, necessarily a matter of chance . . . in intelligence tests of the nature of our detour (roundabout-way) experiments, everything depends upon the situation being surveyable by the subject from the outset'.[6]

Koehler emphatically distinguished between two kinds of behaviour which he labelled 'intelligent' and 'mechanized'. Intelligent behaviour consisted of a performance which is appropriate to the structure of a new situation and in its overall organization is relatively independent of past experience, while mechanized behaviour referred to performance which, initially either intelligent or occurring by chance, has become automatic through repetition. In adult human behaviour it may be difficult to make this distinction and this was one reason why Koehler believed that work with chimpanzees had such potential. 'We must, however, be on our guard against constructing our standard of values for these tests on the basis of human achievements and capabilities; we must not simply cancel what appears to us to be intricate, and leave what appears to us elementary in order to arrive at an ape's capabilities . . . We must avoid such judgements because the primitive achievements we are here investigating have become mechanical processes to humans. Thus the comparative difficulty of achievements may have been quite altered, nay, reversed, by the increased *mechanization* of these processes, the degree in which this has taken place being independent of the original difficulty.'[7]

As for the behaviour of his chimpanzees Koehler found it easy to tell what was mechanized and what was intelligent. The latter is typically preceded by a pause and then 'takes place as a single continuous occurrence, a unity, as it were, in space as well as in time; in our example, as one continuous run, without a second's stop, right up to the objective'. Also, 'if the experiment has not been made often, there is the additional fact that the moment in which a true solution is struck is generally marked in the behaviour of the animal by a kind of jerk . . . Thus the characteristic smoothness of the true solution is made more striking by a discontinuity at its beginning'.[8]

Koehler gave great weight to the pauses occurring just before a successful attempt or before what he termed a 'good error'. He described a visit by a sceptical colleague who believed that chimpanzees were as much limited to trial-and-error methods as Thorndike's cats and dogs: 'But nothing made so great an impression on the visitor as the pause after that, during which Sultan slowly scratched his head and moved nothing but his eyes and his head gently, while he most carefully eyed the whole situation.'[9] Nevertheless, like Hobhouse and unlike Thorndike, Koehler made no attempt to record this aspect of his subjects performance: 'the duration of an experiment (trial) depends on so many accidental and changing circumstances (e.g. futile attempts at solution, lack of interest, depression on account of failure or isolation, etc.) that measures of time would only give the *semblance* of a quantitative method'.[10]

In general the style in which Koehler carried out and reported these experiments was informal and observational, since the tests were intended as provisional explorations which, when a satisfactory theoretical account was developed, would lay the ground for the kind of systematic, quantitative work prized by American animal psychologists; thus, 'the further development of this branch of science depends much more on the clarity and sharpness of our questions (the way of thinking) than on the details of experimental procedures. Why take pains to achieve technical refinement when we do not quite know how to proceed validly in essentials?'[11]

The kind of questions Koehler wished to ask of his chimpanzees seemed best answered by developing the tasks that Hobhouse had devised. To convey the flavour of this work it is worth looking at it in some detail. Two particular kinds of task were used extensively: problems requiring the use of a stick to rake in some object, usually a banana, which was otherwise unobtainable, and ones where the animal had to move a box, or boxes, into an appropriate place for climbing up to an objective. Other tasks involved the use of ropes and it was with a problem of this kind, straight from Hobhouse, that serious testing started in January, 1914.

A basket filled with bananas was suspended from a cord which ran over a pulley with the other end attached to a ring. The ring was hooked to a branch of a tree. The main lessons that emerged was that this was far too complicated a situation with which to begin and that chimpanzees were likely to find solutions other than those they were intended to discover. Sultan, the first animal tested, learned by what appeared a straightforward trial-and-error process that jerking the cord vigorously was effective in upsetting the bananas from the basket and consequently he never faced the problem of unhooking the ring.

By the end of the month Koehler tried the box test for the first time. A banana was suspended from the ceiling of the test area and the only way for a chimpanzee to reach it was by moving a large box into a position underneath. All seven chimpanzees eventually performed well, but Koehler provided a description of what happened when an animal first encountered this test for only two of the chimpanzees. When Sultan was first tested about five minutes elapsed

Fig. 7.3. One of Koehler's chimpanzees, Grande, on an insecure construction, while Sultan looks on

before he succeeded in obtaining the banana. Immediately after a period of restless pacing he seized the box and moved it part way towards the objective, climbed up on the box and sprang to reach the prize. On the second occasion he quickly moved the box directly beneath the banana and jumped up to get it. From then on the performance was repeated smoothly. In the absence of a box others suitable objects, such as a light table or the backs of the experimenter or of other animals, were immediately used as substitutes. The second animal whose performance was described was tested in isolation. The solution occurred after fifteen minutes of the first test period and was preceded by a few tentative movements of the box. In further tests carried out over the next nineteen days the animal rarely touched the box, until he quite suddenly repeated his original performance of dragging the box directly underneath the banana. No lapses occurred after this. The two chimpanzees certainly did not display the kind of gradual improvement described by Morgan and Thorndike in their accounts of trial-and-error learning with accidental success.

Once all the animals had mastered the one-box situation, Koehler tried the next step of requiring that one box be stacked on top of another in order to reach the fruit. One animal, again Sultan, clearly solved the problem without aid. Two other animals eventually performed adequately, but it is not clear to what extent assistance from Koehler or observation of Sultan helped in this. Once all four had become accustomed to the two-box situation, further tests were given in which the height of the bananas was increased and more boxes were made available. All four managed constructions of more than two boxes, but only two chimpanzees, Sultan and Grande, became expert. Illustrations of Grande's achievements are shown in Figures 7.3 and 7.4. Three other chimpanzees showed no sign of solving even the two-box test.

In carrying out the tests the position of the objective was varied. Towers were always built in the appropriate position, except in what Koehler described as a one in a hundred case of building below the spot where the objective had been on the previous test. In addition to the complete failure of three animals, another observation convinced Koehler that the tests involving more than one box were approaching the limits of the chimpanzees' capabilities. The structures erected by the four successful animals were always clumsy and liable to collapse when the builder started to climb upon them. Achieving a stable structure appeared to be a matter of chance and he noted no improvement as the tests were repeated. When rocks were placed underneath the objective, in such a way as to prevent the construction of a stable tower of boxes, none of the four animals attempted to remove the rocks.

The box tests tell us about an animal's ability to find some means of getting to a desired object when a direct approach is prevented. An alternative solution to this kind of problem is to find a way of moving the object nearer. The use by Hobhouse's chimpanzee of, first, his rug and then of various sticks to rake food into his cage provided the inspiration for Koehler's other main set of tests.

Two of the chimpanzees, including Sultan, were already familiar with the use of sticks as rakes before Koehler arrived on Tenerife. Three animals, which

Fig. 7.4. Grande achieving a four-storey structure

had been isolated from the others and which, as far as Koehler knew, had not previously used sticks as implements, were independently given the stick test. A banana was placed outside an animal's cage and could be reached only by the use of one of the sticks as rakes. As in the box tests, the solution emerged suddenly and the technique was subsequently used with other appropriate objects. The time elapsing before the successful movement was made appeared to be dependent on the relative positions of the objective and of the stick. If a stick was likely to be seen when the animal was gazing at the banana, the solution would appear quickly. This claim, important for Koehler's theoretical approach, was again based on impressionistic evidence.

In other tests an effective stick was not im-

Fig. 7.5. Sultan making a double-stick

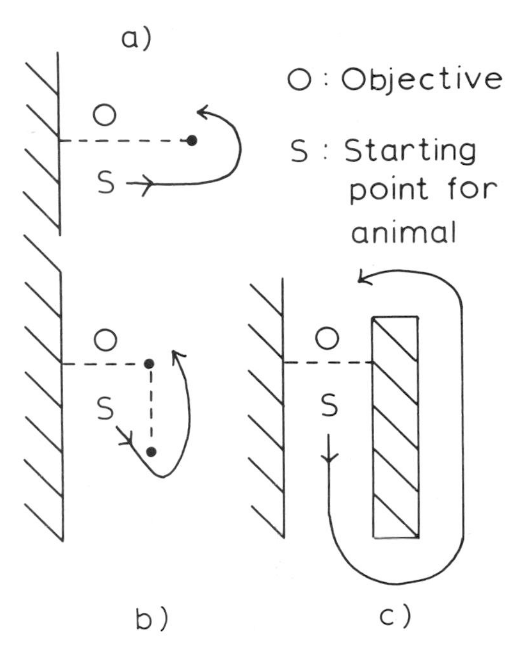

Fig. 7.6. Variations of the detour test, differing in terms of the starting angle which is 90° in a), 125° in b) and 180° in c)

mediately available. These included the one Hobhouse had devised, where a small stick was first used to obtain a larger stick lying outside the cage and the latter then used to rake in the banana; and also one in which the stick could be reached only by climbing onto a box, which first had to be placed in an appropriate position. The first of these was solved by four of the animals, but not by a further two that were tested. The second, involving the box, proved to be even more difficult and the successful solution, by Sultan and two other animals, was preceded by false starts in which the box was moved part of the way towards the stick, but then abandoned.

The most complex problem in this series was apparently solved only by Sultan and one other chimpanzee. Two rods were made available and a banana was placed outside the cage at a position where it would be reached only if a long rake were fashioned by inserting the end of one rod into the other. Sultan's eventful solution was observed by the keeper, and not by Koehler: it occurred when, apparently by chance, Sultan happened to hold the two rods in either hand in such a way that they lay in a straight line. Figure 7.5 shows Sultan working on this problem.

Before describing the final set of stick problems, it will be helpful to consider a simpler kind of situation that Koehler began to use in March. This situation provided him with an important conceptual basis for interpreting the other tests. It has been called in English the 'detour' or 'round-about' test, and the German term used by Koehler, 'Umweg', has also been used in English texts. An animal is placed in a situation where it is close to some desired objective, but direct access is prevented by a barrier, for example, wire netting or bars. To gain the objective the animal has to make its way around the barrier. Situations of this kind may be ordered in terms of the initial direction which must be taken and the extent to which the animal loses sight of the objective while making the detour. Thus, the diagrams in Figure 7.6 illustrate a simple version, a), where the animal has to start at 90 degrees to the direction of the objective, which always remains within sight; an intermediate problem, b), which involves starting at an angle of 125 degrees; and the most difficult, c), where the detour is initially in the opposite direction to the objective, which is out of sight for a considerable part of the detour.

Koehler distinguished between two kinds of performance in a detour test. One consisted of a sequence of random movements of which one, apparently by chance, took the animal to a point from which the objective could be seen and reached directly. Even in a simple version, of the kind shown in a), chickens would typically show this kind of behaviour. At the opposite extreme, a few attempts to get directly at the objective might occur, then the animal would quieten, gaze around and proceed to take the appropriate path in a smooth unhesitating manner. Thus, in a situation of the kind shown in c), provided either that the appropriate route could be seen or that an animal was familiar with the geography of the situation, a dog, a little girl of one year and three months and the chimpanzees showed this second kind of performance.

By this time the animals had become familiar with the technique of raking objects towards them. A number of tests were devised that, although involving the use of sticks, were similar in conception to a detour task. In one test the banana outside the cage could not be obtained if the animal raked it directly towards him, but only if directed towards a gap to one side. In a second, the banana was placed in a three-sided box, placed outside the cage in such a way that the banana had first to be pushed away from the animal before being raked in. This arrangement was like that of Figure 7.6b, except that the objective, rather than the animal, had to move along the appropriate detour path. A final example of this type of test was obtained by suspending a basket of fruit on a horizontal rail above the animal. In this case a stick had to be used to push the basket along the rail, at 90 degrees to its direction from the animal, until, on reaching the end, it fell off.

Performance in the second kind of test, the one with the three-sided box, was described at relative length. In the position shown in Figure 7.6b the problem was solved directly by only two chimpanzees. This pair again included Sultan, although for once his performance was inferior to that of the second animal and Sultan's first solution appeared to depend on a chance roll of the banana towards the exit from the box. Even though all five of the remaining chimpanzees, and a two-year-old boy, were very familiar with the use of sticks and could solve the equivalent detour problem when it involved *their* movement along the detour path, they appeared unable to solve this combination of the two problems.

Two further kinds of observation deserve mention before concluding this account. First, the chimpanzees' use of sticks was not confined to the situations devised by Koehler. He noted that they appeared to discover for themselves how to use sticks as levers, as aids to jumping – though not in the manner of a human pole-vaulter – and as a means of fishing for ants. Fifty years later, in the first intensive study of chimpanzees in a natural environment, Jane Goodall discovered the same use of twigs by wild chimpanzees in catching ants and termites. Sultan had learned to use the jumping stick method to reach objectives high out of reach before he solved the two-rod problem. Once the latter had been mastered in the situation where the double rod could be used as a long rake, he applied it outside of the test situation as a means of fashioning a more effective jumping stick.

The second thing to note is that Koehler was not, as some later texts have implied, trying to demonstrate how clever chimpanzees are. He was as much interested in failure as success. There were a number of situations where the chimpanzees' behaviour appeared to be surprisingly unintelligent by human standards. The very first test, for which the intended solution was to unhook a ring, was repeated in various forms. This kind of problem appeared to be insoluble for the animals, except by trial-and-error methods. Similarly difficult were the problems of uncoiling a rope and of detaching a stick, whose crook or cross-piece had caught on a bar. Koehler's chimpanzees showed no more understanding of how to detach a crook caught in a bar than had Lloyd Morgan's dog, whose attempts had been observed for two minutes by the passer-by.

Insight

Koehler viewed his work on Tenerife as an important part of the mission to save psychology from 'elementism, sensationism and associationism'. Back in Germany his earlier collaborator and mentor, Wertheimer, had made a start in the study of human perception; now Koehler was in a position to do the same for animal psychology. The main enemy here was Thorndike with his claim that animal intelligence can solve problems only in a blind fashion. Koehler was personally acquainted with Oskar Pfungst, whose study of Clever Hans had been completed just before Koehler arrived to study in Berlin, and being familiar with Pfungst's work, he must have viewed it as a local example of the dangerously negative trend in animal psychology that Thorndike had initiated.

From the outset of his work with chimpanzees Koehler was convinced that their performance could not be adequately explained as the result of a process of trial-and-error learning with accidental success. On some occasions such a process did appear to underlie the emergence of the correct solution. Just as chickens were able to reach the objective in a detour situation following a period of random movements, in some other tests a chimpanzee might succeed in an apparently similar manner. However, as noted above, in many tests the solution appeared to occur in a very different way, just as the performance of the dog, the child and the chimpanzees in a detour situation could be clearly distinguished from that of the typical chicken.

Koehler maintained that a new conceptual

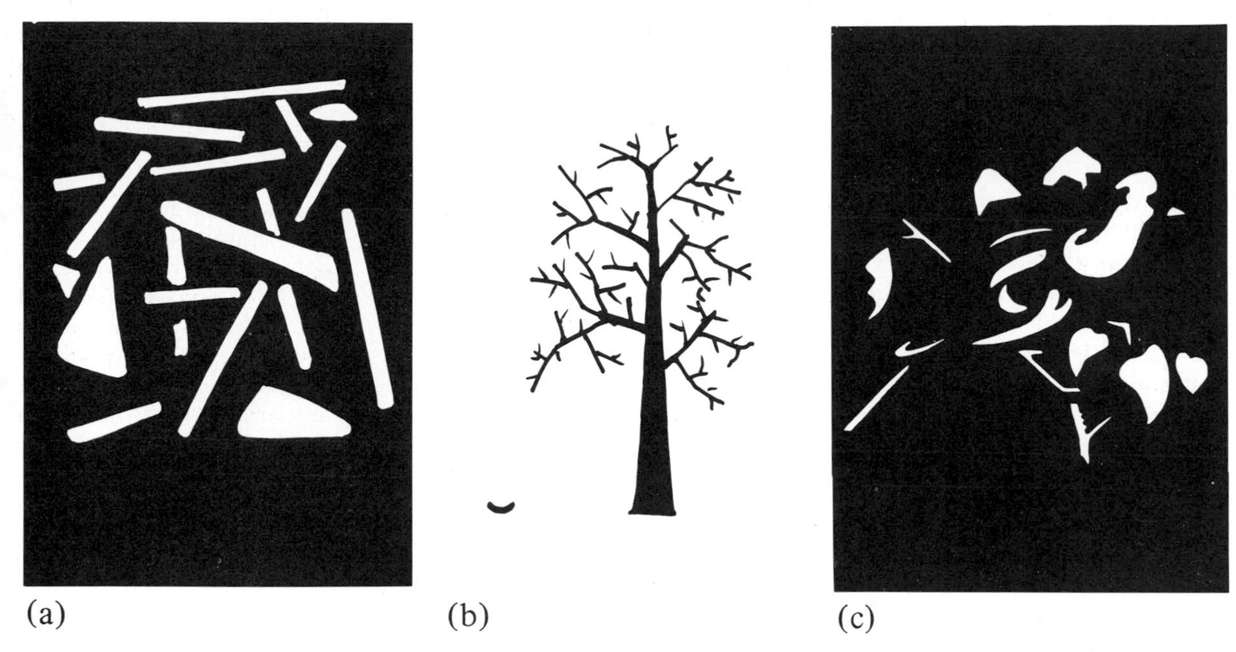

Fig. 7.7. Examples of perceptual re-organization showing, from left to right, a) a hidden digit, b) a branch of a tree and a banana, and c) a visual puzzle (McGill closure test)

approach was required to understand the abrupt appearance in a situation that was novel to the animal of a successful solution or of a 'good error', an attempt which appeared to indicate some understanding of the situation, but happened to fail. Such an approach should also go at least some way towards explaining why some problems were more difficult than others and why the configuration of the situation was important. On this last point it was clear to Koehler that to attribute to the chimpanzees some form of abstract reasoning power would be as inappropriate as attempting to analyse their behaviour solely in terms of associative learning. To credit them with some internal logical processing, in which various alternative strategies and their expected outcomes were reviewed in a systematic manner, would provide no understanding of why a stick lying on the ground was readily used as a rake, but not one forming part of a bush, or of why solutions involving movements at 90 degrees to the line between the animal and the objective were easier than those involving movements at 180 degrees.

The term 'insight', and the related German term 'Einsicht', had been used quite widely by psychologists, both to refer to a presumed form of mental processing and to the subjective experience accompanying the sudden recognition of a solution to some problem: the most ancient example being that of Archimedes' presumed feelings when leaping from his bath. It was sometimes termed the 'Ah-ha! – experience'. But when Koehler described the chimpanzees' solution as arising from the occurrence of 'insight', he used the term in a much more specific way than his predecessors. The crucial idea was that an abrupt change in performance occurred because the animal saw the situation in a different way: a process of 'perceptual restructuring' had occurred. Thus, he described a situation in which the apes, needing a rake, spent a great deal of time vainly tugging at a stout bar firmly fixed to a door, when all that was needed was to tear a branch off a nearby tree. 'The black iron bar . . . stands out *visually* better from the wooden door *as a separate object* . . . To "see" a branch of the tree, so to speak, *as a stick* is much more difficult.'[1]

Such notions are better illustrated than described. Figure 7.7a can be seen as a set of unrelated lines: it can also be seen as the digit '4' with other lines superimposed. Figure 7.7b can be seen as two independent objects; it is seen in a more 'structured' way as a tree with a branch pointing towards a banana. Figure 7.7c can be seen as some unrelated shapes: it can also be seen as a bird. Important features of these kinds of 'perceptual restructuring' are that they tend to occur abruptly and that they are usually irreversible. Once the '4' or the bird has been seen, then it is usually immediately seen again on a second occasion, even if this occurs after some considerable time has passed.

The simple detour situation suggested a way of applying such ideas to the results of other tests. Koehler assumed that the most primitive perception of any problem was simply of the relationship between the animal and the objective; given that the animal desired the objective, this would impel him to move directly towards it, somewhat in the manner of one of Loeb's tropisms. If the direct route is blocked, then either the animal generates random movements in chicken-like fashion until an effective route is found by chance or he achieves a more comprehensive and structured perception of the situation which enables him to 'see' the appropriate route in its entirety before beginning to move.

For Koehler a test that was easy was one in which the essential elements could be readily incorporated into the perception of the problem situation. Some tests were difficult because an element already formed part of some other perceptual organization, as with the stick seen as part of a tree, illustrated in Figure 7.7b. In some cases the solution depended on distinguishing the different elements forming part of some overall pattern. A variety of results suggested to Koehler that a chimpanzee's capacity for this kind of perceptual analysis was inferior to that of man. In a test where a string was attached to a basket of fruit the animals would readily pull on the string if only one were present, but, if two or three additional strings overlaid the first, they showed no immediate perception of which was the one connected to the basket; this could be seen immediately by an adult human being. A ring hooked over a nail was seen as a single perceptual unit, Koehler argued, and the persistent failure of chimpanzees to unhook rings from nails arose because of their failing to see the separate components in such a unit.

The presumed difficulty that chimpanzees have in distinguishing the components of some larger unit and in achieving a perceptual re-organization into separate units, does not explain their poor performance in two further kinds of problems. Koehler suggested that the stick–detour problems were very much more difficult than the simple detour problems because an appropriate perceptual representation can be more easily attained when based on the relationship between the elements and the perceiver than when the perceiver does not serve as a reference point. The idea here resembles the distinction Jean Piaget was to make later between a child's early egocentric perception of the world and the later development of a more objective form of perception.

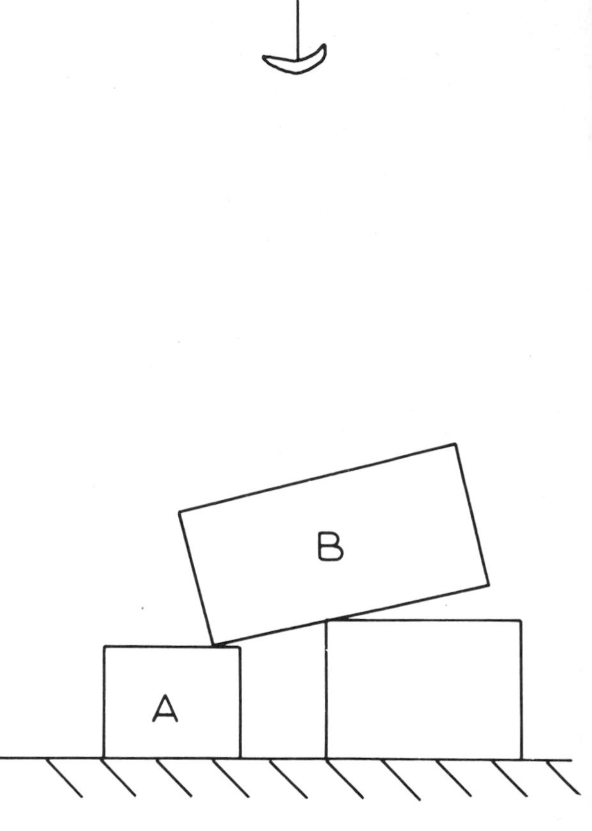

Fig. 7.8. Pillar construction suggesting the chimpanzee's poor grasp of statics

An interesting implication of this discussion of egocentrism is that both problem-solving and the perception of relationships have their roots in an animal's need to move through his environment. As Koehler noted, every time a chimpanzee in some new situation reaches out to some convenient branch in order to swing across to another spot, or adjusts its body so as to negotiate some narrow aperture, it is solving a detour problem.

A second set of difficult problems demanded a different kind of explanation. The inability of many chimpanzees to solve the problem of stacking boxes to reach a high objective and the unimpressive performances of the successful few did not appear to result from failing to see the situation in an appropriate way. A number of observations suggested to Koehler that the chimpanzees lacked even a very basic understanding of statics, of the principles involved when one object provides support for another. Some of these observations were mentioned earlier. Others were the occasional attempt to place a box in a suitable position by pushing it against a vertical wall, as if the box were expected to stick to the wall; or, having erected a construction of the kind shown in Figure 7.8, to remove Box A and attempt to place it quickly on Box B before the construction collapsed. Since the floor of the test area was flat and the sides of the boxes even, a 'visual' solution was sufficient for the one-box test and no knowledge of statics required for this. It is interesting in this context that contemporary work on picture analysis by computer has suggested that an implicit knowledge of statics may serve an important function in the perception of scenes by human beings. This might well be an important difference between the visual processes of man and those of apes.[2]

Koehler's exchanges in Frankfurt with Wertheimer and Koffka prior to arriving in Tenerife clearly inclined him towards an emphasis on the visual nature of problem-solving and the definition of insight as involving a change in the way a situation is perceived. They also affected his analysis of the way his chimpanzees behaved in and out of test sessions. Just as the notion of 'side' only has meaning if a line or edge is part of some shape, like a square, the detailed movements of an animal only have meaning in terms of the overall action of which they form components. Thus, heading off at an angle of 90 degrees or reaching up for a stick are significant only as parts of an integrated sequence defined by the goal it is designed to attain. He believed that to break down behaviour into highly specific movements was as misguided as treating lines or flashes as the most appropriate level at which to begin the study of perception.

'If we change to this method and direct our observation and description to *parts* of these movement complexes, the result turns out to be entirely unsatisfactory: it provides detailed examples of the mechanism of movement and of the pure physiology of muscles and glands. The further we push the analysis in striving for this kind of objectivity, the less we are inclined to call the description one of the "behaviour" of apes, and the more it dissolves into purely physiological statements . . . We suspect a train of thought that leads to such conclusions must contain an error, even if it is not at first clear where to look for it . . . If the subject matter of objective psychological observations disappears as soon as one tries to describe it analytically beyond a certain point, *then there are realities in the animals investigated which are perceptible to us only in these total impressions.*'[3]

One of the 'realities' Koehler referred to concerned the way that his chimpanzees learned by imitation. This showed two important characteristics. The first was that such learning only seemed to occur when they had some understanding of the situation; Koehler came to believe that 'the animal must work *hard* to gain some understanding of the model, before it can imitate it' and found that, even among chimpanzees, it was very rare for any animal to 'manage to imitate a performance enacted before him of which he knew nothing before'.[4] The widespread failures in recent years to obtain much evidence for imitation was attributed by him to the use, especially in cases of a human being demonstrating to another animal, of tasks where it was impossible for the animal to perceive the crucial relationships.

The second characteristic noted by Koehler was that, where imitation did occur, it was not simply a matter of copying some movement made by the skilled animal. One example was an obstacle task in which a heavy cage blocked the use of a stick to rake in a banana; initially only Sultan found the solution of moving the cage to one side, but, after watching Sultan perform, one of the younger animals, Chica, later abruptly solved the problem. Koehler commented that 'if Chica's achievement was performed in imitation of what she had seen Sultan do, then she certainly imitated the *substance* of his actions, and not their *form*, for her movements in displacing the box were quite different from his, though both came under the category of "removal of the obstacle"'.[5] Thus, it appeared that chimpanzees represented the actions of their fellows at a general and interpretative level, one in terms of the apparent purpose of the sequence of movements, and similar to the one most human beings use when describing them. However, it is worth noting that, as with other topics, Koehler made no systematic study of imitation learning; this was partly because he found it was often necessary to carry out tests in a group situation, since the animals were likely to become dejected and unreactive when tested in isolation. Consequently it was very difficult to sort out the role of social learning in an unambiguous way.

In studying topics other than problem-solving Koehler was able to carry out formal experiments yielding more conclusive results. These were reported in scientific papers and not included in his book, *The Mentality of Apes*, which reported the research so far described. Some of his earliest work on Tenerife was on perceptual constancies in chimpanzees; for example, does the apparent size of an object vary with its distance from the observer or, as with human vision, remain constant? In general he found that chimpanzees showed exactly the same constancies as man.[3]

A later set of experiments used a discrimination

procedure and discovered a phenomenon he called 'transposition', which became a key theoretical issue some twenty years later. Chickens were first trained to choose a box painted a medium grey over one painted a dark grey and then were given a choice between the medium and a still lighter grey; Koehler found that they now chose the latter, even though it was novel and their choice meant that they turned away from the box that had always hitherto contained food. Similar findings were obtained when Chica, the chimpanzee, and a three-year-old child were trained in a similar discrimination task and also from an equivalent experiment involving variations of size instead of greyness. These results indicated to Koehler that animals can base their performance on the relationship between stimuli in such choice tasks, learning to go to the lighter or to the smaller of the two stimuli, as well as on the absolute properties of a particular stimulus.[7]

Speculation on what constitute the major differences between the human and ape mind led Koehler to study one further topic. As already noted, he believed that two major differences involved perception: the chimpanzee's lesser ability to distinguish the component parts of some situation – to use a term becoming very familiar in Europe at that time, it is as if this animal lives in a world in which objects are much more highly camouflaged than in the human world – and its relatively poor grasp of the elementary physics of common objects, such as support relationships, which means that its solutions to problems remain at a visual level. Lack of language provided the third difference; Koehler concluded from his many years of close contact with chimpanzees that their calls and gestures never serve to designate objects or describe states of the external world, but can only express their moods and emotions. He said little on this subject, beyond suggesting that the lack of articulate speech could not be ascribed to anatomical factors.[8]

The fourth, and final, major difference was that in Koehler's opinion '"the time in which the chimpanzee lives" is limited in past and future'.[9] He suspected that it is unable to anticipate future events to anything like the same extent as man and that this is the main reason for the absence of anything that could be described as even a primitive form of culture; he thought it totally unlikely that a chimpanzee could prepare a tool for use on the following day. However he was unable to devise any tests that satisfactorily examined this aspect of his animals' intelligence.

Their ability to take into account past events was much easier to study. Koehler knew of Walter Hunter's research on delayed reactions and regarded this as the first real study in animals of 'something like real "remembering" rather than the reproduction of former behaviour in a like situation'. He used a simplified version of Hunter's technique, which consisted of burying fruit in the clear, sandy yard outside the home cages in full view of a chimpanzee, but well outside its reach. At some later time the animal was either given a stick with which to reach the hidden treasure or allowed out into the yard. It turned out that memory for where the fruit had been hidden remained very accurate; even after a delay of up to sixteen hours they would simply head straight for the right spot and start digging. Since during the preceding retention interval they had been involved in various activities and in many cases had also slept for the night preceding a morning memory test, there was even less possibility than in the case of Hunter's raccoons or children that performance depended on some kind of orienting response or maintained mediating behaviour. Various control procedures confirmed that they were not using smell or subtle visual cues to guide them to the hiding places.[10]

This work started too near to the end of Koehler's stay to allow an extensive study of memory. When he returned to Germany in 1920, the Tenerife station was closed and the chimpanzees transferred to the Berlin Zoo. Koehler's productivity had confirmed Stumpf's original opinion and in 1922 he became Stumpf's successor as the Professor of Psychology at the University of Berlin and its institute of psychology became a vigorous centre of Gestalt psychology. The institute in Berlin, housed in the former palace of the Kaiser, now attracted students of psychology from Germany and from abroad in the way that Leipzig had done in earlier years.

Very little research on animal psychology was carried out after this by Koehler or by any of the other Gestalt psychologists in Germany.[11] Koehler's own research followed his belief that the most fundamental problem in psychology was that of understanding perception and that this was most effectively pursued by using human observers. The approach to the study of problem-solving that he had developed on Tenerife was applied and developed in studies with human subjects by some of his students, notably Karl Duncker. Also, Wertheimer had written on this subject before Koehler had left Frankfurt and he continued these studies. But Wertheimer's interest was in how children and adult human beings solve problems, and not at all in animals.[2]

In 1925 Koehler made his first visit to the United States as a visiting lecturer at Clark University. The first English translation of *The Mentality of Apes* appeared in the same year. Gestalt ideas began to make a major impression on the way Americans studied perception. However, although Koehler's work became very well-known and his book widely read in the English-speaking world ever since, his ideas had only a limited impact on animal psychology in America. Gestalt psychology displayed the now pre-eminent virtue of generating experimental research, but Koehler's ideas crossed the Atlantic at a time of growing hostility to any form of nativism, of which Koehler was judged guilty, and to any kind of theory that was not ground on the laws of association. As these two aspects of his chimpanzee studies have continued to receive attention, some comments on them are appropriate before concluding this section.[13]

Koehler was not particularly interested in the way that past experience contributes to the solving of problems, beyond the very important point of emphasizing that this does not occur simply as the expression of a habitual response earlier acquired under slightly different circumstances. He was quite sure that what an animal has previously seen and done has a great effect on its performance in a novel task; he did not believe, as some later commentaries have suggested, that problem-solving of the kind he studied reflected some special inherent capacity which develops in autonomous fashion. He might have speculated on whether his chimpanzees' poor grasp of support relationships was due to their lack of the extensive manipulation of objects and of piling up things which most young children display. He did not do so, but in any case, since little was known about the infancy of his subjects and there was no opportunity for studying chimpanzee development, such speculation would have been empty.

The other aspect of Koehler's work that perturbed many Americans was its uncertain theoretical status. Did he *explain* how chimpanzees solve problems or was his account merely descriptive? In what ways did it differ from the loose anthropomorphic tales of the Victorians? Koehler knew that many of his readers would have accepted the strictures of Lloyd Morgan or of Wundt, would have been impressed by Thorndike's research or by Pfungst's analysis of Clever Hans, and would not have been convinced by Hobhouse's studies. Against the trend supporting a rather low estimate of any animal's intelligence, performances interpreted as indicating complex mental processing were likely to be received with scepticism. Aware that on two counts, style and content, he might be classed as a latter-day Romanes, Koehler was anxious to defend himself against possible charges of 'mentalism' at various points in the *Mentality of Apes*; thus, 'I must explicitly warn my readers against the mistake of thinking that I am implying any supernatural mode of interpreting behaviour'.[14] The problem was to extend the general Gestalt rejection of the mechanical by attacking the connectionism of Thorndike and others, and yet to avoid any appeal to the non-material forces of the vitalists.

This problem seems to have been an acute one for Koehler during his stay on Tenerife. The second book he wrote there, *Physical Gestalten*, was an attempt to lay the foundations for an explanatory system in psychology which was related to physical concepts, but not to 'mechanics'. The areas in physics he concentrated upon were those to do with field effects. He suggested that a direct relationship might be discovered between psychological phenomena and the properties of electric, magnetic or gravitational fields.

As for his discussion of insight Koehler saw this as the most fruitful way of approaching the study of thought, but not as an *explanation* of the way his chimpanzees behaved. A theory of problem-solving in chimpanzees would have to await a satisfactory general theory of perception. For Koehler the attitude of a sceptic demanding hard evidence and a precise theory before taking any interest was one that missed the point of scientific enquiry. The principle of parsimony, Lloyd Morgan's canon, is not a sufficient guide. One should also employ the 'principle of maximum fertility' by examining a set of phenomena from a variety of viewpoints, even if these cannot be specified very vigorously.

In later years Koehler decided that explanations of perceptual phenomena can only be sought at the level of neurophysiology. He came to believe in direct parallels between visual effects and changes in electric fields operating across the surface of the brain. Subsequent developments in neuroscience made these beliefs seem as unlikely as those of Pavlov's on the physiological substrates of conditioning phenomena, but Koehler hung on to them – even to the extent of insisting in one study that quite high voltages be applied to his scalp in an attempt, which was repeatedly unsuccessful, to see the changes in vision that should be caused by the resultant distortion of his brain fields.[15]

Few American psychologists accepted Koehler's idea on field theory. Much later their doubts seemed justified following experimental work by Karl Lashley, one of the American animal psychologists most sympathetic to Gestalt theory, who manipulated the electrical fields on the surface of monkeys' brains and found no change in their ability to perform on visual discrimination tasks.[16] The general notion that psychological phenomena are to be explained in terms of field forces was largely rejected. The unconvincing physiology meant that Koehler's third way to explanations of behaviour, one that provided an alternative to both S-R links on one side and mentalism on the other, was widely seen as a cul-de-sac.

Robert Yerkes' studies of apes

All of Robert Yerkes' early research was on the performance on simple tasks of turtles, frogs and his dancing mice. Around 1914 he devised his multiple-choice method for studying what he began to call 'ideational behaviour', using this first with human subjects and later with pigs and crows. As his interests turned to larger creatures, Yerkes' earlier ambition to establish an institute for the study of comparative psychology changed into an ambition to establish one for the study of primates. During the decade or so since Thorndike had given his low estimate of monkey and ape intelligence, a series of experiments by American psychologists had failed to settle whether he was correct or not.

The major problem, as Yerkes saw it, was that studies of monkeys or apes were restricted to the occasional and imperfect opportunities offered by zoos or by the odd pet animals psychologists gained access to. Given that the major point of comparative psychology was to provide an understanding of the human mind, it seemed to him absurd that, simply on grounds of cost and convenience, so little effort should be expended on the study of primates and so much on animals more distantly related to man. Consequently in 1913 he was delighted by the news that a primate station had been set up in Tenerife and by the invitation to visit it. A sabbatical leave from Harvard, due in two years' time, would provide the opportunity. The possibility of maintaining it as a joint Prussian–American venture was discussed, but the outbreak of the First World War ended these plans.[1]

Fortunately, Yerkes received an invitation from Gilbert Hamilton, his former student, to spend the sabbatical in California at the private laboratory the latter had established near Santa Barbara. Hamilton also generously offered to cover all the expenses of any research Yerkes wished to carry out. There were already ten macaque monkeys at the laboratory and now an orang-utan was purchased for his visit. For over six months – about the same amount of time as Koehler had spent on his tests, which Yerkes knew nothing about when he started – Yerkes worked with two of the monkeys and with the five-year-old orang-utan, Julius. With little respect for the previous 'incidental, casual and qualitative' approaches to the study of a primate's mind, he was determined that his research would be intensive, quantitative and systematically maintained throughout the period. What he planned was an exact repeat of the procedures he had used earlier with pigs. But before starting he had to wait until the large scale apparatus this required was ready.

Fig. 7.9. Robert Yerkes with the orang-utan, Julius, in California

Meanwhile he adopted the approach taken by another of his Harvard students, Mark Haggerty, who some years earlier had carried out some tests of imitation learning and problem solving, first with monkeys and later with two orang-utans in the New York Zoo. Haggerty was fiercely opposed to Thorndike's 'sense-impulse' theory and in his attempts to demonstrate the superior abilities of primates tried some of the Hobhouse tasks.[2] Yerkes had supervised this work and also used some of Hobhouse's methods.[3] While waiting to start the main experiment Yerkes decided to try some of these tests himself, since they did not require any special equipment.

Yerkes was astounded by the performance of Julius. He began with what Koehler had already discovered was a difficult form of the box-test. A banana was suspended at a sufficient height so that it could be reached only if two boxes were stacked one on another. Julius did not solve the problem until Yerkes himself had demonstrated the solution, but nonetheless his first reaction in the situation had been to move the largest box immediately to a spot beneath the banana. Yerkes jotted in his note book: 'Despite all that has been written concerning the intelligent behaviour of the orang-utan, I was amazed by Julius's behaviour this morning, for it was far more deliberate and apparently reflective as well as more persistently directed towards the goal than I had anticipated. I had looked for sporadic attempts to obtain the banana, with speedy discouragement and such fluctuations of attention as would be exhibited by a child of two to four years. But in less than ten minutes Julius made at least ten obvious and well-directed attempts to reach the food.'[4]

Yerkes was impressed by what he saw as the ingenuity displayed by the orang-utan in various other attempts. When given the stick-test, Julius immediately used the stick, and subsequently various substitutes, to rake in fruit from outside the cage. Julius failed to solve one of the problems that Watson had tried with his monkeys, that of poking out a banana from the middle of a long, thin box. Yerkes' own two monkeys in general performed no better than those of Watson in the Hobhouse tests. However one that Hamilton had previously classified as 'feeble-minded' showed such a startling ability to use a hammer and a saw that Yerkes was convinced that this monkey possessed an 'instinct for mechanical ability'; he wrote that this monkey 'has importantly modified my conception of genius'.

Fig. 7.10. Julius working on a two-box stacking problem; this may be compared with the performance of Koehler's chimpanzee, Grande, shown in Figures 7.3 and 7.4

By this time the sizeable and elaborate building specially commissioned for the main experiment was ready for use. It contained a row of nine compartments, each with a door that could be locked or unlocked for a given trial; at any one time only a subset of the nine doors was unlocked and the number and particular combination of openable doors varied from trial to trial.

The three primates were each given the same sequence of problems. For Problem 1 choice of the extreme left door of the available subset was correct; for Problem 2 the second door from the right was correct; for Problem 3 the correct door was alternatively on the extreme right and the extreme left; and for Problem 4 the middle one of the available doors was correct. If an animal entered an incorrect door he was usually confined in the compartment it led to for ten or twenty seconds; choice of the correct door led directly to a piece of banana or carrot.

Yerkes was convinced of the efficacy of his multiple-choice method for investigating intelligent behaviour and naturally assumed primates to be more intelligent than pigs. Given the ability that Julius had already displayed in the Hobhouse tasks, it was natural to predict that he would perform exceptionally well in the main experiment.

The results surprised Yerkes. On the first problem, choice of the extreme left compartment, the two monkeys took two and three times as many trials to reach the criterion as the pigs. Julius took six times longer, a total of three hundred trials to reach the same criterion of ten correct trials in succession.

The way Julius performed was completely different from the monkeys. He was much easier to work with and got through his daily sessions of ten or so trials much more promptly than the monkeys; Yerkes described him as 'always gentle, docile and friendly' in contrast to the often infuriating monkeys who at best were 'stealthy, furtive and evidently suspicious of the experimenter as well as of the apparatus'. Nonetheless he showed no improvement whatsoever until quite abruptly he stopped making any but the occasional error. 'The curve of learning plotted from the daily wrong choices . . . had it been obtained with a human subject, would undoubtedly be described as ideational, and possibly even as a rational curve; for its sudden drop from near the maximum to the base line strongly suggests, if it does not actually prove, insight. Never before has a curve of learning like this been obtained from an infrahuman animal. I feel fully justified in concluding . . . that the orang-utan solved this simple problem ideationally.'[5]

Since Julius took twice as many trials to learn this first problem as the slowest, 'feeble-minded' monkey, Yerkes commented that 'where very different methods of learning appear, the number of trials is not a safe criterion of intelligence'. It is worth noting that Yerkes' main objective was to find a satisfactory *measure* of intelligence, as compared to Koehler's concern with the *nature* of intelligence.

The second problem, second compartment from the right as correct, took even longer to solve. The two monkeys eventually reached the criterion, but the control tests suggested that neither had learned the rule, 'second from right', in any sense, but had learned particular responses to particular settings. Even after 1400 trials Julius showed no improvement and had long begun to find other activities, such as playing with the sawdust on the floor, frequently more attractive than entering a compartment. The use of a whip, which made the animal whine with fear, did not improve his performance; a finding that Yerkes might perhaps have predicted from the Yerkes–Dodson Law.

The complete failure by Julius to solve this second problem, which the monkeys appeared to learn by the trial-and-error development of various habits, suggested to Yerkes that 'in this young orang-utan ideational learning tended to replace the simpler mode of problem solving by trial and error. Seemingly incapable of solving his problems by the lower grade process, he strove persistently, and often vainly, to gain insight.'[6]

Fig. 7.11. Julius being led by a worker on the estate; the latter may have served as a subject in the earlier experiment by Gilbert Hamilton which was described in Chapter 6

By this time the sabbatical was coming to an end. There was little time to study the final two problems. One monkey completed the third problem, alternating far left and far right, in about the same number of trials that he had required for the second problem, but showed no more sign than the pigs of solving the fourth, that of selecting the middle compartment in each setting. The second monkey was still performing at a chance level on the third problem when Yerkes had to leave California to return to Harvard.

In reporting his work in California Yerkes suggested that the results from the few informal tests were at least as interesting as those from the multiple-choice experiment on which he had spent so much time. But because the Hobhouse tests were carried out in a much less systematic fashion and because the animals' behaviour was difficult to describe in precise, objective terms – let alone quantifiable ones – he felt dissatisfied with them. Indeed it was almost an embarrassment that they should have proved so interesting. In the summary to his report he proudly trumpeted the virtues of his multiple-choice method and of well-controlled, quantitative experimentation. 'Never before . . . has any ape been subjected to observation under systematically controlled conditions for so long a period as six months. Moreover my multiple-choice method had the merit of having yielded the first curve of learning for an anthropoid ape . . . so far as one may say by comparing it with the curve for various learning processes exhibited by other mammals, it is indicative of ideation of a high order, and possibly of reasoning.'[7] That so little had been gained from the huge amount of effort invested was not discussed. Nor was the earlier admission that good performance on one of the problems did not necessarily mean that the animal had perceived the appropriate relationship, had got the 'idea' of what was required, since various unintelligent response strategies could also generate quite accurate performances.

Yerkes used the term 'insight' to describe some aspects of Julius' behaviour, but he did not try to specify what he meant by this. Experimenting left too little time to develop his thoughts on such matters while he was still in California. He did not work out how the conclusions he drew from the Hobhouse tasks could be substantiated by firmer evidence to convince tough-minded critics such as Watson, who edited his report. Instead Yerkes relied on the authority of his personal reputation as a skilled and highly experienced researcher.

Yerkes had objected strongly to Watson's 1913 paper on behaviourism. Although in later years the terminology he employed became progressively more behaviouristic, Yerkes never abandoned his respect for Titchener's approach to human psychology or his critical attitude towards Watson's behaviourism. In turn Watson did not conceal his opinion that Yerkes' work in California was of poor quality. In a letter commenting on a harsh review by Hunter of Yerkes' book Watson included the following remarks. 'I am going to punch you one under the fifth rib. It will make you a little mad but I think we have known each other long enough for you to cuss me out in the way I deserve, and then settle down into the old grooves which are based along lines of sincere friendship . . . I have just gone over your book on the apes again and I confess if I had to review it I would say some pretty mean things. You have made statements which are based on such flimsy and anthropomorphic evidence that for a while I seriously questioned your scientific spirit . . . The papers on the multiple-choice stand for a loose type of theoretical interpretation which has been absolutely foreign to the spirit of behaviour for the last fifteen years. I agree thoroughly with Hunter who says that the only thing that can be brought out by the multiple-choice method is reaction tendencies. Anything more which is brought out through interpretation is completely gratuitous and reacts strongly against both the method and the interpreter . . . I don't understand this momentary lapse, nor do any of the behaviour men I have talked with.'[8]

In reply Yerkes wrote a lengthy defence. But their friendship did not survive this exchange. Yerkes did not carry out any further animal research for the next few years. When he returned to Harvard from his Californian sabbatical he became involved in other matters, one of which was the continuation of work on a version of Binet's intelligence test.

Two years after his Californian sabbatical Yerkes became President of the American Psychological Association. With the United States' entry into the First World War he and other American psychologists became concerned with the contribution that psychology might make to their country's efforts. Since Yerkes happened to be the president of the Association and also had some knowledge of mental testing, he was chosen to head the massive Army testing programme that resulted from these deliberations.[9]

It was an entirely new experience for Yerkes and he discovered that he was a good organizer and enjoyed committee work. When the war was over he decided to stay on in Washington to supervise the analysis and publication of reports on the huge volume of data accumulated by the programme. He had long been interested in questions concerning heredity. Various animal projects he designed to investigate the inheritance of behavioural characteristics had previously been disrupted by various accidents. More recently he had become involved in the eugenics movement and he saw the army testing programme as an unparalleled opportunity for gaining an understanding of the relationship between race and intelligence.

Yerkes also realized that remaining in government service and serving on various committees of the new National Research Council might be the best way of achieving his major ambition, the founding of a primate station and the launching of full-scale studies of apes both in captivity and in the wild. At the end of his report on his Californian research he had made detailed proposals for such work and had estimated that a satisfactory station would need an initial endowment of at least a million dollars.[10]

Yerkes appears to have been highly effective in promoting the professional development of psychology in America and in ensuring that psychological research received a reasonable share of the funds allocated by the research council. His involvement extended well beyond the usual boundaries of academic psychology. Thus, for many years he chaired a committee funding sex research which, among other things, supported Kinsey's massive and epoch-making survey of sexual behaviour. Another example, one less delicate and presumably much easier in political terms, was support of the first expeditions to Africa whose sole aim was the study of chimpanzee life in the wild.

Yerkes' activities in Washington left little time for research. The one study of apes in which he became directly involved consisted of an attempt to teach two chimpanzees to talk. Before the war the first intensive project ever to test la Mettrie's claim had failed miserably in that five years' work with both chimpanzees and orang-utans had progressed only to the stage where one chimpanzee could say 'Mama' and an orang-utan could produce two or three recognizable words. The investigator, William Furness, wrote in 1916 that it seemed 'well nigh incredible that in animals so close to us physically there should not be a rudimentary speech center in the brain which only needed development. I have made an earnest endeavor and am still endeavoring, but I cannot say that I am encouraged.'[11]

The pair of animals studied by Yerkes and his co-worker were easy to work with and one was exceptionally fast at mastering all sorts of tasks, except learning to speak. Neither chimpanzee made any progress at all. 'Although I admit surprise in this outcome of my effort at training the animals to speak, I am not yet convinced of their inability.'[12] Later he was less confident and was particularly struck by their lack of vocal mimicry. 'Evidently, despite possession of a vocal mechanism which closely resembles the human, and a tendency to produce sounds which vary greatly in quality and intensity, the chimpanzee has surprising little tendency to reproduce other of the sounds which it hears than those characteristic of the species, and very limited ability to learn to use new sounds either affectively or ideationally.'[13]

Like Koehler, Yerkes was impressed by the rich use his chimpanzees made of gestures. In complete contrast to their lack of verbal mimicry they readily

Fig. 7.12. Yerkes in the summer of 1923 carrying the two young chimpanzees, Chim and Panzee, who made no progress in learning to speak

copied human hand gestures and body movements. Yerkes suggested that 'perhaps they can be taught to use their fingers, somewhat as does the deaf and dumb person, and thus helped to acquire a simple, non-vocal "sign language"'.[14] Unfortunately he never followed up this idea and no one else seems to have noticed it. Research forty years later has shown that it was an excellent suggestion.

In the immediate post-war years there was little work on primates anywhere in the world. The one exception used a method first employed with an animal by Hamilton, who in 1908 had described it in a paper that did not attract attention even from Yerkes, despite the latter's general interest in methodology and his high regard for Hamilton's other ideas. An uninformative title, unclear presentation and inconclusive results from a single dog seem to have guaranteed that Hamilton's use of an ingenious procedure remained entirely unnoticed. After it was reinvented twelve years later the procedure continued to be used widely and is known now as the matching-to-sample task.[15] By the time this re-invention occurred Hamilton had given up his attempts to continue animal studies, but, although he became a highly successful psychiatrist in private practice, he managed to continue with other kinds of research. In the course of investigating the sex life and problems of married people he met Eugene O'Neill and managed both to cure the latter's alcoholism and, by encouraging O'Neill to recall his childhood, to provide the initial inspiration for *Long Day's Journey into Night*.[16]

In his 1908 experiment Hamilton put his bull terrier in a box from which it could escape as soon as it pressed the correct one of four pedals. On a given trial the correct pedal was indicated by marking it with a stimulus that was identical to one on a large board mounted to the side of the pedals. Hamilton explained that 'an adequate reaction to the situation required the animal first to seek the sign board, then to inspect it, and finally, to strike the pedal bearing the only card that would offer him the same odor or visual stimulus that he got from the sign board'. Hamilton decided that at times the dog's choice was influenced by what was displayed on the sign board, but this was based on a very generous reading of the results. In general there was no sign that the animal learned the task, or would have done so even if training had continued beyond the six hundred or so trials already given.[17]

The performance of Hamilton's terrier is in complete contrast to that of a chimpanzee, named Ioni, who in the early 1920s was given training on an equivalent task by Ladygin Kohts, a psychologist working at Moscow University. To begin with Ioni was trained to pick out from a set of small discs the one that was the same colour as the sample she was shown at the start of a trial. Her eventually high level of performance with this procedure meant that it could be used as a very efficient technique for testing Ioni's perceptual abilities, which was Kohts' main interest. She used a large variety of colours, shapes, sizes and even letters of the alphabet in this work, which complemented Koehler's studies in yielding accurate information on how well chimpanzees can see.

Two kinds of observation persuaded Kohts that her chimpanzee was displaying more than rapid learning of a complex task and that in some sense she understood the general principle of the matching-to-sample procedure, that of identifying stimuli that are similar to each other independent of their particular sensory properties. One was the ease with which Ioni could be switched from one set of stimuli to another; thus, after she had had considerable experience involving discs that differed only in colour, her

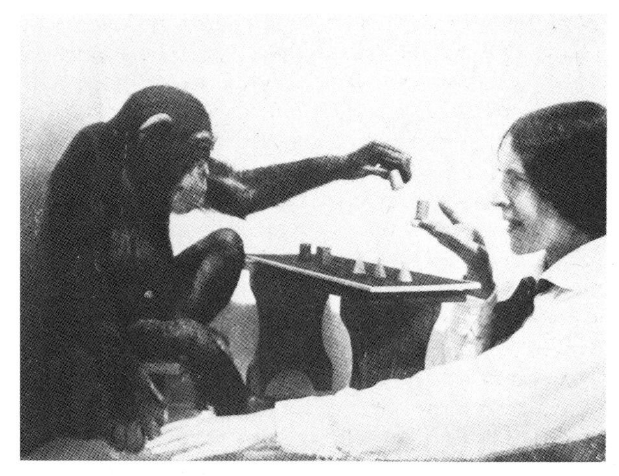

Fig. 7.13. Ladygin Kohts testing her chimpanzee, Ioni, on a matching-to-sample problem using stimulus objects varying in shape

performance was little affected when these were replaced by discs that differed in shape or size. Furthermore, she displayed a remarkably good ability to choose an unseen object from a bag on the basis of touch alone, after being allowed only visual inspection of a sample. Incidentally, this last observation rules out the possibility of an experimenter-cueing effect, since there was no way that an unintended gesture or movement of Kohts' eyes could signal which was the correct object to pull out of the bag.

However, when Ioni was given training across sensory modalities that included sound, she showed no more ability than had the apes to which Furness and Yerkes had tried to teach some speech. Despite extensive training Ioni showed no inclination to associate characters from the Russian alphabet with particular sounds. Kohts decided that this was because hearing, unlike vision, does not play a fundamental role in a chimpanzee's life.

Finally, like Koehler, Kohts also decided to test her subject's memory ability. By interposing a delay between the presentation of a sample and the presentation of the choice stimuli Kohts was also able to use the procedure as a test of short-term memory that provided an interesting alternative to the kind of procedure employed by Hunter. These experiments showed that, if more than fifteen seconds elapsed, Ioni's score fell to a chance level.[18] There do not appear to have been any further studies of apes in Russia that continued this pioneering research by Kohts.

In 1924 Yerkes returned to academic life when he was invited to become a research professor at Yale. He visited a private colony of primates that had been set up in Cuba to get further information on what was involved and then set up a temporary laboratory at Yale for work on chimpanzees. He wrote a series of both popular and scholarly books about apes and also during this period carried out the first assessment of gorilla intelligence. The one member of this species that he tested performed very poorly on the Hobhouse tasks in which both chimpanzees and orang-utans could be so impressive.[19] Despite his prestigious new position, his very considerable research reputation and his unrivalled knowledge of the way Washington dispensed research money, it took five more years before a massive grant was obtained to realize his plan for a permanent primate laboratory.

In 1929 the Yale Laboratories of Primate Biology were set up near Jacksonville in the sub-tropical climate of Florida to serve as a breeding, as well as a research, centre. This came as the culmination of his efforts on behalf of animal psychology that had begun thirty years earlier. By his books, editorial labours, committee work and unflagging correspondence, as much as by his scientific achievements, Yerkes had done more than anyone to keep animal studies alive in America and based in psychology departments, as well as to initiate the study of apes. He was the first to admit that his had been an organizational rather than conceptual contribution. Shortly after the primate laboratories opened Yerkes described himself in the following way. 'Endowed with a mentality in many respects ordinary, I have always had the advantage of a few wholly extraordinary abilities. Love of work and the power to tap new reservoirs of energy seem to have been paternal heritages which the circumstances of my life greatly strengthened. From childhood I have been able to work easily, effectively, and joyously, even when associates whom I considered my superiors physically and intellectually faltered or failed. This I attribute more largely to exceptional planfulness, persistence, sustained interest, and abiding faith in the values of my objectives, than to unusual gifts or acquisitions.'[20]

Concluding discussion

The studies described in this chapter were the first real tests of Descartes' claim that apes cannot think. The chimpanzees tested in various ways by Hobhouse, Koehler and Kohts and the orang-utan and chimpanzees tested by Yerkes sometimes appeared capable of responding in an appropriate fashion when confronted with a novel situation. On the other hand these pioneering studies suggested that Descartes was correct in stressing the unique properties of human language and that, like von Osten's belief in his horse, Clever Hans, la Mettrie's idea that an ape could be taught to speak was mistaken. After years of close contact with his colony of chimpanzees Koehler was unable to detect any form of communication that resembled human language, while explicit attempts to train apes to talk, first by Furness and then by Yerkes, were miserable failures, even though both investigators had started with considerable sympathy for la Mettrie's view.

In discussing the studies of problem-solving reviewed here Bertrand Russell once observed that 'animals studied by Americans rush about frantically, with an incredible display of hustle and pep, and at last achieve the desired result by chance', whereas those 'observed by Germans sit still and think, and at last evolve the solution out of their inner consciousness'.[1] He did not find the situation as discouraging as this observation might suggest, since it seemed that national characteristics mainly influenced an experimenter's choice of task and thus only indirectly the behaviour of their animals. It is reassuring that, while different attitudes to science were reflected by the methods favoured by Hobhouse, Yerkes and Koehler, when they used the same tests, they obtained the same results. On being given the Hobhouse tests, the orang-utan in California behaved in much the same way as the chimpanzees in Tenerife. All three men agreed with Romanes and disagreed with Descartes, in concluding that under appropriate conditions apes can display more than blind habit and show evidence of rudimentary thought.

As to the nature of their thinking and the way apes differ both from other non-human species and from man, Yerkes did not really explain what he meant by ascribing to his orang-utan and to the chimpanzees he tested years later the capacity for 'ideation'. Hobhouse and particularly Koehler had more to say on this and both concurred with Romanes' suggestion, implicit in many of the pertinent terms used in everyday language, that thinking and understanding are intimately related to visual perception. The superior abilities of apes to solve problems in comparison with other animals was not because the solutions well up from unusually rich inner consciousness, but appeared to stem mainly from seeing the world in a more articulated way.

Yerkes would dearly liked to have obtained much more dramatic and unambiguous results from the multiple-choice apparatus in which he required his animals to rush about. The reason he attached so much importance to this approach was that at least no one could argue about the precision of the data it produced; if he reported that an animal made eight correct choices out of a possible twelve in a day's session, then no one could object that the results were projections of the experimenter's beliefs. To claim that Julius moved a box with the intention of placing it beneath the banana or, in the absence of a box, attempted to drag the experimenter towards the appropriate spot, was, Yerkes knew full well, to invite disbelief and disapproval from the majority of his peers.

Perhaps because he had fewer powerful and articulate critics of this kind close at hand, Koehler faced this problem more directly than Yerkes. He decided that Hobhouse and, by implication, Driesch were right to emphasize the purposive nature of at least some types of behaviour, in particular the characteristic of potential flexibility with respect to a fixed end-point. Consequently, the dangers of subjective interpretation seemed to Koehler a necessary price to pay in order to ensure that what an animal does is not viewed in a meaningless way. For Koehler and for Hobhouse to decide that some action on the part of the animal reflected a specific intention did not mean that the animal possessed a non-material mind of a Cartesian kind or was imbued with Driesch's *entelechy*. Both believed that living organisms work according to the same laws of chemistry and physics that govern the physical world and that the difference lies in the extraordinary complexity of their structure. They were not at all sure how the structure of an organism and the organization of its behaviour were to be understood. Some of Hobhouse's remarks suggest that he would have been pleased by the later application of control theories developed for man-made self-regulating systems. For his part Koehler looked to field theories and his ideas developed into what was an unsuccessful physiological model of brain function.

The details of Koehler's work became widely known among American psychologists in 1925 when the first English translation of *The Mentality of Apes* was published. This was two years before the English version of Pavlov's *Conditioned Reflexes* appeared. Pavlov's ideas and discoveries were quickly absorbed, albeit with some distortion, by animal psychologists in America and the general study of conditioning became the mainstream research tradition. In contrast, the impact of Koehler's work on apes was quite limited. This was partly due to the greater attraction of Pavlov's more familiar kind of theory and of his objective methodology. But perhaps, more important, was the fact that it was easy to see what kind of empirical questions needed to, and could, be answered within the framework offered by the conditioned reflex; it was not at all clear what kind of research could develop naturally and productively out of the early studies of problem-solving.

Over twenty-five years later Koehler commented on the unreceptive treatment given to his ideas on insight and expressed his belief that American-style research on such issues was more feasible than it appeared at the time. 'The American tradition is averse to non-experimental forms of observation. It would therefore have been advisable not to insist too much on the simple kind of evidence from which the concept of insight had been derived. The conditions under which insight occurs, and the consequences which it may have, can surely be investigated in perfectly orthodox experimentation . . . Clearly it must be possible to combine the American insistence upon precise procedures with the European tendency first of all to get a good view of the phenomena which are to be investigated with so much precision.'[2]

There were other reasons for a cautious attitude towards Koehler's ideas. One has been alluded to throughout this chapter, namely the suspicion that reference to 'Gestalten' and 'field theories' might be a loose cloak for vitalism. Clark Hull reacted so negatively to lectures given by Koehler's comrade-in-arms, Koffka, that he dedicated himself to developing explanations for Gestalt phenomena that were to be based solely on principles of conditioning.[3]

Another reason has been referred to less often. This was the hereditarian bias that Gestalt theory appeared to display. Any claim regarding the superiority of ape intelligence to that of other species implied that this was inherited via the ape's genes. In an unreflecting way, this was seen as related to the claim that one branch of the human race is innately more intelligent than another. Koehler's work happened to become well-known in America when opinion was sharply shifting from the attitude that such a claim was obviously true towards the belief that it was just as obviously false. Some of the reasons why this shift occurred are discussed in the following chapter.

Before concluding it is of interest to note that both Koehler and Yerkes provide further examples, adding to those of Thorndike and Watson, of men who became highly influential in the professional world of psychology – Koehler as professor in Berlin and Yerkes first in Washington and then as a professor at Yale – and whose reputation in psychology was first established by their work with animals.

8
Nature and nurture

'Of all vulgar modes of escaping from the consideration of the effect of social and moral influences on the human mind, the most vulgar is that of attributing the diversity of conduct and character to inherent natural differences.'

John Stuart Mill: *Principles of Political Economy* (1848)

During the nineteenth century there was considerable confusion concerning the concept of instinct. The term was used a great deal by biologists and psychologists as if it had some precise, technical meaning. Yet what one writer meant when talking of a particular instinct often had very little in common with the meaning attached to such phrases by someone else.

One of the older sources of ideas on the subject was the extensive discussion of instinct provided by an otherwise obscure professor of philosophy from Hamburg, named Herman Reimarus, who believed that instincts were to be understood in terms of the purposes that they appeared to subserve. Thus, a 'maternal instinct' did not refer to some specific reaction that an animal might display towards its eggs or young, but to some general tendency to act in various ways so as to promote the welfare of its offspring; such an inner tendency was supposed to be non-physical and to operate without the animal's awareness. Reimarus attempted to categorize such tendencies and, in doing so, produced the first systematic list of instincts, which totalled forty-seven distinct types.[1]

In Lamarck's theory of evolution a central place was given to a concept of instinct similar to that of Reimarus. Many later evolutionists agreed with Lamarck on the inheritance of tendencies to behave in certain ways, accepting the view that the innateness or ready acquisition of an action by some living creature must be the result of arduous practice on the part of its ancestors. However, they did not necessarily also accept other aspects of Lamarck's theory.

A notable example was Herbert Spencer, who believed in Lamarckian inheritance as the principal explanation for the origin of instincts, but emphatically did not share Lamarck's view of what constituted an instinct. For Spencer an instinct was simply an innate reaction pattern more complicated than a reflex, but best understood as a chain of reflex actions or a 'compound reflex'. His friend, George Lewes, also criticized the view that instinctive action is guided by some central, immaterial purpose, but accepted Lamarck's view on inheritance. Lewes popularized his theory under the label 'lapsed intelligence': instincts were actions which had initially involved some perception of their consequences or were in some other sense rational, which had then become habitual and which over a number of generations had become to some degree innate.

Charles Darwin's views on the matter at first marked a complete break with earlier thinking. One of the most revolutionary ideas in Darwin's *Origin of Species* was that natural selection acts upon patterns of behaviour as well as upon bodily organs. He pointed out that various complex, yet completely innate, reactions found in certain species of social insect, such as bees and ants, could not be the product of Lamarckian inheritance, since they were displayed by neuter individuals without offspring. Like Spencer, Darwin believed that such reactions were different only in level of complexity from simple reflexes.

The simple contrast between Darwinian and Lamarckian theories of evolution became obscured in the 1870s and 1880s as first Darwin and then his successors, like George Romanes, became convinced of the importance of Lamarckian inheritance. Romanes came to conclude that there were two kinds of instincts: primary ones, produced directly by natural selection, and secondary ones, corresponding to Lewes' 'lapsed intelligence'. At the same time he rejected Spencer's analysis of instincts as compound reflexes and felt sympathy for Reimarus' original view that an instinct is a *set* of activities and reactions working towards some common goal.[2]

The variety of views held by instinct theorists a hundred years ago is still reflected in colloquial English. An instinct can refer to some set of related activities, as in describing sex or aggression as instincts; to some specific, innate reaction, as in referring to an animal's tendency to right itself when first falling from a height or to swim when first immersed in water; or to some reaction so well practised that it has become automatic, as in claiming that the driver of a car 'instinctively jammed on the brakes'.

This last kind of usage, meaning a well-learned habit that can occur without conscious reflection, was until about 1890 essentially the only way in which the term was applied to human behaviour. During the last century most of the arguments over instinct were developed in the context of non-human animals, for traditionally their behaviour was thought to be guided largely by instinct, while man alone was guided primarily by reason. 'Nothing is commoner', wrote William James in 1890, 'than the remark that man differs from lower creatures by the almost total absence of instincts.'[3] In characteristic fashion James then tried to persuade his readers that the received view was entirely wrong, and he did so by exploiting in a seemingly deliberate way the ambiguity surrounding the meaning of instinct. He concluded that a higher animal 'appears to lead a life of hesitation and choice, an intellectual life; not, however, because he has no instincts – rather because he has so many that they block each other's path'.[4]

Five years later Lloyd Morgan summed up the by now total confusion surrounding the subject. 'Instinctive activities are unconscious (Claus), non-mental (Calderwood), incipiently conscious (Spencer), distinguished by the presence of consciousness (Romanes), accompanied by emotions in the mind (Wundt), involve connate ideas and inherited knowledge (Spalding); synonymous with impulsive activities (James), to be distinguished from those involving impulse proper (Hoeffding, Marshall); not yet voluntary (Spencer), no longer voluntary (Lewes), never involuntary (Wundt); due to natural selection only (Weismann), to lapsed intelligence (Lewes, Schneider, Wundt), to both (Darwin, Romanes); to be distinguished from individually-acquired habits (Darwin, Romanes, Sully and others), inclusive thereof (Wundt); at a minimum in man (Darwin, Romanes), at a maximum in man (James); essentially congenital (Romanes), inclusive of individually-acquired modifications through intelligence (Darwin, Romanes, Wallace).'[5]

There are plenty of words which continue to be used in various and often contradictory ways without ever upsetting anyone very much. In the case of instinct, however, the rapid shift of ideas on inheritance occurring around the turn of the century meant that biologists and psychologists either had to agree on a more consistent definition of instinct or discard the term altogether from their technical vocabulary.

The crucial changes in views on heredity consisted of the rapid acceptance of Weismann's arguments against Lamarckian inheritance and the subsequent rediscovery of Mendel's Laws, which together laid the foundations of modern genetics. This had two major consequences for psychology: one was the need to clarify the distinction between the learned and innate behaviour of animals, and Morgan was the first to show how this might be done;[6] the other was that it brought to the forefront questions concerning the inheritance of human mental abilities.

For at least two centuries beliefs about the extent to which man is bound by his nature have swung slowly to and fro. During certain eras, often ones of economic expansion followed by a rapid social or political change, there has been widespread agreement that a person's ability and temperament are largely fixed by the way he or she has been nurtured and by the more immediate influences of the physical and social surroundings of everyday life. Such an attitude was common in the latter part of the eighteenth century, particularly in France, and also in the early Victorian England of John Stuart Mill and Alexander Bain.

In other periods an individual's potential has been viewed as severely constrained by a set of factors called 'human nature' and differences between individuals have been seen as largely a result of differences existing at birth. When the pendulum is passing through this latter position it has been common to assume that variations in characteristic patterns of action across cultures, classes or sexes reflect inherent constitutional variation. At such times it has seemed natural for a man to do certain things or pursue certain interests, but unnatural for a woman to do so; obvious that one social class produces born leaders, while another consists of people capable of only manual work; and beyond question that some races show an innate sense of rhythm and others inherit a talent for shopkeeping.

Such beliefs provided a convenient ideology for justifying, and administering, the vast colonial empires that many European nations had acquired during the nineteenth century. They became particularly strong in Great Britain, perhaps because the whole colonial experience loomed so large in British life and possibly also because of the development in British India of rigid, detailed class distinctions that mirrored the intricate caste system of the native culture. But whatever the reasons were, from around 1885 nativistic beliefs became more pronounced in Western Europe and, to a lesser extent, in North America than they had been for centuries. When men of outstanding intellectual reputation and moral authority, like James, concluded that man, after all, possessed a great many inherited tendencies to act in certain ways or, like Francis Galton, stated that human achievements were mainly the result of good selective breeding, the stamp of scientific authority, with all the prestige that this now enjoyed, was put upon the hereditarian outlook.

General views of this kind provided the context for two of the most lively developments in British psychology during the decade or so before the First World War. One was the invention by men such as Karl Pearson and Charles Spearman of statistical techniques specifically designed to analyse the inheritance of mental abilities. The other started by taking seriously James' claim that there are a great many human instincts and tried to understand their properties by comparing them to the instincts displayed by other species. The most persuasive advocate of this approach was William McDougall (1871–1938). He decided that Darwin's successors in comparative psychology had placed far too much weight on studying the relative *intellectual* abilities of different species; in doing so they had neglected Darwin's discussion of the links between emotional states in different species, including man.[7] McDougall believed that the most important contribution the study of animals could make to human psychology was to provide an understanding of human emotion and the springs of human conduct. It was largely due to McDougall and to his compatriot, Wilfrid Trotter (1872–1939), that during the early years of this century psychologists and sociologists on both sides of the Atlantic took a great deal of interest in the question of instinct. Subsequently, when a reaction to instinct-based theories of human behaviour gathered momentum in America around 1920, McDougall's work became a principal focus for attacks on the hereditarian outlook in psychology.

William McDougall, Wilfrid Trotter and human instincts

McDougall was born in 1871 to a wealthy family in the North of England. He spent most of his boyhood in a suburb of Manchester and showed precocious intellectual ability. Many aspects of his later career seem to have been guided by the motive of maintaining the childhood conviction that he was cleverer than anyone else. Since his father had a northern industrialist's suspicion of Oxford and Cambridge, and since these institutions still did not offer the scientific education that McDougall's reading of Huxley, Spencer and Darwin impelled him to obtain, he studied science, concentrating on geology, at the local University of Manchester. He began this at the uncommonly early age of fifteen, so that four years later he obtained a first-class degree at a time in life when most boys of his background were about to leave school.

By this time McDougall had developed an interest in physiology and this subject had become well-established at Cambridge. Since there was no urgent need to earn a living, he now took a further four-year course of study in Cambridge which led to another first-class degree in 1894. This was followed by three years of medical studies at a London hospital; he saw this training less as a preparation for a career in medicine than as a useful step towards his goal of election to a Cambridge fellowship. It was at this stage that the future course of his career was decided when he read William James' *Principles of Psychology*.

In 1898 contacts with people in Cambridge led to an invitation to join a notable anthropological expedition to the Torres Straits in Indonesia. His main duty on the expedition was to assess the sensory abilities of the natives of this area. Afterwards he spent some months studying the lives and customs of various primitive tribes in nearby Borneo. On returning to Europe his education finally ended with a year spent in Germany, learning about experimental methodology in the psychological laboratory of G. E. Mueller in Goettingen.

McDougall's fourteen years of training left him ideally equipped for a profession which hardly existed in England. The best appointment he could find was a part-time teaching post at University College, London where he was responsible for the psychological laboratory and where he carried out some experiments on visual perception. This research was inspired by his violent antipathy to currently prevailing theories, an attitude that was to characterize his later work in other areas of psychology. He spent four years in London and then in 1903 was appointed to the Wilde Readership in Mental Philosophy at Oxford. Since the terms of this appointment were designed to deter the holder from attempting to engage in any empirical work, McDougall had to carry out research in a room loaned to him by the physiological department.

The ideas that brought McDougall fame did not arise from his research on vision nor from other work

Fig. 8.1. William McDougall as a student at Cambridge University

in physiological psychology, but from the lecture course he was required to give at Oxford. 'Lecturing one day in 1906, I found myself making the sweeping assertion that the energy displayed in every human activity might in principle be traced back to some inborn disposition or instinct. When I returned home I reflected that this was a very sweeping generalization, one not to be found in any of the books; and that, if it was true, it was very important. I set to work to apply the principle in detail, becoming more and more convinced both of its truth and of its importance; and my *Social Psychology* emerged.'[1]

This book was published in 1908 bearing as the full, and somewhat misleading, title, *An Introduction to Social Psychology* – misleading, because it had little to say about social psychology as usually conceived either then or now, but rather was intended as a preliminary volume to a series of books on the topic, of which only one was ever written. The main concern of the book was described by McDougall as being with 'the department of psychology that is of primary importance for the social sciences, that which deals with the springs of human action, the impulses and motives that sustain mental and bodily activity and regulate conduct; and this, of all the departments of psychology, is in the most backward state, in which the greatest obscurity, vagueness, and confusion still reign'.[2]

McDougall's major claim was stated in the following way. 'The human mind has certain innate or inherited tendencies which are the essential springs or motive powers of all thought and action . . . These primary innate tendencies have different relative strengths in the native constitutions of the individuals of different races, and they are favoured or checked in very different degrees by the very different social circumstances of men in different stages of culture; but they are probably common to the men of every race and of every age.'[3] He argued that such tendencies can be divided into two classes: 'the specific tendencies' or instincts, and the 'general or non-specific tendencies'.

The early and most influential part of the book is devoted to establishing a general definition of instinct and to deciding what are the basic human instincts. McDougall regarded Spencer's definition and Morgan's subsequent modification of this as inadequate because they concentrated on only one aspect of an instinct. 'Instincts are more than just innate tendencies or dispositions to certain kinds of movement' because they also involve distinctive emotions and distinctive ways of perceiving the world. Thus, to describe a salmon guided by instinct to cross thousands of miles of ocean and battle its way up some mountain river, persistent in its efforts to surmount the many waterfalls and rapids on the way to the remote tributary where its eggs are laid, as simply displaying a series of reflexive movements was plainly not enough.[4] McDougall believed that each basic instinct contained three essential elements: a specific cognitive set, a specific emotion and a 'conative' aspect, a striving towards some specific end.

He defined an instinct 'as an inherited or innate psychophysical disposition which determines its possessor to perceive, and to pay attention to, objects of a certain class, to experience an emotional excitement of a particular quality upon perceiving such an object, and to act in regard to it in a particular manner, or at least, to experience an impulse to such action'.[5] Many earlier definitions had insisted that instinctive actions are necessarily performed in an unconscious manner or that they are little influenced by past experience, but neither of these factors was important in McDougall's concept of instinct. He was content to allow animals, as well as people, to be aware of their instinctive actions and to modify them to a great degree. To take the example discussed by Wallace, the way that a bird builds a nest may well be greatly influenced by the way it saw nests built when it was a fledgling, may well improve hugely with practice and may well show great flexibility according to the situation and available material; nevertheless, according to McDougall, the bird would still be displaying a nest-building instinct.

He was well-aware of pitfalls in the way of understanding human instincts. Even in the case of animals he regarded earlier attempts to analyse instincts as completely useless: 'attribution of the actions of animals to instinct . . . was a striking example of the power of a word to cloak our ignorance and to hide it even from ourselves.'[6] And as for human behaviour, 'lightly to postulate an indefinite number and variety of human instincts is a cheap and easy way to solve psychological problems and is an error hardly less serious and less common than the opposite error of ignoring all the instincts'.[7] Accordingly, McDougall did not accept as true instincts many that were suggested by his contemporaries. For example, he denied that people are endowed with a religious instinct or an instinct for rivalry. Similarly, he decided that it was misleading to label imitation or play as instincts; these should be regarded as general tendencies.

The major difference between man and other species is that 'man has an indefinitely greater power of learning, or profiting by experience, of acquiring new modes of reaction and adjustment to an immense variety of situations'.[8] This means that in man innnate tendencies are deeply overlaid and it accounts for the unsatisfactory nature of previous discussions of human instinct. From a psychological point of view the greatest value to be gained from studying animal behaviour stems from the possibility of more easily discerning the basic instincts, since in animals they are far less distorted by learning. The other way of approaching the problem of human instincts and their emotional components, one that is complementary to animal psychology, is to study psychopathology where, as it were, instincts rise to the surface in exaggerated form.

This is how McDougall explained his approach to the subject. In fact *Social Psychology* does not contain a single example of working systematically from sets of studies of animal behaviour to the analysis of a particular human instinct. Instead its list of primary instincts contains as high a ratio of speculation to specific evidence as any of its predecessors. The only starts in the direction that McDougall claimed to be taking occur when apparently comparable behaviour in animals is briefly described. There is no sign of questioning whether he might be selectively searching for examples to confirm beliefs already firmly held; or whether descriptions of animal behaviour might be distorted by the observer's preconceptions about human psychology. He knew of the work of Morgan and of Hobhouse, but did not absorb their critical rigour. And in what McDougall has to say about animals there is nothing of the shrewdness and the direct impact of first-hand observation that Morgan and Hobhouse both convey.

A peculiarity of McDougall's treatment of instinct was that he kept to his three-fold definition in a rigid manner so that sometimes what might on other grounds be regarded as a prime candidate for the status of a basic instinct was rejected because it was not associated with a distinctive emotion. He described the following seven primary instincts and their associated emotional quality: flight and fear; repulsion and disgust; curiosity and wonder; pugnacity and anger; self-abasement and subjection; self-assertion and elation; and finally, the parental instinct and the 'tender emotion'. He also mentioned other instincts in which the emotional tendency was less well-defined, such as the gregarious and acquisitive instincts, and other emotions compounded from primary ones, such as sexual jealousy and female coyness.

A notable feature of the early editions of *Social Psychology* is that sex is itself coyly referred to as the 'instinct of reproduction' and receives only a passing mention. As McDougall later admitted, this was odd in that 'consideration of sexual experience and conduct affords the clearest illustration and the most obvious support' for his theory.[9] Indeed, it arguably provides the best model for a general treatment of human instincts along the lines proposed by McDougall. The main reason for omitting any extensive discussion of sex was because McDougall felt this might be offensive to the general reader for whom the book was intended. It is a good example of the 'rosewater' attitude of psychology in that era, particularly since McDougall was in many respects an unusually courageous and unconventional person.

The extent of Freud's subsequent influence and of a rapid shift in convention is illustrated by the fact that in 1914, six years after *Social Psychology* was first published, McDougall added a lengthy and relatively frank discussion of sex. Nevertheless, despite his early and unfulfilled promise to bring evidence from psychopathology to bear on the question of human instincts and despite a close interest in psychoanalysis, McDougall resisted the general conclusions Freud drew from the analysis of his patients. 'I incline strongly to the view that they (the psychoanalysts) have extended to normal individuals generalizations which are true only of a certain number of persons of somewhat abnormal constitution.'[10] He fiercely rejected the possibility of childhood sexuality, noting that infants who masturbate may 'belong to the minority of abnormal innate constitution'.[11]

McDougall's interest in Freudian theory was one of many things that he held in common with an English contemporary, Wilfrid Trotter, who also wrote a popular book about human instincts. In 1908 McDougall had argued that the study of animal behaviour was one of the most fruitful ways of gaining an understanding of human motives, which in turn was needed in order to understand people's behaviour in any social setting. In the same year Trotter made a similar claim, except that he appealed to comparative psychology to concentrate on species of animals that displayed some degree of social organization.

Trotter obtained his medical degree from University College in London in 1897, the same year and place as McDougall. However, he did not have McDougall's financial resources and so, despite an equally deep interest in psychology, he continued with medicine and remained associated with University College for the rest of his working life. His medical career was highly successful and reached its peak in 1928 when he performed an operation which saved the life of King George V. Well before this he had become famous for his views on human social behaviour.[12]

Fig. 8.2. Wilfrid Trotter at about the time he first wrote on the herd instinct

Trotter's first contribution to the discussion of human instincts claimed that there have been two sudden and striking developments in the course of evolution that overshadow all others: one was the emergence of multi-cellular from uni-cellular organisms and the second, and in some respects parallel, advance was the emergence of social animals. He was impressed by the high degree of social organization displayed by certain insects, whereby a hive of bees or a colony of ants seem often to act as a single organism, their individual members having very specialized functions and being incapable of survival on their own. In contrast, among mammals social organization is rare and, where it exists, as in wolves or sheep, it is of a relatively simple kind, the one great exception being man. For Trotter the extraordinary aspect of man that marks him off more than his adaptability or intelligence from all other mammals is his degree of gregariousness, which exceeds even that of the social insects. Two generations later he might have argued that reaching the moon was more a tribute to the complexity of collaborative effort that human beings can achieve than to the intellectual capacity that made specific technical achievements possible.

Trotter felt that, although it might well be satisfactory to consider the behaviour of most species in terms of instincts serving self-preservation, nutrition and sex, this was clearly inadequate for any gregarious animal where such instincts are often effectively checked by social pressures; to condemn a person to solitary confinement can be a harsher punishment than curtailing the food he receives or inflicting physical pain. Nineteenth-century evolutionists like Huxley, Spencer or Haeckel, who had found it so difficult to include human morality within their biological theories, made the mistake of concentrating entirely on the individual. Or, at least, this was Trotter's opinion and he singled out Karl Pearson as the one theorist who had fully appreciated the way in which the existence of a social organization modifies natural selection: 'the so-called ethical process, the appearance, that is to say, of altruism, is to be regarded as a directly instinctive product of gregariousness, and as natural, therefore, as any other instinct'.[13]

Such a quotation makes Trotter sound as if he anticipated the recent development of sociobiology, but he provided no developed examples to back up his claim that the study of social animals is crucial to an understanding of man. The occasional references to the bee, the wolf or the sheep were, if anything, even vaguer than the passing nods to natural history provided by McDougall. The main reason why Trotter is interesting here is that, like McDougall, he believed that the prevailing introspective psychology was sterile, that it had nothing to offer social psychology or sociology and that what was needed was an application of the objective approach that animal psychologists had begun to develop. In less forthright terms than those used later by Watson he described 'the sense of unimaginable complexity and variability of human affairs', which arises when we furnish explanations of our own conduct, as a kind of anthropomorphism. He contended that 'a reaction against this in human psychology is no less necessary therefore than was in comparative psychology the similar movements the extreme developments of which are associated with the names of Bethe, Beer, Uexkuell and Nuel . . . it is this anthropomorphism in the general attitude of psychologists which, by disguising the observable uniformities of human conduct, has rendered so slow the establishment of a really practical psychology.'[14]

Trotter's ideas became widely known when he combined two earlier papers with new material in a book of 1916 entitled *Instincts of the Herd in Peace and War*. This included an extended discussion of Freudian theory; Trotter was one of the first British scientists to attend a meeting of the psychoanalytic society and his close friend and brother-in-law, Ernest Jones, became the leading figure in the psychoanalytic movement in England as well as Freud's biographer. Trotter was far more sympathetic to Freudian theory than McDougall, but nevertheless worried that no biological context was provided: 'it seems to feel no need of bringing its principles into relation with what little is known of the mental disabilities of the non-human animals'.[15] He argued that consideration of the social instincts provided this necessary context.

Although the treatment of psychoanalysis no doubt added to the book's appeal, what made it notorious and of huge interest to the British wartime reader was its description of national differences. Among the varieties of social organization Germany, it claimed, displays the aggressive characteristics of the wolf-pack, while English society shows the advanced gregariousness of the bee. For example, he maintained that among Germans there was 'deliberate cultivation by superiors of a domineering harshness towards their inferiors, of habitual cruelty towards animals, and indeed of the conscious deliberate encouragement of harshness and hardness of manner and feelings as laudable evidence of virility'. Even the German musical tradition and love of singing were cited in support of the argument, since 'the wolf, then, is the father of the war song, and it is among peoples of the lupine type alone that the war song is used with real seriousness'.[16] In a hive 'decisions of policy of the greatest moment appear, as far as we can detect, to arise spontaneously among the workers, and whether the future is to prove them right or wrong, are carried out without protest or disagreement' and English society is marked by a similar lack of central direction; he wrote to his fellow countrymen of 'the deep, still spirit of the hive that whispers unrecognized in us all'.[17]

Trotter claimed that this nonsense was serious analysis and it followed that, like a dog or wolf, Germany needed to be given a sound thrashing. When the war was over he admitted that some of this was prejudice, but excusable because of the time it was written. By then the harm had been done; to invoke scientific theory in support of chauvinism is in the long run as harmful to the theory as appeals to 'God on our side' are to religious belief.

Trotter wrote no more on the subject and he was soon forgotten. Although his contribution was a minor one, his book was important at a critical time for spreading very widely the opinion that an objective psychology, which was based on the study of animal behaviour, avoided introspection and was allied to Freudian theory, would help cure the ills of society.

Like Trotter, McDougall was much affected by the war, but his influence on psychology continued long after it was over. Following the German invasion of Belgium in 1914, McDougall immediately volunteered as a private in the French army. Some months later he transferred to the medical corps of the British army and for the first time in his life became deeply involved in applying his knowledge of psychology, in particular to the treatment of victims of shell-shock. This experience increased his interest in psychoanalysis and, when the war was over, he visited Jung to submit himself to analysis at the hands of an expert. He came away 'enlightened but not convinced'. On returning to his academic work at Oxford an invitation arrived from Harvard to take the vacant chair of psychology. Although there was more interest in psychology at Oxford than before the war, the low esteem in which the subject was held there and at other British universities was still discouraging. McDougall accepted the invitation and remained in America for the rest of his life.

In the few years before his departure many of what had been tentative opinions became more confident and extreme, and in a direction that was usually opposed to the general trend of the time. If self-assertion was the primary instinct behind his career, then the second was certainly pugnacity. 'Whenever I have found a theory widely accepted in the scientific world, and especially when it has acquired something of the nature of a popular dogma among scientists, I have found myself repelled into skepticism'.[18] He began to take a sympathetic interest in Lamarckian inheritance where earlier he had decided that it was probably correct to reject it. In his *Social Psychology* of 1908 he was less dismissive of the mental abilities of non-Europeans than many of his generation, but by the time he left for America his opinions had become much more racist.

There were many members of the psychological community in America who in 1920 shared McDougall's interest in Lamarckism and his beliefs about racial and sexual differences. Amplified by his haughty, unfriendly manner, what distanced him from many of his new colleagues was what they saw as his abandonment of a scientific outlook. Although eight years earlier he had defined psychology as the science of behaviour, this did not endear him to the early American behaviourists, since what McDougall meant by 'behaviour' differed from Watson's meaning. McDougall believed that behaviour had to be defined in purposive terms and that the actions of a living organism are inherently different from the movements displayed by any kind of physical object. For example, he compared the movements of a billiard ball with those of a timid guinea pig, frightened by being removed from its hole and persistently trying to return there by one way or another, despite obstacles placed in its way, until it succeeds or its energy is exhausted.[19] 'When we survey the whole world of material things accessible to our perception these are seen . . . to fall into two great classes, namely (1) a class consisting of those things whose changes seem to be purely physical happenings, explicable by mechanical principles; and (2) a class of things whose changes exhibit the marks of behaviour and seem to be incapable of mechanical explanation, but rather to be always directed, however vaguely, towards an end – that is to say, are teleological or purposive.'[20]

In 1912 he had admitted it just possible that in the distant future science might succeed in 'establishing that all seemingly purposeful action is mechanically explicable', but by the time he left for Harvard he was convinced that behaviour is impelled by a special kind of non-physical force which he later called 'hormic energy'. His outlook had become identical with that of Hans Driesch, whom McDougall never directly cited, but who nonetheless was the main inspiration for his now unabashed vitalism.

McDougall's *Social Psychology* had been well-received by American psychologists as well as by the general public; by 1920 the book had gone into its fifteenth edition. But he seemed like a figure from an earlier era when he arrived in America, a speculative philosopher with no solid background of experimental papers on the topics on which he was supposed to be expert; although this was more a reflection on the state of British psychology than on McDougall. 'In America I was known as a writer who had flourished in the later middle ages and had written out a list of alleged instincts of the human species.'[21]

Once settled, McDougall pursued new interests in a way that might have been designed to provoke further antagonism in view of his influential and prestigious position as professor of psychology at Harvard. As one historian put it, 'his name became almost synonymous with theories and practices regarded by most American psychologists as remnants of exploded but still dangerous superstitions – with animism, vitalism and teleology; with nativism in the discredited form of Lamarckism; and with shady ventures into psychical research and extrasensory perception . . . Perhaps more than any other individual, McDougall became a symbol of what American psychology has most heartily set itself against'.[22] The result was that even to take an interest in purposive behaviour, instincts, and the very real problems of motivation that McDougall attempted to grapple with, came to be viewed with great suspicion.

The Psychological Darwinism of race and instinct

The discussion of instinct just prior to the First World War by McDougall, Trotter and other writers was just one aspect of a general approach for which the term 'Psychological Darwinism' provides a useful label, even though it contained many elements which Darwin himself would have rejected. Other important aspects included the belief that there are several pure 'types' of human being possessing well-defined physical characteristics and 'neophrenology', the belief that there are significant differences in brain structure that give rise directly to psychological difference between individuals.

Such beliefs were well-entrenched in Europe, and

not just among British scientists. On the continent Ernst Haeckel continued as an influential figure right up until the war, using his immense authority as the leading evolutionist in German-speaking countries to promote views on inherent racial differences. Haeckel endorsed the anti-semitism that erupted in late nineteenth-century Germany, proposing, for example, that Christ's merits stemmed from the fact that he was only half-Jewish, since his true father was a Roman officer who had seduced Mary.[1]

Haeckel also associated evolutionary theory with specific social and political views. His professional career coincided with the growth of the Second Reich and his lifetime hero was Bismarck, the chief architect of the German Empire. Haeckel argued that Germany's expansionary policies and its autocratic form of government could be justified on Darwinian grounds. With the outbreak of war Haeckel, like many other German scientists, felt betrayed by the British; referring to Germany's 'treacherous, murderous English brother', he stated that England had polluted Europe by bringing into battle on her side 'the inferior races of the Empire' and by allowing them to fraternize with the racially superior Europeans.[2] Such sentiments led Trotter to reply in kind. Across the Channel it was solemnly reported to the French Association for the Advancement of Science in 1917 that 'the German race suffers from polychesia (excessive defecation) and bromidrosis (body odour) . . . Their advantage to the enemy in wartime is that they serve to detect infallibly the spies of the German race masquerading in France as Alsatians'.[3]

The reason for mentioning such jingoistic outbursts here is that they may help to account for the fierceness of the criticism to which Psychological Darwinism was eventually subjected on the other side of the Atlantic. Underlying the post-war debates in America over issues of instinct and race, and often not far from the surface, was the mood of rejecting an alien ideology, one that is irretrievably embedded in the mesh of feuding nationalism, undemocratic politics and antiquated social structures of Europe.

The American attack on Psychological Darwinism came from two directions. One was the experimental tradition of rejecting speculative theory and concepts not based on hard evidence; as discussed in earlier chapters, such an approach had already discredited Lamarckian inheritance and also Haeckel's key 'biogenetic law', the principle of recapitulation of phylogeny by ontogeny. Around 1920 the concept of instinct was subjected to similarly critical examination. The second direction of attack was from the social sciences, and formed part of their struggle to assert the independence of their subject matter from the 'biologizing' approach and evolutionary framework that founding fathers such as Herbert Spencer had imposed. One focus for this part of the attack was the question of race and its main theme was the cultural relativity of social and psychological norms. The debates over race and instinct, and the related controversy over mental testing, had a profound effect on American psychology, leading to the dominance of an environmentalist ideology and encouraging a subsequent emphasis on the study of learning.

In the early years of the century the attitudes of American scientists towards racial and ethnic differences in general resembled those common in British or German science. Wide concern over the tide of immigration from southern and eastern Europe led to frequent cries of alarm from experts who discussed the dilution of American genetic stock, using phrasing as racist as any to be found in a book by Galton or Haeckel. But at the same time there were developments beginning within anthropology that eventually led to the overthrow of such beliefs, to the extent that by 1925 few established psychologists or social scientists would express in public views on race that had been held as 'scientific' facts less than ten years earlier.

The major figure in this development was the anthropologist, Franz Boas. In 1883, as a twenty-five year old student at Berlin, Boas had gone to live among the Eskimos of Baffinland and from this experience became increasingly sensitive to the relativity and arbitrariness of human customs. This attitude was strengthened by subsequent expeditions to study the Indians of British Columbia. In 1889 he emigrated from Germany to take up a teaching post at Clark University where, partly stimulated by G. Stanley Hall and by a close friendship with Henry Donaldson, he became heavily involved in the then popular activity of amassing measurements of various bodily dimensions. When physiological psychologists working at the turn of the century, like Henry Donaldson, began to measure the head, brain and spinal cord of rats and when the occasional student they supervised, like John Watson, tried to relate those measurements to an animal's intelligence, they were implementing in experiments with animals what physical anthropologists were attempting in surveys of human populations.[4]

From 1880 until the end of the century a vast amount of data on human head shapes and sizes had been gathered, largely for the purpose of clarifying differences between human races. Boas' concern was different in that, instead of the usual attempt to examine what were hoped to be stable characteristics, he studied changes in bodily form with the aim of discovering what conditions modify the effects of inherited dispositions. An early expression of this interest was an extensive programme started in 1891 for obtaining various measures of body size from schoolchildren in Worcester. There was an uproar in the local press and it became one of a number of incidents that led the citizens of Worcester to view Clark University with as much suspicion as those of Baltimore viewed Johns Hopkins. The positive outcome of this survey was that it yielded the first substantial body of results on human physical growth.

In 1908 Boas' skills as a physical anthropologist were deployed in a study supported by the U.S. Immigration Commission, whose purpose was, for Boas, to study the effect of 'change of environment upon the physical characteristics of man' and, for the Commission, to find out 'whether the marvellous power of amalgamation which has worked so well in assimilating immigrants from north-western Europe would continue to operate on the more remote types entering the country from southern and eastern Europe.'[5] In measuring children of recent immigrants Boas came upon some 'very striking and wholly unexpected results', which showed very rapid changes in various body measurements, including, most critically, those used to classify headform which had been regarded as one of the most stable features for use in distinguishing various races. These changes were most highly related to the time elapsing between the arrival of the mother in the United States and the birth of the person who had been measured.

Like most of his contemporaries Boas accepted Lamarckian inheritance for much of his career, but he had been gradually persuaded to adopt a Mendelian view of genetics by 1911, when his ideas on race reached a wide public through his book, *The Mind of Primitive Man*.[6] In the context of the new theories of heredity one conclusion that could be drawn from his research was that, as environmental conditions become equal, the physical differences between human groups become minimal, even when for countless generations the groups have experienced highly contrasting ways of life.

There were other arguments against the idea of fixed racial types. A linear ordering of physical characteristics from ape to northern European was repeatedly described by the Psychological Darwinians and often represented in textbook illustrations drawn so as to suggest a strong correlation with intelligence; one such example from Haeckel is shown in Figure 8.3. Boas pointed out that such scales depended on a completely arbitrary choice of measures. Thus, in terms of hairiness or the proportion of his limbs the European tended to be the most 'animal-like' of all human types. Also, the belief that there was a clear separation between distinct races was shown to be at variance with a huge degree of overlap between the distributions of any bodily dimension that one might choose.

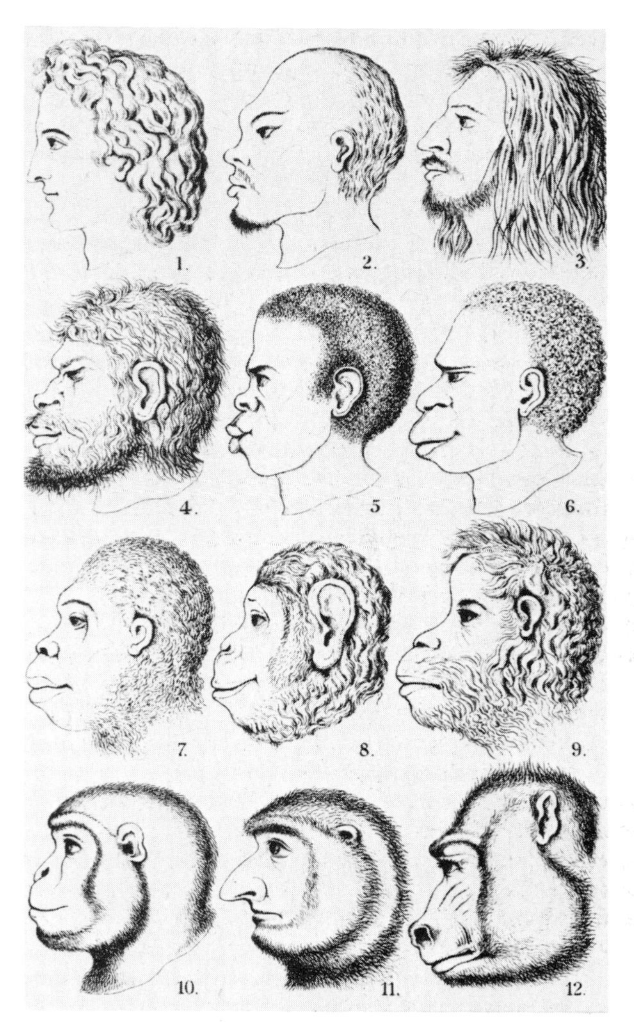

Fig. 8.3. Heads of different types of primate in ranked order from the first book on human evolution by Ernst Haeckel (1868)

Possibly the most important argument concerned the issue of interbreeding. An obvious objection to the claim that there are human races representing distinct biological species is the absence of any inherent barrier in the way of sex between members of different races. One standard rejoinder to this objection was that children of mixed parentage are 'degenerate' and less fertile. Boas found no evidence for this when he studied 'half-blood Indians' of mixed European and North-American Indian origin. If anything, the average half-blood woman bore more children than a mother of 'pure Indian stock' and these children were taller.

Such discoveries led Boas to a position of considerable scepticism towards any claim about inherent racial differences. By the beginning of the war some of the younger anthropologists and sociologists influenced by Boas' work went beyond this to belief in complete equality of intellectual or cultural potential across different races. This was the view, for example, of William Thomas, whose study of changes occurring in traditional customs among immigrants from Poland paralleled the study of physical changes carried out by Boas.[7] It was some years before any psychologist publicly endorsed an equally environmentalist outlook, even though many had close personal ties with the anthropologists and were familiar with recent work in the field. Thus, Watson later wrote that the book that had influenced his whole outlook more than any other was William Sumner's *Folkways* of 1906, which described the overwhelming variety and complexity of human mores and customs, particularly those to do with marriage and child-raising.[8] Within anthropology the critical examination of entrenched hereditarian assumptions had confronted the central concept of race, but within psychology such questions were at first avoided because of the exciting possibilities offered by the development of mental testing.

One major effect of the war on American psychology was to reduce sharply the remaining feeling of deference towards the European traditions which had inspired its origins. A second effect served as an equal boost to American psychology's self-esteem; this stemmed from Robert Yerkes' success in mobilizing his country's psychological expertise to assist the army in selecting and training its personnel. From almost every American university, psychologists left to join the Division of Psychology, which to Yerkes' chagrin was designated part of the Sanitary, and not the Medical, Corps. Some were allocated to training or testing schemes related to specific military functions, but the great majority worked on the development and administration of the general Army tests.

It is doubtful whether the effectiveness of the American army was improved by this massive exercise. Testing was often haphazard and was frequently prevented or ignored by sceptical officers; one noted that he 'would be just as much helped by a board of art critics to advise me which of my men were most handsome, or a board of prelates to designate the true Christians'.[9] In any case, the Armistice was signed shortly after full-scale testing was under way. Even Yerkes admitted that it showed how efficiency *could* have been improved and millions of dollars *could* have been saved by using the test results to assign men to appropriate postings or to reject those deemed incapable of benefiting from the most simple training. The important thing for psychology was that nearly two million men had been tested and the potential of this new kind of psychology had received official recognition; 'if psychology had not in fact contributed significantly to the war, the war had contributed significantly to psychology', for, as Yerkes noted, 'wartime publicity accomplished what decades of academic research and teaching could not have equalled'.[10]

When the War was over and the lengthy, but eagerly anticipated, task began of analysing all these data, the theoretical assumptions made by Yerkes and his colleagues were strongly hereditarian. The leading members of the American eugenics movement had played a central role in developing the Army tests and this influence became very apparent in the interpretations given to some of the major findings emerging from the results. For example, the fact that black Americans from northern states gave average scores higher than those from southern states was seen as a result of selective migration northwards by the most able members of this population. Again, the marked positive correlation in the results from recent immigrants between high test scores and length of time spent in the United States was taken as proof that, as widely feared, the mental calibre of immigrants had been steadily declining.

The authors of the report do not seem to have entertained at all seriously the possibility that the tests might measure familiarity with mainstream American life. The types of question asked, the conditions under which the tests were given and the high incidence of zero scores indicating that many of the men were totally confused by the whole procedure, these and other considerations strongly suggest that the average mental scores assigned to different groups simply indicated the extent to which their members had become assimilated to the dominant culture.

However, even the idea of discussing *cultures* in the plural was still unusual. The word normally occurred only in the singular form to denote the accomplishments, interests and styles of elite groups within Western society. The new, technical use of 'culture' was being adopted by anthropologists and social scientists influenced by Boas, but was not yet widespread.

An attack on the hereditarian bias of the mental testing movement began by concentrating on its social and political implications. This had been heralded by an address to the American Psychological Association given in 1916 by John Dewey, who discussed in very general terms the connection between structural approaches in psychology and conservative attitudes towards social problems. 'The ultimate refuge of every standpatter in every field, education, religion, politics, industrial, and domestic life, has been the notion of an alleged fixed structure of mind. As long as mind is conceived as an antecedent and ready-made thing, institutions and customs may be regarded as its offspring. By its own nature the ready-made mind works to produce them as they have existed and now exist. There is no use in kicking against necessity.' He went on to welcome 'the advent of a type of psychology which builds frankly on the original activities of man and asks how these are altered, qualified and reorganized in consequence of their exercise in specifically different environments'.[11]

A more detailed version of Dewey's argument came later from outside the profession when in 1922 the journalist, Walter Lippmann, began a series of articles in the *New Republic*. He examined current claims on behalf of intelligence tests and expressed alarm that the promise of Binet's work was 'in danger of gross perversion by muddleheaded and prejudiced men'.[12] Lippmann was primarily concerned with education: 'the danger of intelligence tests is that in a wholesale system of education, the less sophisticated or the more prejudiced will stop when they have classified and forget that their duty is to educate. They will grade the retarded child instead of fighting the causes of his backwardness.'[13] He argued that correlations between social status and intelligence were 'hardly an argument for hereditary differences in the endowment of social classes. They are a rather strong argument for the traditional American theory that the public school is an agency for equalizing the opportunities of the privileged and the unprivileged.'[14] Lippmann's articles were well-informed, closely argued and – in contrast to Dewey's elaborate and difficult style – very forceful. They ended with the plea that 'psychologists will save themselves from the reproach of having opened up a new chance for quackery in a field where quacks breed like rabbits, and they will save themselves from the humiliation of having furnished doped evidence to the exponents of the New Snobbery'.[15]

Just at this time intelligence was also attracting wide interest in another context, the public debate over immigration that came to the fore in 1922 when a congressional committee began hearings in preparation for new legislation. There was a strong movement in favour of much more restrictive quotas drawn up on racist grounds. Many of the mental testers with an interest in eugenics gave public support. However, these psychologists were not so much leading lights as Johnnies-come-lately to a movement which had campaigned against further immigration from southern and eastern Europe long before anyone thought to cite eugenics arguments or intelligence test data in its support. There were a variety of banners to wave other than what one supporter of restricted immigration dismissed as 'highbrow Nordic superiority stuff'.[16]

Results from the Army testing programme were used on behalf of the campaign for selective immigration just as this campaign began to arouse deep revulsion among other American intellectuals. In 1924 the new and highly restrictive law was passed, but by then the kind of belief that had helped its passage was seen as disreputable among American scientists. Jacques Loeb's personal experience before the war of racial discrimination based on his Jewish background had long turned him into an enemy of racial theories.[17] By the mid-1920s he was joined in public by other pillars of the biological establishment, including his former adversary over the behaviour of lower organisms, Herbert Spencer Jennings, who was now one of the world's leading geneticists; Jennings attacked the eugenics movement for the way it distorted genetic theory in order to provide a spurious scientific basis for its claims.[18]

Within psychology many who had been involved in the eugenics movement or in mental measurement began to change their minds. Cattell, the father of mental testing in America, confessed that he now found 'the Army tests sadly inadequate as scientific measurements'.[19] In 1923 Carl Brigham published with Yerkes' encouragement a popular summary of the results of the Army programme that incorporated a highly racist interpretation under the title *A Study of American Intelligence*; in 1930 he declared that 'the study, with its entire hypothetical structure of racial differences, collapses completely'.[20] Unlike Brigham, Yerkes continued to believe in racial differences in mental ability, but he had been impressed by Lippmann's arguments concerning the problems of mental testing and anyhow considered that the climate of opinion had changed so drastically it was too dangerous a topic to study or discuss.[21]

The immigration debate was a turning point for American anthropology and sociology. Until then the dominance of the Psychological Darwinians in the social sciences had been prolonged by events connected with the war. Most of them were of British descent and Protestant tradition, were born in the North-East and associated with the Ivy League universities; in general they had favoured American entry into the war. Among the group of anthropologists associated with Boas and among the Chicago sociologists were many of Jewish or of other 'non-Aryan' tradition and they were strongly suspected of being at best neutral or, in some cases, pro-German in their politics. They were certainly not as conspicuous in their eagerness to rush in to help the war effort as Yerkes, Thorndike and their fellow psychologists had been. Bitterness created by wartime differences persisted for a few years and kept the Boasians at a distance from the establishment. But in the mid-twenties it was gradually ceded that biology did not provide the most appropriate training or framework for the study of society or of human customs. The long era in which evolutionary ideas dominated American social sciences ended at last. Terms like 'social heredity' faded away as others like 'cultural transmission' took their place.[22]

Some three years before Lippmann's critique of intelligence tests and the hearings on immigration, a debate erupted within psychology over the issue of instinct. Although this began by considering the logical validity of appeals to instinct, the relative importance of nature and nurture in human and animal behaviour quickly became one of the main issues examined by the stream of papers on instinct that were published in the early 1920s. As a result of this, the instinct debate was as important as the criticisms of racial theory and mental testing in accelerating the environmentalist trend of American psychology.

By the time the arguments began, McDougall's list of seven primary instincts in his *Social Psychology* of 1908 looked exceedingly modest in comparison with the suggestions of later instinct theorists. One particularly thorough review showed that, in the first twenty years of the century, four hundred authors of books or articles had proposed nearly six thousand classes of instinct encompassing over fourteen thousand individual cases. Among the miscellaneous examples discovered by this survey were: within the aesthetic group, the instinct of a girl to pat and arrange her hair; within the altruistic group, the desire to liberate the Christian subjects of the Sultan; as a social instinct, that of socialists towards international relationships; as an anti-social instinct, that of upper-class Mexicans to consider themselves a people apart from the lower classes; within the intellectual class, the politicians' unfailing instinct for exhausting every wrong device before trying the obviously right one; in the religious group, the English instinct to begloom Sunday; and as a self-abasement instinct, that of women that they all worship strength in whatever form, and seem to know it to be a child of heaven.[23]

The polemics were begun by Knight Dunlap, Watson's colleague at Johns Hopkins, who wrote a paper in 1919 with the provocative title, 'Are there any instincts?'[24] Dunlap was worried about the conceptual confusion surrounding the topic and proposed that two points should be made clear. One was the need to distinguish carefully between the term 'instinctive reaction', used at a physiological level to refer to an unlearned response, and the 'teleological' use of 'instinct' by theorists like McDougall to refer to classes of activities serving some common purpose. The second point made by Dunlap was that, at the teleological level, instincts should be regarded as no more than labels for convenient groupings of activities, serving to classify behaviour much as documents are sorted out in a filing system in whatever way is currently most useful. What Dunlap objected to was the tendency in social psychology to assume that, once an activity was described as an example of some instinct, it was possible to reach some conclusion about the nature of this activity; 'having posited a "pugnacious instinct" for example, one writer proceeds gravely to infer that war is forever a necessity, as the expression of this instinct'. Such inferences would be justified only where the underlying psychological processes were understood. Since Dunlap did not know of any case in which this was remotely true, he concluded that in this sense there were no instincts.

Dozens of papers on instinct followed within the next four or so years. The main accusation was expressed by one leading critic in the following, and relatively mild, way. 'In many cases it is a sort of catch-all for vague and indefinite ideas about the causes or relationships of activities. Writers, unable to account clearly for the occurrence of a particular behavioristic phenomenon on a purely objective basis, bring in the term instinct and use it as a charmed word, thus sidetracking further responsibility for an explanation. Race has been a similar term to conjure with, a stop-gap to a complete explanation of social phenomena in terms of scientifically-determined facts.'[25] The kind of examples used to illustrate this aspect of the problem were to explain that a man jumped off a cliff because he had a suicidal instinct or Trotter's claim that wolves behave as they do because they possess the lupine form of the herd instinct. A related form of criticism was that appeals to instinct were masked attempts to retain within psychology explanations based on non-physical forces. As suggested earlier, McDougall's defence of vitalism and increasing interest in beliefs that most colleagues dismissed as ones involving the supernatural encouraged such attacks.

While an initial group of critics followed Dunlap in concentrating on the logic of instinct and its status as a technical term within psychology, in 1921 a second group began to raise different and more substantial issues. One of the first and most vociferous was Zing Yang Kuo, who had left southern China to study in America and was still an undergraduate at the University of California when he wrote his first contribution to the instinct debate.[26] He argued that even the limited kind of usage allowed by Dunlap, whereby innate patterns of behaviour were called 'instinctive acts', should also be abandoned since all behaviour is a product of interactions between the genes and the environment that begin with the fertilization of an egg.[27]

The most novel aspect of Kuo's discussion concerned what fifty years earlier Douglas Spalding had called an imperfect instinct and what more recently has been known as imprinting. As discussed in the first chapter, Spalding found that a certain amount of exposure to the mother duck or hen was needed at an early period in life in order for a chick or duckling to develop the usual tendency to follow its parent. Almost everyone who had subsequently discussed phenomena of this kind, including the more behaviourist writers like Harvey Carr and Walter Hunter, had made the assumption implied by Spalding's label, that appropriate experience allows some pre-ordained developmental process to become complete; thus, when some other experience occurs instead of the usual one, as for a two-day-old chick that only sees a moving human hand, 'modification of instinct' is said to occur.

Kuo rejected this view, basing his argument upon an example of what has become known as sexual imprinting, which had been reported a year earlier. This was the discovery that the choice of a mate by male pigeons on reaching sexual maturity is strongly influenced by the company they have kept when young. Thus, 'a male passenger-pigeon that was reared with ring-doves and had remained with that species was ever ready, when fully grown, to mate with any ring-dove, but could never be induced to mate with one of his own species'.[28] In further examples from this study, early experience later

Fig. 8.4. Zing Yang Kuo

caused male pigeons to attempt to mate with other males, and females with females. Kuo maintained that only in a statistical sense was such behaviour less 'normal' or 'natural' than usual heterosexual preferences within the species: when some environmental condition experienced by the vast majority of individuals causes development to proceed in a certain direction we tend to view the behaviour as 'instinctive' and development in any other direction as distortions of what nature intended. 'There is no sex instinct in the sense that it necessarily involves coition between two opposite sexes. The fact that mating always takes place between two opposite sexes of the same species is because the members of the same species always live in the same community where the heterosexual habit is normally developed . . . The point I am driving at', Kuo continued, 'is this: that all our sexual appetites are the result of social stimulations. The organism possesses no ready-made reaction to the other sex, any more than it possesses innate ideas.'[29]

Kuo's anti-hereditarian stand was also adopted by psychologists and sociologists who had little interest in animals, but who wanted theories of human action appropriate for social psychology. These writers were sensitive to the findings from cross-cultural studies and attacked instinct theory on the grounds that no list of instincts, no matter how well-defined, subtle or numerous, could account for the diversity of human customs; what was required from psychology was an analysis of the processes by which aspects of a culture are maintained by a social group.[30]

In concluding this section, it is worth noting one crucial difference between the arguments over race and instinct. The racial issue was to some degree an empirical one, in which the various protagonists advanced certain kinds of evidence or criticized others, whether the findings were from an anthropologist's field-work or from the Army testing programme. Thus, the swing away from biological determinism and from the racism of the evolutionists was encouraged by the discovery that people changed after arriving in the United States to an extent that had not been thought possible.

This was not the case with the debate over instinct, which remained largely conceptual and to a surprising degree free from the consideration of evidence. In fact, it was just this absence of relevant evidence that made most of the younger psychologists so suspicious of the concept. From the time of James, chapters on instinct continued to be included in psychology textbooks, but they remained empty of the bustle of experimental data for or against this or that theory which accumulated in other chapters. Even a committed experimentalist like Kuo only began to make an empirical contribution to the study of such matters long after the debate was over, as in a celebrated study on the development of rat-killing by cats where rats and cats were reared together.[31]

When McDougall first discussed instinct he took pains to insist that such an approach did not deny the plasticity of behaviour displayed by many species. He was not troubled by the occurrence of sexual imprinting in pigeons nor by what were to him familiar findings that both nest-building and bird-song can be similarly affected by early experience, which were the only other pieces of evidence entered into the instinct debate. However, his later espousal of racist atttitudes promoted an identification of analyses of behaviour based on instinct with a hereditarian attitude towards human differences which critics like Kuo were happy to encourage. Consequently, given the general swing towards environmentalism it was easy for psychologists to accept implicitly the fallacious logic whereby, since instinct is an unscientific concept and since it implies hereditarian beliefs, hereditarian beliefs are themselves unscientific.

The number of examples drawn from the study of animals' behaviour to be deployed in the instinct debate may have been small, but, when argument concentrated on the specific issue of human instincts, there was only one source of directly relevant evidence. This was a study of the behaviour of newborn babies that Watson began in 1916. This research and Watson's part in the nature–nurture controversy are described in the next section.

The final part of John Watson's career

For good or for ill, since the 1960s there has been a massive invasion of behavioural methods into the practice of clinical psychology. Variants of conditioning theory have inspired specialized forms of therapy for individuals suffering from crippling phobias, obsessions or addictions and have suggested major innovations in the way that institutions for the mentally ill or retarded are organized. It appears as a natural development of the behaviourist viewpoint to consider abnormal patterns of behaviour as habits or conditioned reactions that have been acquired according to the same principles of learning that produce the acceptable and adaptive actions of those regarded as perfectly sane and rational. The puzzle is that this development came so late.

Watson was well-aware of the potential application of his ideas to clinical problems. His interest in the importance of conditioning for psychology had first been aroused by Bechterev's *Objective Psychology* and one of Bechterev's main hopes for this new kind of psychology was that it would transform the field of psychiatry in which he was one of Russia's leading authorities. In a paper of 1916 Watson expressed a similar view, paying tribute to Freud's ideas on the pervasive influence of childhood experience on adult life, but arguing for the use of an objective terminology based on the concept of habit.[1] Using as a hypothetical example the favourite animal in Russian laboratories, Watson described a dog that, in displaying all sorts of curious and unpredictable responses to familiar objects, could appropriately be labelled neurotic. The point was that we are quite willing to accept the dog's behaviour as a product of training by a human master, but we are loath to believe that comparable human reactions might arise as habits produced in a disastrous childhood environment. At the time the paper was written Watson was on the point of moving into the Phipps Psychiatric Clinic at Adolf Meyer's invitation so that he was in an excellent position to develop such ideas. So why did behaviour therapy take so long to develop?

The last event may offer a clue to the answer. Watson sent a pre-publication copy of his paper to Meyer, who was infuriated by what he regarded as the cheap ridicule cast upon other approaches, by its lack of any informed appreciation of what mental illness is like and, above all, by the evident failure to grasp Meyer's own views on the subject, even though the two men had now been fellow professors at Johns Hopkins for over six years. Meyer decided that 'the paper is merely another of the half-cocked pioneer schemes, devoid of any serious attempt to do justice to others and attempting to handle a field concerning which you are obviously too uninformed'.[2]

Watson was taken aback at Meyer's reaction. He protested that Meyer had overlooked the modesty with which his tentative suggestion had been made and that he was not trying to butt in on Meyer's preserve. Meyer was unmollified. 'Your "neurasthenic" dog is a permissible product of imagination. Your statement of what the physician would do with it is however a farce . . . Your temperament as shown in your work is not unlike Loeb's. You have to shut out everything that might confuse your outlook. It always is entertaining, but, as far as convincing him, useless to debate with Loeb. It is most satisfactory to take him for what he gives and not to ask for any assimilation of one's own viewpoint. So it may be with you . . . My forefathers have been free of the dogma of exclusive salvation since 1521; and I never had any need of eliminating a whole sphere of life interests as you did when you shed the Baptist shell . . . You may have much more up your sleeve than your paper showed. You will find me receptive for all but one thing: intolerance and the dogma of exclusive salvation.'[3]

Meyer was pleased to have Watson in his clinic, but the message was clear: if Watson wished to keep his laboratory space and promote his own ideas on psychopathology, then he had better make sure first that he had the kind of clinical experience and knowledge of the subject that Meyer had already gained; which was a tall order, given Meyer's intellect, integrity and twenty hard-working years in the field. Meyer suggested that Watson examine thoroughly three to five cases of neurosis or psychosis and then see how promising the conditioning model looked.

Once settled at the Phipps Watson was diverted from this challenge by an equally promising field in which to develop behavioural research and one which would avoid the strain of frequent collisions with Meyer that were inevitable if he continued with psychopathology. So instead of psychiatric patients in the clinic, Watson studied the reactions of newborn babies at a nearby maternity hospital.

The aim of this work was to obtain empirical evidence on the issues that James and McDougall had discussed and to demonstrate that the behaviourist could study productively the key subjects of instinct and emotion. For Watson an emotion was a distinctive, hereditary pattern of response, 'involving profound changes of the bodily mechanism as a whole, but particularly of the visceral and glandular systems' and which 'throws the organism for the moment at least into a chaotic state'.[4]

Watson and his students looked at babies within a month of birth and decided that only three such patterns could be distinguished: fear, rage and love. These conveniently corresponded exactly with those he had described some years earlier, before ever contemplating empirical work on such a topic. The more productive and interesting aspect of the study came from attempts to discover what types of stimuli elicited these reactions at this early stage of life. Three main kinds of event were found to produce strong fear reactions: sudden loss of support, a sudden shake or pull of the blanket when the baby was nearly asleep, and loud sounds. The only way of producing a rage reaction seemed to involve hampering the baby's movements, as by holding the head or constraining the arms or legs. Finally the love reaction was obtained when the baby was tickled, gently rocked or given another form of 'stroking or manipulation of some erogenous zone'.[5]

For twenty-five years psychologists had discussed the wide variety of emotional reactions believed to be innate, and in particular the numerous 'natural' fears to be expected during a child's development, which Stanley Hall and his students had described. Consequently Watson was as much intrigued by the negative results from his study as by the positive ones. He could, for example, find no sign of distress when the babies were brought into close contact with various animals and, to take an example closely related to his own experience, they showed no fear of the dark.

The work on emotion was accompanied by some initial tests that were intended to focus on the subject of human instincts. Watson adopted Loeb's definition of instinct as a system of chained reflexes, which in turn came from Herbert Spencer; this meant that a start should be made by finding out what reflexes are present at birth or soon after. The preliminary study revealed a large number of specific and well-developed reflexes, such as hiccoughing, sneezing, yawning and eye-movements. This surprised the student who was supervised in this work by Watson and she concluded the first report by noting that 'the reflex and instinctive equipment of the child at birth is more complex and advanced than has hitherto been thought'.[6] Watson himself was especially intrigued by the grasp reflex and the babies' ability to support their own weight, which disappeared as they grew older; Figure 8.5 shows him carrying out a test of this kind.

An early opportunity to talk about this new work and the practical uses of behaviourism arose when Watson was invited to deliver one of a series of public lectures sponsored by the Joint Committee on Education, a body which, among other aims, was to examine what 'new light on the subject of public schools might be obtained from modern science'.[7] The four main participants included Jennings, who represented biology and talked about health and heredity, and Meyer, who represented psychiatry and discussed its implications for intellectual and moral development; only the sociologist, William Thomas, was from outside the Hopkins group. Whereas the other three remained either very general or non-controversial in tone, Watson characteristically had concrete and memorable suggestions to offer.

He began with examples like the grasping reflex or the question of how right-hand preferences arise in order to illustrate what he was doing and then went on to consider the meaning of his discovery that only a very limited number of stimuli elicit emotional reactions in young babies. Two years earlier with Karl Lashley he had used the conditioned reflex as a method for studying perceptual problems, but now, as in his paper on mental illness, he presented it as a process of learning which could explain how emotional reactions developed to a wide range of situations. 'If we do possess, as is usually supposed, many hundreds of emotions, all of which are instinctively grounded, we might very well despair of attempting to regulate or control them and to eradicate wrong ones. But according to the view I have advanced it is due to environmental causes, that is, habit formation, that so many objects come to call out emotional reactions.'[8]

The focus of the conference was education in the schools. Since Watson's results had confirmed him in the belief that the first few years of life are all-important in shaping the emotional life of the child, it followed that 'we can rapidly improve matters by making the positions of the early grade teachers the most desirable and the best paid ones in our schools', provided that they received good training in psychology and psychopathology; 'I do not wish to cast stones, but I do wish to decry the tendency in our American schools to think that any teacher is good enough to teach young children'.[9] In general his advice was cautious, promising that education might expect a great deal from future behavioural studies, but admitting that there was not much on offer yet. In this context he echoed Morgan's appeal for an institute for the study of comparative psychology and Yerkes' more recent call for one in which apes might be studied by explaining the need for 'an experimental nursery where fifteen to twenty children can be brought up during the first five years of life . . . under strict experimental conditions . . . I have something in mind far more scientific and far more important than any material which can be gathered from the use of the so-called scales of measuring "intelligence", however useful such scales may be'.[10]

Watson's new emphasis on nurture in human development and the possibilities for change was also seen in his discussion of some of the animal research carried out by his students. One issue was the now familiar one of whether spaced practice is more effective than massed practice: in addition to results showing that rats learned to open a puzzle box in fewer trials when given one, rather than many, trials a day, Watson cited a parallel study by Lashley yielding similar results from human subjects learning to use the English long-bow.

The other issue referred back to James' pessimistic conclusions on the crystallization of habits by the age of thirty. A Hopkins student, Helen Hubbert, had compared the times to learn a complicated maze required by rats of different ages. She found that, although the younger ones were quicker, even those of eighteen months or so – corresponding to human old age – learned quite rapidly. In his lecture Watson suggested that 'these experiments should give those of us who have passed the first bloom of youth a good deal of hope. Many of us in that too often unfortunate condition say that we do not know how to dance, to skate, and to play games because we did not learn such things when we were young; but this excuse is no longer valid. We now have experimental evidence to show that the contention of William James concerning the non-plasticity that is supposed to go with old age, which has been so unanimously accepted, is completely unfounded . . . Convention has more or less frowned at middle age putting on so-called youthful habits: we look askance at a middle-age individual who is trying to learn such acts . . . I should say that here our conventions are wrong; that middle age and early old age would be much more exciting periods for all of us if we would only become willing to scorn such

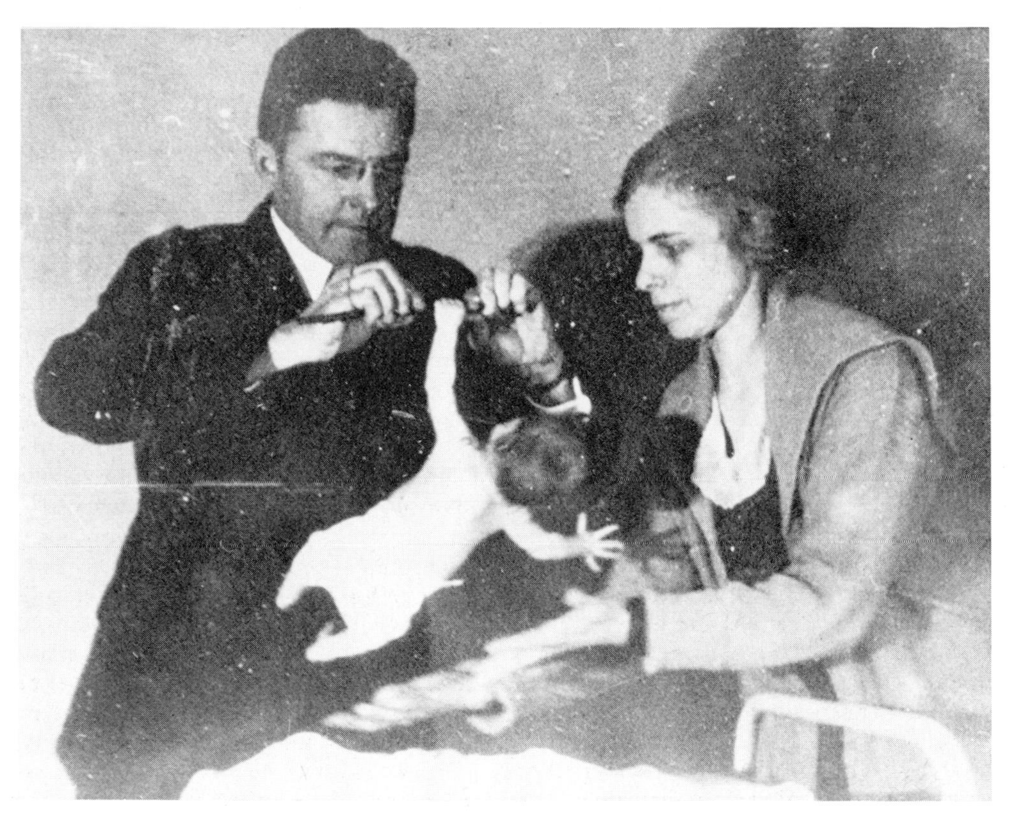

Fig. 8.5. Watson testing the grasp reflex of a new born baby; this is a still from a film Watson made in 1919, hence the poor quality of the print

conventions and dare to learn whatever we please to learn.'[11]

The study of babies had been going for just over a year when America's entry into the war led to a temporary interruption of the work. Presumably because of his known lack of experience and of enthusiasm for mental testing, Watson was not involved in the general Army testing programme. When in August, 1917 he was eventually commissioned, he was assigned at Thorndike's suggestion to the aviation board. His responsibilities included that of devising tests for selecting men most suited for training as pilots and later, when joined by Dunlap, studying the effects of oxygen deprivation on flying skills.

Watson's wartime experience was not important in terms of the results obtained from his various projects. What was significant for him was the realization, in common with many other of his fellow academics, that he had worthwhile skills to offer the outside world and the ability to use them as effectively as any professional. In Watson's case such feelings of confidence in his practical value were combined with growing contempt for the typical army officer. 'The whole army experience is a nightmare to me. Never have I seen incompetence, such extravagance, such a group of overbearing, inferior men.'[12]

With his general attitude and his scorn for the kind of diplomatic skills that enabled Yerkes to function so effectively in the army, it was predictable that Watson should get into trouble with his superiors. As a result of one clash he was sent on his first-ever trip to Europe. Its purpose was to use questionnaires which Thorndike had designed to interview British airmen fighting in France. Even if they had not been otherwise engaged, it is difficult to envisage much success from this venture and Watson eventually returned to America with little to show for his travels.[13]

Another clash followed when he let his opinion be widely known that some test results demonstrated the complete worthlessness of a selection procedure favoured by his superior officer. Watson was assigned to front-line duties in France and there were strong hints of a hope that he would not return. However, the Armistice was signed before he left and by December, 1918 he returned to academic life in Baltimore.

The Phipps Clinic once more became a busy

centre for research. The work with babies started again and there were also new projects, some of them extending scientific method into areas which made the university administration nervous. One was a project planned during the war in which Watson and Lashley investigated the effectiveness of some instructional material on venereal disease designed to emphasize the perils of promiscuity. While evaluating an Army film called *Fit to Fight,* they gave a showing in a small town in Pennsylvania, where no advanced warning had been given about the content and explicit nature of the footage, and narrowly escaped arrest by the sheriff. Of equal potential for unwelcome publicity in this new era of Prohibition was research into the effects of alcohol on human performance. Watson was a drinking man, whose survey of the evidence revealed nothing to indicate that 'the consumption of alcohol in small amounts when taken after the working period of the day is over produces any evil effect either upon the individual or upon his progeny'.[14] He was eager to obtain a supply of whisky adequate to sustain experimental work likely to confirm this conclusion. Both projects took a great deal of time, but they contributed their share towards making the laboratory a lively and often hilarious place for Watson and his small group to work in.[15] Somehow in the middle of all this Watson managed to finish his second book, *Psychology from the Standpoint of a Behaviorist,* which was published in 1919.

The book began with a clear statement of his now familiar views on psychology, followed first by some standard textbook information on the physiology of the senses and the rest of the nervous system and then by a series of topics reflecting the research interests concerning human behaviour that he had begun to explore since leaving the study of animals. The outlines of a behaviourist programme, oriented towards practical problems, were now quite clear: it was to include the development of emotional reactions and the integration of reflexes into skilled habits; the basic laws of habit formation, especially those bearing on educational practice; and the effect of factors, ranging from alcohol, caffeine and climate, to sex, age and time of day, on various aspects of human performance. Watson repeated the proposal to study thinking and problem-solving by monitoring responses of the larynx, but there was still nothing in the way of substantial research to report in this area.

Any reader who noticed that the author was affiliated to a psychiatric clinic or who had read Watson's paper on mental illness three years earlier might well be surprised to find that the subject of psychopathology was given only the barest treatment in the final two pages of the book. Watson simply endorsed Meyer's rejection of appeals to hypothetical organic lesions as causes of neurosis or psychosis, suggesting that in most cases it was much more productive to view a patient's problem as resulting from a 'diseased personality' caused by 'habit distortion' and 'long continued behavior complications'. Prior to publication the book was discussed chapter by chapter in weekly seminars attended by Meyer. His many other interests had not left Watson time to work with any of the patients at the clinic and one presumes that the memory of Meyer's challenge over mental illness was kept fresh.

It would have been natural to treat the question of 'habit distortion' in more depth, since a central theme of the book was the development in the child of what Watson now called 'conditioned emotional reactions'. In presenting this idea he had appealed to plausible examples, such as the fear of lightning developing because it is often followed by a loud noise. The only direct illustration he could quote was that of a six-month old baby, who previously showed no hesitation in approaching animals, but showed signs of fear when a small dog was tossed into his pram. However, around the time of the book's publication Watson carried out a demonstration of his ideas on emotional conditioning that has remained his most famous experiment.

An eleven-month-old infant, named Albert B., had been observed since birth and found to show no fear of anything except sudden, loud sounds. When various animals were brought close, he would readily reach out and touch them. A conditioning procedure was used by Watson and his assistant which consisted of presenting Albert with a white rat and, as he stretched his hand towards it, a steel bar was sharply struck immediately behind his head. Two such trials were given and then a week later the rat was presented again. Albert's reaction was now much more cautious and, after five more trials in which reaching for the rat was followed by the loud noise of the bar, he began to cry and move away when he saw the rat. In later tests there was some sign that the fear reaction transferred to other objects, since Albert showed distress when a rabbit, dog or fur coat were put in front of him.[16]

Forty years later, when modern behaviour therapies were developed, this experiment with little Albert became widely cited as an early piece of evidence in support of conditioning theories of phobias. It is therefore worth noting that from Watson's initial report it is clear that to label Albert's reaction as a 'rat phobia' is a misleading exaggeration; even after the final conditioning trial he 'began to fret'

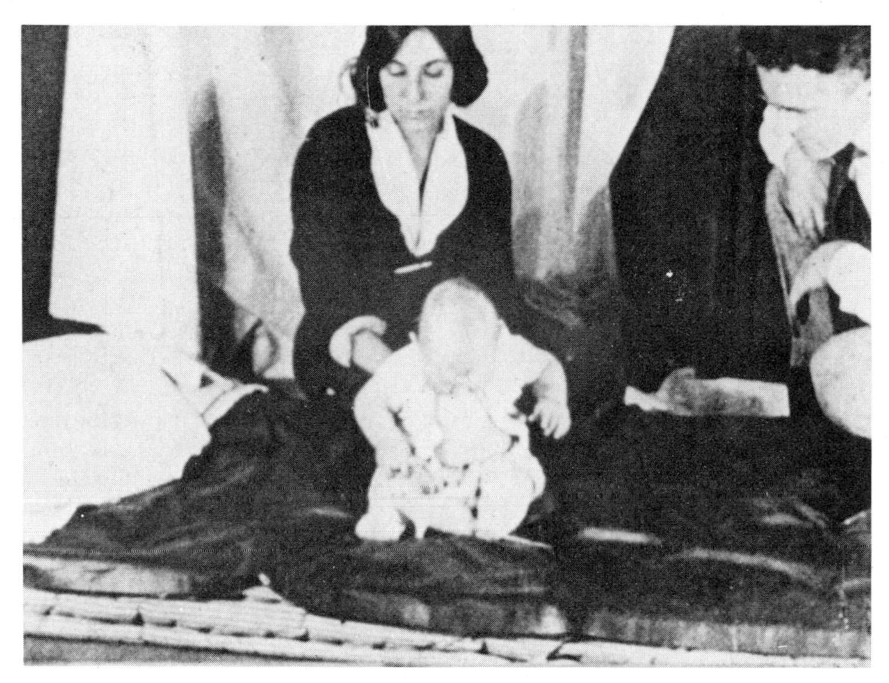

Fig. 8.6. Testing a baby's fear of animals by placing a white rat in his lap; a still from the film referred to in Figure 8.5

only when the rat was placed on his arm or crawled about his chest. It may be that the mildness of Albert's reaction accounts for what is so curious in the light of subsequent interpretations of this study, including Watson's own later accounts: namely, that no effort was made to rid Albert of his fear of the rat before he was removed from the hospital.[17]

The absence of any attempt to 'de-condition' Albert is comprehensible in terms of the research tradition to which Watson belonged. As already noted, comparative psychologists in England and America had concentrated on the problem of how animals solve problems and how intelligent behaviour is acquired. From Thorndike onwards there was not a single experiment from a North American laboratory that examined the *disappearance* of a previously learned response. This was not true of Pavlov, who from the beginning viewed this as a central problem, but in 1919 there was still little to read even about Pavlov's early work and, although Bechterev could be read in French or German, it was not an aspect of conditioning that he said much about. Even if Watson had not been constrained by Meyer's insistence that informed clinical experience precede pronouncement on matters of psychiatry, he would have had no experimental analogies to suggest ideas as to how a neurotic patient might shed a distorting habit.

Two graduate students had joined Watson after the war: Curt Richter and Rosalie Rayner. The latter assisted him with the baby work and was involved in the experiment with Albert. Soon after her arrival he had concluded a paper with a passage that intriguingly combined his belief in the importance of studying emotion and his rejection of James' pessimism over the possibility of escaping the trap of thirty years of habit training. Watson wrote of an individual whose 'social relations at home and on the outside are on the same dead level. His emotional attitudes are stereotyped – there seems to be a wall around these people. Is there no way of breaking through this wall and getting the individual to reach a higher level of achievement? Emotionally exciting stimuli occasionally seem to accomplish it . . . may break through the stereotyped and habitual mode of response and arouse the individual to the point where he can accept and profit by intensive training . . . and eliminate his errors, work longer hours and plan his work in a more systematic manner.'[18] So what are these 'emotionally exciting stimuli'? Watson listed wealth, rearing a family, strong rage or fear; the omission of an obvious further example, of falling in love, seems to have been deliberate. By the time these comments on emotion appeared, Rosalie Rayner had begun to play an important part in his life.

Rosalie was an intelligent, attractive, well-educated and lively young lady from a rich Baltimore family and with an uncle who had been a senator for Maryland. By the early part of 1920 she and Watson were seen so often in each other's company that newcomers to Hopkins often assumed that they were man and wife. Watson's marriage had started badly many years earlier in Chicago and the relationship with his wife, Mary, had been very cool for some time. For a while it seemed that Mary might tolerate this close friendship for Rosalie. But they separated in April and it turned out that Mary had previously stolen some love-letters from Rosalie's house that her husband had written. Her lawyer and a brother who had always distrusted Watson prompted her to make the most of this evidence.[19]

By August Watson knew that there was a strong chance that he could lose his job. He was in New York and he wrote to Meyer for advice, confiding to him the details of his married life and his present dilemma. 'I can find a commercial job. It will not be as bad as raising chickens or cabbages. But I frankly love my work. I feel that my work is important for psychology and the tiny flame which I have tried to keep burning for the future of psychology will be snuffed out if I go – at least for some time . . . I think since the first year of my marriage I have always been a fatalist – I've said always that Mary and her brothers would finally force me out. Possibly this has accounted for some of my impatience – attempting to force things and to get as far as I could before the storm overtook me.'[20]

Meyer was sympathetic and hoped that a solution could be found, but admitted to earlier disquiet over Watson's disregard for convention and over his belief that psychology could be developed into a science without regard for ethical principles. 'Many passages in your book gave me a feeling of taking, for example, great freedom on your part in the illustrations from the sex-life, and the references to Freudism . . . always made me fear danger to you or to the cause or both – not because of any timidity as to what others might say, but owing to doubts as to whether there was scientific proportion in the statements and safety in the control of the effects among the readers. It may be that I did not want to be taken for a prig.'[21] The only way out, Meyer believed, was to convince the university president that the affair was completely finished and that no scandal could leak out. What parents could send their offspring to an institution known to tolerate flagrantly immoral behaviour on the part of its faculty?

Somehow or other copies of the love-letters reached the president and, in any case, Watson would not accept permanent separation from Rosalie. Meyer advised the president that 'without clean-cut and outspoken principles in these matters we could not run a co-educational institution nor could we deserve a position of honor and responsibility before any kind of public nor even before ourselves'.[22] Watson resigned and left Baltimore. Later, in November, divorce proceedings were begun in his absence and, by supplying journalists with personal details of their marriage and copies of the letters, Mary ensured that the resulting publicity would end Watson's chances of obtaining an appointment at any other university for at least several years. His career at Hopkins had begun with the crisis over Baldwin's visit to a brothel and was now ended at the age of forty-two by one over his own sexual behaviour.

On his return from the war Watson had felt increasing dissatisfaction with academic life and, as always, found it difficult to live on his professorial salary. He had toyed with the idea of applying his talents in some business and had begun to put out feelers to various firms. Upon resigning he contacted the J. Walter Thompson advertising agency and was soon given a trial appointment designed to acquaint him with the basics of the world of commerce. By the time his name featured prominently in the *New York Times* and the Baltimore papers he was travelling up and down the Mississippi valley testing the market for rubber boots. In December he returned to New York and early in 1921, now that the divorce was final, he and Rosalie were married. Both the marriage and his new career were successful and brought him a great deal of contentment and satisfaction.[23]

A behaviourist approach and respect for empirical evidence turned out to be great assets in the advertising world. Watson promoted the then almost unknown concept of market research. His psychophysical expertise proved very useful, as in some tests to determine whether blindfold smokers could identify their favourite brand of cigarette. To Watson's surprise these tests revealed that his subjects' marked preferences were not based on any intrinsic property of the cigarette, such as its taste, but entirely on the associations that went with the brand name. He sold Johnsons' Baby Powder, Pebeco Toothpaste, Ponds Cold Cream, Maxwell House Coffee and Odorono, one of the first deodorants. 'It can be just as thrilling to watch the growth of a sales curve of a new product as to watch the learning curve of animals or men.'[24] By 1924 he was made one of the four vice-presidents of what was now a booming agency.

It says a great deal for Watson's commitment and energy that he managed to maintain an active interest in psychology despite the considerable demands of his job. Before leaving Hopkins he had told Meyer that 'he was fearful of cutting loose and going into business because it seems to produce the same effect as does the eating of the poppy', but for ten years his determination that this should not happen was successful. On the more popular side he produced a stream of magazine pieces on psychological topics and became known to a wide public as 'Mr. Behaviourism'. Many of his former colleagues were offended by the simplistic messages and occasional self-puffery in these articles; one virulent critic complained that, like Dadaism in art and jazz in the world of music, behaviourism was becoming the mindless symbol of 1920s modernism in psychology.[25]

There was also a professional side to Watson's psychology. He continued to serve as a journal editor for several years and to give and arrange series of evening lectures. He also searched for ways of continuing the research with babies. A year or so after the divorce he wrote to a former Hopkins colleague that he 'would never go back into academic work as such – I could be tempted by an experimental farm for babies. I hate to leave the babies hanging in the air as I did, but the probability of my ever getting this under suitable conditions is almost *nil*'.[26] However, in 1923 he did succeed in obtaining some support for a research project on the emotional development of children and he supervised the young psychologist, Mary Cover Jones, who carried out most of the work. This research included a study in which techniques for ridding disturbed children of irrational fears were at last tried out and thus it represents the first example of work in behaviour therapy.[27]

The debates over race and instinct had not impinged strongly on Watson at first, partly because of his general attitude that solid empirical data, rather than more discussion, were needed and partly because of the developments in his private life that followed soon after Dunlap's paper sparked off the controversy over instinct. The results of his baby research and his emphasis on the importance of conditioning placed him on the nurture side of such debates, but at first he voiced no strong opinion on these issues that could compare, for example, with Kuo, who was accused by McDougall of out-Watsoning Watson.[28]

In his evening lectures he found that the attitude of suspended judgement was constantly being challenged by students who wanted to know: what is inherited and what is not inherited?[29] In his 1919 book Watson had suggested in a tentative manner a modest list of human instincts. Unlike Kuo, he continued to believe that instinct was a perfectly serviceable concept to use in the analysis of animal behaviour, a view strengthened by his experience with birds. But the cumulative results of his baby research weakened his belief that there are any human instincts. The congressional hearings on immigration were in progress; in his Hopkins research he had not found any marked difference between the occasional black baby born in the hospital and more usual white ones, and now he came to decide that no one else had any evidence either to support the assertions about innate racial differences being made to the committee. His plan for a 'baby farm' came to include the idea of having different racial groups so that at least it would be possible to establish whether under identical environmental conditions children of different races still displayed characteristic psychological traits.

Watson's lecture series provided the basis for a third book, *Behaviorism*, which was published in 1924. The content did not differ very much from that of the one six years earlier. What stood out were the style, reflecting the deliberate appeal to a wide readership, and the central emphasis now given to an environmentalist approach to human behaviour. In short, emphatic sentences, with pithy examples from everyday life in New York, he presented the argument that everyone has an equal psychological inheritance and that the differences between individuals which become apparent in childhood are the result of very rapid learning which occurs early in infancy and perhaps even before birth. In opposition to the familiar claims of the eugenicists Watson stated, for example, that he 'would feel perfectly confident in the ultimately favorable outcome of careful upbringing of a *healthy, well-formed baby* born of a long line of crooks, murderers and thieves, and prostitutes'.[30]

Stress on the enormous importance of learning during the first few months of life was one way in which Watson's environmentalism differed from earlier versions. Another was its sensitivity to the way that subtle differences between the life experience of two individuals, which to the outsider may not be at all obvious, may have far-reaching consequences. He discussed the dynamics of family life and the way brothers and sisters may receive very different treatment from their parents and also examples of the following kind, an actual case he knew about. 'Two girls, aged nine, live in adjoining houses. They have the "same" training (mothers are close friends and bring up children according to the same rules). One day they took a walk. The girl on the left looked at the street and saw only street activity, the one on the right looked towards the houses and saw a man exposing his sex organs. The girl on the right was considerably troubled and disturbed and reached equanimity only after months of discussion with her parents.'[31] Adults may never know about many similarly important events in a child's development, especially at an earlier age, and so, when seeing differences develop between children from apparently identical environments, have traditionally attributed these to inheritance.

For most of the assertions, there was usually the cautionary note: that behaviourism was still in its infancy or that little direct evidence was available yet on some issue. The central argument was that James and McDougall, who described the natural instincts and emotions, or the eugenicists, who described inherited traits of intelligence or criminality, had no evidence to support their claims; not that there was strong evidence for the opposing view. However, it was the bold claim that was memorable, and the more modest admission that often accompanied it easily forgotten. Thus, the following extract from *Behaviourism* has been quoted a thousand times: 'I should like to go one step further now and say, "give me a dozen healthy infants, well-formed, and my own specified world to bring them up in and I'll guarantee to take anyone at random and train him to become any type of specialist I might select – doctor, lawyer, artist, merchant-chief and, yes, even beggar-man and thief, regardless of his talents, penchants, tendencies, abilities, vocations, and race of his ancestors".' Usually omitted is the sentence that immediately follows: 'I am going beyond my facts and I admit it, but so have the advocates of the contrary and they have been doing it for many thousands of years'.[32]

Many of Watson's former colleagues were offended by the aggressive overstatement, the brushing aside of complexities or conflicting evidence, the persistent use of examples involving sex and the way many of their own views were cheerfully insulted. They saw in the book's brashness a sign of Watson's new profession. It is undoubtedly true that five years of writing advertising copy enabled him to achieve the aim of getting his ideas across to the public at large more effectively than if he had taken up some other trade.

This aspect of Watsonian behaviourism infuriated McDougall. In 1924 the two men took part in a public debate. McDougall's pugnacious wit and verbal felicity, of the kind much prized in British universities, did not win the same appreciation from American psychologists. He should have known that his attitude of intellectual superiority and the argument that, since European psychology had shown no interest in behaviourism, it was highly suspect, would not help his cause. Nonetheless, when a vote was taken at the end, it was a close thing; McDougall concluded that 'when account is taken of the amusing fact that the considerable number of women students from the University voted almost unanimously for Dr. Watson and his Behaviorism, the vote may be regarded as an overwhelming verdict of sober good sense against him from a representative American gathering'.[33] Consequently, when commenting on this exchange three years later, he was bitter at the continued attention paid to behaviourism by the press and disturbed at its refusal to disappear from American psychology. 'Dr. Watson knows that if you wish to sell your wares, you must assert very loudly, plainly, and frequently that they are the best on the market, ignore all criticism, and avoid all argument and all appeal to reason . . . The susceptibility of the public to attack by these methods in the purely commercial sphere is a matter of no serious consequence. When the same methods make a victorious invasion of the intellectual realm, it is difficult to regard the phenomenon with the same complacency.'[34]

Fig. 8.7. William McDougall when a professor of psychology in the U.S.A.

There is considerable truth in McDougall's accusation. Whatever the validity of his opinions on

Watson's style and lack of scholarly virtues, there is certainly a compelling analogy to be drawn between Watson's advertising and the effect of *Behaviorism*. Just as he created a glamourous image for Ponds cream by persuading genuine queens to appear in advertisements for the product, so behaviourism became intimately associated with the new environmentalism and identified, by its language and by its appeal to Jefferson's statements on equality in the Declaration of Independence, as a truly American kind of psychology.

Watson's message must have given comfort to a great many parents whose family line seemed to have more than its fair share of criminals or failures. At the same time it placed a heavy weight of responsibility on parenthood. Watson offered this as an explanation of the hereditarian outlook's perennial appeal. When children prove wayward, parents can always say: 'If these tendencies are inherited we can't be much blamed for it.' This was certainly the kind of reassuring message conveyed by Stanley Hall's picture of development as a series of unfolding instincts and his ideas had by then been widely absorbed by child-guidance experts. With the rapid social changes following the war and decreasing confidence in traditional patterns of child care, more and more parents looked to science for advice.[35] To meet this demand Watson and Rayner wrote *The Psychological Care of Infant and Child* in 1928. In the long run this book probably had a more profound effect on more people than anything else Watson wrote.

After his departure from Johns Hopkins the tone of the authoritative scientist who knows best became increasingly evident in Watson's writing. His pet scheme for bringing up babies under tightly controlled conditions had a sinister ring to it from the very beginning. In 1916 he had explained that this kind of project was necessary in order to obtain the kind of evidence that would make it possible to give parents 'advice of a scientific character; only a charlatan would presume now to give "expert advice"'.[36]

In the following twelve years his experience had been limited to two years of intensive work with babies in Baltimore, to acting as a very much part-time consultant to Jones' work in New York and to some contact with the two children of his first marriage, plus rather more with the two boys that Rosalie bore. On this basis he presumed to advocate a particularly controlled style of child care that was strangely grim for someone who in his own life could be so spontaneous and sociable and display a lively sense of fun. 'Treat them as though they were young adults. Dress them, bathe them with care and circumspection. Never hug and kiss them, never let them sit in your lap. If you must, kiss them once on the forehead when they say goodnight. Shake hands with them in the morning.'[37] Rosalie Rayner was no doubt the saving grace in her two children's lives. In an article titled 'I am the mother of a behaviourist's sons' she wrote: 'In some respects I bow to the great wisdom in the science of behaviourism, and in others I am rebellious . . . I secretly wish that on the score of (the children's) affections they will be a little weak when they grow up, that they will have a tear in their eyes for the poetry and drama of life and a throb for romance . . . I like being merry and gay and having the giggles. The behaviorists think giggling is a sign of maladjustment.'[38]

The book was dedicated to the first mother of a happy child. A happy child was described as one who was, and would remain, self-reliant, 'so bulwarked with stable work and emotional habits that no adversity can quite overwhelm him', as well as 'a problem-solving child' and one that 'puts on such habits of politeness and neatness and cleanliness that adults are willing to be around him, at least part of the day'. The care needed to achieve this result was spelt out to a surprising level of detail, including precise specification for the times of meals and of sleep and for the level of the bathwater. Despite some earlier expression of progressive views on the status of women, Watson was no advocate of major changes in traditional parental roles. In discussing the period before bedtime he suggested that 'this is a good time to give the father his half-hour. It keeps the children used to male society. Then, too, they have their chance to ply him with questions.'[39]

The aspect of current practice in child care that most troubled him was the incidence of 'too much mother love', the title of the book's third chapter. It was strongly reminiscent of the contempt James had expressed, when writing about habits almost forty years earlier, for the 'nerveless sentimentalist and dreamer'. Watson believed that a mother who showed more than a minimum of affection – one who was perhaps not careful to ensure that the 'bath should be a serious but not gloomy occasion' – was likely to produce a dependent child, who would grow up unable to cope with life and prone to 'invalidism', a tendency to find the aches and pains of his body of far more interest than events in the external world.

It is difficult to assess the impact of Watson and Rayner's advice. However, at the very least it must have encouraged the trend towards inflexible patterns of child care that appeared in North America and Britain during the 1930s. Books offering the same kind of advice as Watson and Rayner remained popular until the appearance of Benjamin Spock's *Baby and Child Care* in 1946.[40] This rejected Watson's good-habits approach and, influenced by Freudian ideas, encouraged parents to be flexible, to trust their own common sense and to make 'natural loving care' the most important ingredient of parenthood. The advice offered by Watson and Rayner seems well calculated to minimize the impact of a child upon its parents' lives. What its other effects might be, or what difference it made when a later generation of parents switched to the methods encouraged by Spock, is one of those questions of paramount importance, which are open to guesswork that is little more informed now than was Watson's.

A few years later Watson admitted that he was not at all satisfied by the book, but by then he had given up trying to remain both a psychologist and a businessman. In 1930 he bought a farm in Westport, Connecticut and commuted from there to his New York office. Lashley came to visit him occasionally when he had important issues to decide about, but otherwise Watson had no further contact with the world of psychology.

Rats, mazes and learning theory

By the time Watson abandoned his effort to maintain an active interest in psychology, the only rats he had handled over the previous fifteen years had been those shown to Little Albert. Following his original paper on behaviourism of 1913 Watson had supervised some students carrying out research using rats as subjects, but otherwise neither he nor any other faculty member of the Johns Hopkins University was directly involved in animal psychology. The subject was in no more flourishing a condition at other American universities. The animal laboratory at Clark University had long since closed down and at Harvard animal experiments stopped with Robert Yerkes' departure for the army and never resumed when, after the war, he went on to his research council post in Washington. When Watson left Johns Hopkins University in 1920, his graduate student, Curt Richter, kept the rat laboratory going, but on a very modest scale, while Karl Lashley left Baltimore for Minnesota.

The one place where an unbroken succession of graduate students continued to gather thesis material by running rats through mazes was the University of Chicago. Harvey Carr had remained there since 1908, when he had taken over Watson's duties as the latter left for Johns Hopkins. In the first thirty years of the century Chicago produced more doctorates in psychology than anywhere in the world and was

Fig. 8.8. Harvey Carr

rivalled only by Columbia University in New York. A very large number of these theses owed a great deal to the support and encouragement provided by Carr. This time-consuming and self-effacing effort was no doubt one reason why his own research career never became outstanding.[1]

Like many of his generation Carr had helped to pay for his college education by teaching in a country school. This stimulated a lasting interest in education and a considerable proportion of the theses he supervised were concerned with learning. Most of these experiments used human subjects, for even at Chicago it was believed during this era to be wise to avoid the animal laboratory; as he later described the situation, 'many of our students expressed an aversion to choosing a thesis topic in this field for fear that they would become known as comparative psychologists, and that this would be detrimental to their profession, placement and advancement'.[2] Nevertheless a sufficient number of students were willing to take this risk to allow the maze-running tradition to continue at Chicago.

In the early 1920s the situation began to change. The public university system of the mid-western and western states had been steadily expanding since well before the First World War. Psychology departments took part in this expansion, but the kind of theoretical or philosophical issues that had earlier given psychology its appeal in the older north-eastern universities cut little ice west of the Appalachians. The pressure was much greater for psychology to show that it could impart skills with some practical value, particularly for teachers. A major function of the big state universities was to provide personnel for the educational system. The rapid expansion of primary schooling that had occurred towards the end of the nineteenth century had been followed by a corresponding development of the secondary sector. New high schools were built at a rate of over one a day during the first twenty years of the century.[3]

There was a large influx of graduate students into psychology after the war and a great many of those who after three or four years left with their Chicago or Columbian Ph.Ds found jobs that were related to education. Many discovered, as had one of Carr's pre-war students, Walter Hunter, at the University of Kansas, that they were expected to give courses in the psychology of learning to students majoring in education. At first Edward Thorndike's three-volume *Educational Psychology* was the only textbook that could be used; the many traditional texts on psychology had little to say about learning.[4]

The Progressive Era was marked by widespread faith in the solution of problems by calling in the right kind of expert. For twenty years Thorndike had been arguing that the kind of expertise needed by educators was a knowledge of statistics, mental testing and experimental studies of learning. The mid-western psychologists concentrated on studies of learning. Their investigations of variables affecting how people or rats learned a particular task had a precision, thoroughness and professional quality that gave them scientific authority and yet, at the same time, the results were easily assimilable. There was another consideration that was no doubt at least as important in making the psychology of learning attractive to the educational establishment: it was essentially conservative. It promised technical improvements that would enable children to learn much more at school, but posed no challenge to the basic system or criticism of the existing syllabus in a way that applications based on other developments in the social sciences might have done.[5]

The approach adopted in these studies had little interest in theory and in later years this tradition earned the title of 'dust-bowl empiricism'. Considerable effort went into ensuring that comparable aspects of learning by rats and people could be studied. In some early studies these went as far as using full-size human mazes, either specially constructed ones or, in at least one case, utilizing one already available in the amusement park.[6] However, most human studies used the stylus maze, where blindfolded subjects were required to trace a route through slots cut in a thin board, or tasks involving simple motor skills or lists of verbal material. Even when some fairly realistic problem was studied it was often quite easy to find some apparently comparable problem to set a rat in a maze. For example, the question of whether a piano piece was best practised in parts or as a whole and the question of how prior learning of task A would affect someone's progress in learning task B prompted equivalent questions that could be studied in the animal laboratory.

In the reading assignments for courses on learning, Thorndike's *Educational Psychology* was joined by Watson's *Behavior* of 1914, mainly because it was still the most convenient book to provide a survey of comparative psychology and not because those giving the courses necessarily saw themselves as behaviourists. Watson's emphasis on applied psychology was welcome and in practice they went along with his dismissal of introspection; there were lots of learning curves, but little discussion of what subjects said about how they learned.

These students of learning also put to one side the problem of whether all learning or knowledge is revealed by immediate changes in behaviour. Watson had dismissed as a 'mythological conception' the view that some experiences can influence future behaviour without producing overt effects in the meantime. 'It is true that I can give the verbal stimulus to you "Meet me at the Ritz tomorrow for lunch at one o'clock". Your immediate response is "Alright, I'll be there". Now what happens after that? We will not cross this difficult bridge now, but may I point out that we have in our verbal habits a mechanism by means of which the stimulus is reapplied from moment to moment until the final reaction occurs, namely going to the Ritz at one o'clock the next day.'[7] This was one of Watson's promises in the opening chapter of *Behaviorism*, but very few of his former colleagues believed that he had, or ever would, cross this bridge. Most of the maze experts were many hundreds of miles from the Ritz and what mattered to them was that habits were very important, not the question of whether habits provided an answer to every question in psychology.

Using Thorndike's and Watson's books in the same course did bring one glaring theoretical difference into focus: their opposing views on how habits are formed. Thorndike consistently repeated his original claim of 1898 that habits are based entirely on connections between stimuli and responses and that such connections are strengthened by the occurrence of a pleasurable or satisfying event immediately following the response. Watson agreed with the first part of this claim, but just as consistently denied the second, the Law of Effect. Following Loeb, he had dismissed appeals to subjective concepts like 'pleasure' and he repeated Loeb's derision for the idea that a moth approaches a candle because it likes the light. Watson argued that the mysterious retro-active effect such events were supposed to have on an S-R connection had no place in a science of behaviour. In place of the Law of Effect Watson maintained that the principles of frequency and recency were the only ones needed to explain how learning occurred; that is, the probability or speed with which a particular response occurred in a given situation is unaffected by its previous consequences and only by how often and how recently it has previously been made in that situation.[8]

This was another case in which Watson stimulated great interest, but failed to convince. Everyone knew, of course, that no rat ever ran more and more rapidly to the place in a maze designated by the experimenter as the 'goal', unless there was a good chance of food or some other reward being there when it arrived. Watson suggested an explanation of why this should be that avoided introducing reward into the learning process and the last paper he published on animal research argued that it made no difference if the introduction of reward into the goal box was delayed.[9] A series of theoretical papers argued against his account. However, the issue failed at first to stimulate further experimental work with animals, except in one place. This was at the University of California at Berkeley, where in 1918 Zing Yang Kuo arrived from China as an undergraduate and Edward Tolman arrived as the new instructor in psychology. They were attracted by the question of how reward for a correct choice and punishment of incorrect choices affects a rat's behaviour. Rats were bought, mazes constructed and within a few years the Berkeley laboratory became one of the most productive centres for the study of learning and, above all, the place where experiments were directed at fundamental questions.

Tolman was no mid-western country boy like most other rat psychologists. He came from a wealthy, urban New England background which he described as 'upper middle or lower upper'. His childhood was spent in Newton, just ouside Boston, and was strongly affected by a family Quaker tradition of plain living and high thinking. At school he was good at mathematics and so went to study science at the Massachusetts Institute of Technology, obtaining his degree in 1911 at the age of twenty-five. By then an interest in psychology had been inspired by summer school courses taught by two Harvard men, Ralph Perry and Robert Yerkes, and, inevitably, by reading William James. Tolman decided to transfer from MIT to Harvard for graduate work in philosophy and psychology.

In the lecture course he was taught that psychology was the study of subjective experience, but in the laboratory he found that subjects' accounts of what they felt or thought were given little weight. Tolman was troubled by this lack of consistency. 'This worry about introspection is perhaps one reason why my introduction in Yerkes' course to Watson behaviorism came as a tremendous stimulus and relief. If objective measurement of behavior and not introspection was the true method of psychology I didn't have to worry any longer.'[10]

Although Watson brought general relief, more specific attitudes to psychological issues were imparted by people closer at hand. One of the most influential was Perry who, despite his interest in psychology, never became involved in research and has been remembered only as a philosopher. Perry and other Harvard colleagues welcomed Watson's views on the methodology and scope of psychology, but rejected his reductionism. They found Watson's treatment of habits simplistic and Perry in particular was much more sympathetic to the views expressed in Leonard Hobhouse's *Mind in Evolution*. Perry's general ideas were very close to the ones Koehler was developing in Tenerife at about this time; he argued that the purposeful nature of behaviour is its central characteristic and that this can be analysed without retreating into mysticism; Watson's 'muscle-twitch' approach was pitched at the wrong level and held no monopoly on an objective approach to psychology.[11]

As a graduate student Tolman worked on a range of topics, but did not become involved in any animal research. When he arrived at Berkeley, he was asked to suggest a new course and, remembering Yerkes' course at Harvard, he chose to give one on comparative psychology. 'It was Watson's denial of the Law of Effect and his emphasis on Frequency and Recency as the prime determiners of animal learning which first

Fig. 8.9. Edward Tolman, *ca* 1911

attracted our attention. In this we were on Watson's side. But we got ourselves – or at least I got myself – into a sort of in-between position. On the one hand I sided with Watson in not liking the Law of Effect. But, on the other hand, I also did not like Watson's over-simplified notions of stimulus and response . . . According to Thorndike, an animal learned, not because it achieved a wanted goal by a certain series of responses, but merely because a quite irrelevant "pleasantness" or "unpleasantness" was, so to speak, shot at it, as from a squirt gun, after it had reached the given goal box or gone into the given *cul de sac*.'[12]

The 'we' included in this quotation must have had Kuo in mind, since he published the first major study of this issue to come out of the Berkeley laboratory. Kuo used a maze situation in which four different routes all led to a food reward. One was a short and direct path, a second involved a long circuitous route, a third involved confinement in a chamber for a time before the rat could resume its journey and choice of a fourth meant that the animal received electric shock on its way to the goal. To no one's surprise eventually the animals showed strong preferences according to the above order: short was preferred to long, which was preferred to a delay, which was preferred to shock. The important point was that detailed analysis of their performance before these preferences emerged showed that the abrupt dropping out of a particular choice could not be predicted from the recency and frequency with which that choice had been previously selected. Kuo's results went against Watson's position and provided support for Thorndike, at least to the extent of indicating that the consequences of making some choice affected whether it would develop as a habit.[13]

Tolman's own interest was not so much in evaluating the relative merits of Thorndike and Watson, but in showing that both were wrong. In a series of articles he argued that the whole S-R framework in which this dispute had been conducted was inappropriate and that the problem of learning had to be approached in another way.

The underlying problem, as Tolman saw it, was that, in adopting the language of the reflex, S-R theorists had misled themselves into thinking that what they meant by a response was the same as the concept used in physiology. 'Our conclusion must be that Watson has in reality dallied with two different notions of behavior, though he himself has not clearly seen how different they are. On the one hand, he has defined behavior in terms of its strict underlying physical and physiological details . . . We shall designate this as the *molecular* definition of behavior. And, on the other hand, he has come to recognize, albeit perhaps but dimly, that behavior, as such, is more than and different from the sum of its physiological parts. Behavior, as such, is an "emergent" phenomenon that has description and defining properties of its own. And we shall designate this latter as the *molar* definition of behavior . . . A rat running a maze; a cat getting out of a puzzle box; a man driving home to dinner; a child hiding from a stranger; a woman doing her washing or gossiping over the telephone; a pupil marking a mental-test sheet; a psychologist reciting a list of nonsense syllables; my friend and I telling one another our thoughts and feelings – *these are behaviors* (*qua molar*). And it must be noted that in mentioning no one of them have we reference to, or, we blush to confess it, for the most part even known, what were the exact muscles and glands, sensory nerves, and motor nerves involved. For these responses somehow had other sufficiently identifying properties of their own.'[14]

By the mid-1920s Tolman had gathered around him an energetic group of students and between them they began to translate various ideas, which many others had expressed before and often more clearly, into experimental research which showed how one could analyse learned behaviour following a 'molar' approach. In that era it was particularly true that to run experiments meant much more than the explicit function of gathering empirical information; it served as a badge of scientific respectability, as a means of making one's ideas precise and coherent, but, even more importantly, as a mode of communication. Generations of psychology students have since learned about many of the experiments run at Berkeley during this period, not because startling results were obtained – in many cases few contemporary animal psychologists would be able to say exactly what the results were from these familiar studies – but because they illustrate a particular set of ideas with such clarity. This is why Tolman, and not Perry, is still remembered.

The main thrust of the early experiments was to show that, despite Watson's claims, reward, or more generally the consequence of a response, is important, but, despite Thorndike, it does not operate directly on the process of learning. Tolman drew a distinction between learning and performance; reward was viewed as unnecessary for learning, but essential for motivating performance. A key experiment on this point was included in the 1925 Ph.D. thesis of H. C. Blodgett and concerned a phenomenon that became known as 'latent learning'. One group of hungry rats was allowed to run through a maze once a day for six days, but without any food reward in the box at the end of the maze. On the seventh day they found food there and within two trials were running as rapidly and with as few errors as a control group which had received food reward since the first trial. This showed that the experimental group had learned a great deal during the first six days, but only when food was provided as an incentive was this learning translated into performance.[15]

Another experiment in this group concerned a second phenomenon which was also to feature prominently in later disputes over the nature of learning. This was carried out by M. H. Elliott for his M.A. thesis of 1928. For nine days a group of rats were given one trial a day in which they found bran mash in the goal box and then, on the tenth day, they found sunflower seed, a much less attractive food to them than mash. This resulted in a sharp increase in the time they took to run the maze and in the number of errors on the next few trials. Here was a case in which on the tenth day the correct response was rewarded and yet its probability of occurring of the next trial was reduced. This result was incomprehensible in terms of Thorndike's Law of Effect.[16]

Other experiments from Tolman's laboratory were directed towards what Perry had termed the 'docile' nature of behaviour and what Hans Driesch had called 'equifinality'. In one experiment run by a student named D. A. Macfarlane, rats were first trained to swim through a maze and, once they had learned where to go, a false bottom was inserted so that they could now reach the goal box by wading to it. This produced some transient disruption, but thereafter the rats made no more errors than before.[17] Such experiments provided concrete evidence on the flexibility of goal-directed behaviour to supplement the hypothetical examples described by Driesch over thirty years earlier. The performance of the swimming rats indicated the inadequacy of the claim by Carr and Watson, arising from their 'kerplunk' experiment. They found that highly trained rats collided with the wall when the layout of their maze was altered and concluded that the only thing learned by a rat in a maze is a chain of responses integrated by kinesthetic stimuli, that is, internal feedback from muscle and joint receptors. In Macfarlane's experiment the switch from swimming to wading should have changed completely the pattern of kinesthetic stimuli received by his rats as they made their way through the maze. The finding that they were not much affected by this change strongly suggested that what is learned in a maze is *where* the reward is located and not *what responses* are needed in order to reach it.

By the end of the 1920s increasing interest in problems to do with learning the location of reward produced experiments that bore a strong resemblance to the detour experiments devised by Hobhouse and developed by Wolfgang Koehler. The classic study in this series was carried out by Tolman and C. H. Honzik and reported in 1930. It used the maze shown in Figure 8.10. As can be seen, the animals were faced with a choice between a direct (Path 1), an intermediate (Path 2) and a lengthy route (Path 3) to the food box. After the rats had been given ample opportunity to become familiar with the maze, tests were given to find out what they would do if the direct path was blocked either at A or at B. Since Path 2 was second in preference to Path 1, it was not of great significance to observe that on finding a block at A the animals turned back to the choice point and chose Path 2. The crucial question was what would happen when a block was placed at B. Provided that they could remember or directly perceive the layout of the maze and were capable of making the appropriate inference, they should return to the choice point and select Path 3,

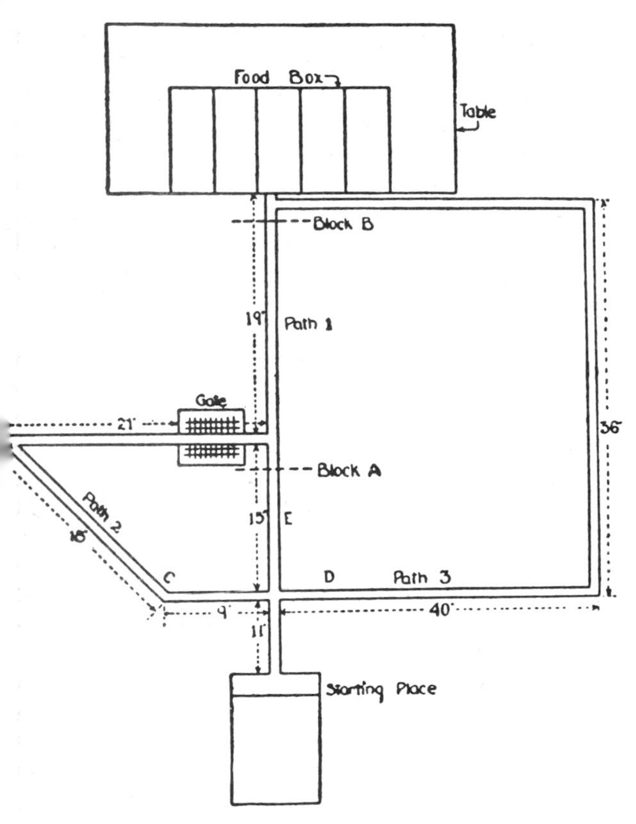

Fig. 8.10. Maze used by Tolman and Honzik in 1930 to study insight in rats

which now provided the only route to the goal. As long as an open, elevated maze, and not an enclosed one, was used, the majority of rats were found to make this choice.[18]

In the early 1920s the similarity between Tolman's ideas on the purposeful nature of behaviour and those of Koehler was probably a reflection both of the common influence of Hobhouse and of Tolman's early interest in Gestalt theory. The latter became more accessible to American psychologists with the appearance of an English translation of Koehler's *Mentality of Apes* in 1925 and shortly afterwards of systematic expositions of their general views on psychology by both Koehler and his former Frankfurt colleague, Kurt Koffka.[19] Koehler made an extended visit to the United States in 1925 when he was based as a visiting professor at Clark University and made several lecture tours. His own experiments with chimpanzees and subsequent work from the Berlin department on memory and perception had generated a lot of interest in America. This was increased by the generally favourable impression and respect created by personal contact. However, the theoretical ideas of the Gestalt psychologists continued in general to be regarded with puzzlement or scepticism. Tolman became one of the few ardent proponents of the Gestalt approach, but somehow in advocating an analysis of spatial inference that followed Koehler's lead, he lost the central idea of insight as a form of perceptual restructuring.

One of Koehler's lecture tours took him to Berkeley, where he spent some time helping with some research related to his own work on memory in chimpanzees. This study was run by a graduate student named Otto Tinklepaugh and its first phase was simply an extension of Hunter's delayed reaction test to monkeys. The second phase was more interesting and illustrates the general experimental approach developed in Tolman's laboratory. With some exceptions, studies of learning in animals up to this time had concentrated on the learning curves provided by individuals or by groups of animals and attempted to answer questions about the underlying processes by examining the details of these curves or by comparing those obtained under a number of different conditions. In contrast, a common method used in Berkeley laboratory was to let an animal master some task and then introduce novel test conditions to find out what specifically had been learned.

Tinklepaugh used the 'direct' version of the delayed reaction test, whereby the monkey could watch as food was placed under one or the other of two containers. She was held by a leash until some specified delay interval had elapsed and when released could go directly to the place in which she remembered food to have been left. Like the children in Hunter's original study and Koehler's chimpanzees, but unlike rats, Tinklepaugh's monkeys performed well over long delays, even when distracted or led out of the room for a while in order to prevent the use of some kind of orienting or other mediating behaviour.

In the second phase Tinklepaugh attempted to find out what the monkey knew at the time of a retention test. Given that she could choose the correct container, did this mean that she remembered only that this was the place to go on the particular trial, or did she retain a precise representation of what had been seen some minutes or even hours earlier? The method chosen by Tinklepaugh resembled, and probably inspired, the one used by Elliott who, as described above, changed the reward for his rats from bran to sunflower seed. The substitution method used by Tinklepaugh was to allow a monkey to see a highly prized banana being placed under a container, but not to let her see it being exchanged for a lettuce leaf before she had an opportunity to make her choice. Under normal circumstances the monkeys would treat lettuce as an acceptable, though unexciting, food. When finding a lettuce leaf where they had seen a banana placed, they consistently rejected the lettuce while they searched around the area of the container, sometimes making angry shrieks. Tinklepaugh concluded that their memory of what they had seen included a representation of the banana in that place.[20]

The influence of Gestalt theory on animal research in America was not limited to work in the Berkeley laboratory. At other universities attempts were made in the late 1920s to find out whether rats could solve problems or demonstrate transposition in a discrimination task. The results were not conclusive, but they were encouraging enough to sustain interest in such questions.[21]

In many of these studies it was a perverse choice of species to test the poorly sighted rat on problems which Koehler had regarded as requiring excellent visual abilities. At Harvard University a new graduate student, B. F. Skinner, proposed that squirrels might be more suitable than rats for tests of insight in mammals other than primates.[22] However, by then there was a further factor to encourage increasing emphasis on the rat. In addition to the huge amount that had been learned since the beginning of the century about this animal's general physiology, a lot of precise and detailed information had been gathered about the structure of its brain. For the long-term aim of understanding behaviour in terms of neurophysiology, to concentrate on the rat had some very clear advantages. One series of studies in particular helped to ensure that the rat secured a pre-eminent place in the kind of physiological psychology based on the animal laboratory that started to expand in America after the First World War. This series also served as an unexpected source of evidence casting further doubt on the belief that learning consists of the formation of specific S-R connections.

As a student at Johns Hopkins University Lashley made a major contribution to the chapter on learning in Watson's *Behavior: an Introduction to Comparative Psychology* of 1914. When he went to Minnesota in 1920 he began to examine the effect of brain lesions on the behaviour of rats in order to understand the neural basis of learning and memory. For fifty years books and articles on psychology and neurology had included illustrations showing connections between one area of the brain and another, including connections through the 'association' cortex that had been established during the lifetime of the individual whose brain was illustrated. Even before Thorndike, much of the appeal attached to the idea of habits based on simple, uni-directional neural pathways came from the belief that such a physiological basis of learning was not only plausible, but actually known to exist. There had been critics of this belief, some armed with logical arguments, some with clinical data from human patients and a few with the results of experiments that observed the behaviour of animals given brain lesions; for example, there was the dispute between Pavlov and Bechterev described in an earlier chapter. But no research on this issue was at all comparable in scale or quality to Lashley's.

Fig. 8.11. Karl Lashley in the mid 1930s

First at Minnesota and then from 1926 at Chicago, Lashley ran experiment after experiment to look at the retention of maze or problem learning and the acquisition of new skills as a function of lesions of various sizes in various parts of the rat's brain. He found very marked effects whereby the degree of impairment was systematically related to the size of the lesion and to the difficulty of the task, but no sign that the *location* of a lesion made any difference. In a

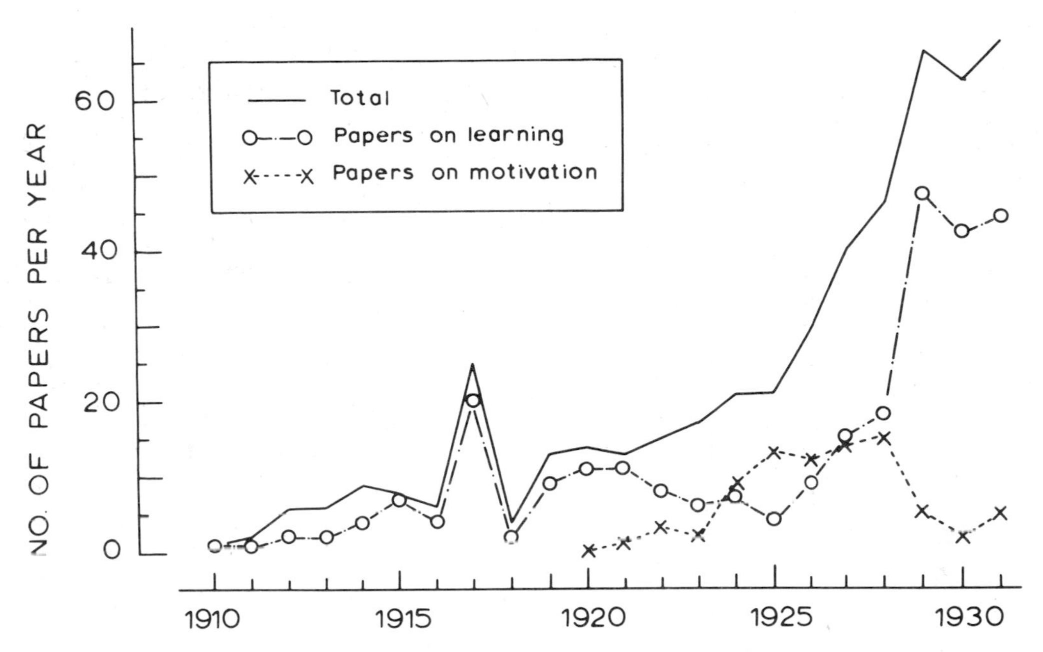

Fig. 8.12. Number of experimental papers per year on the behaviour of the rat from 1910 to 1931; note the temporary increase in papers on motivation during the period 1924–1928 and the very steep increase in papers devoted to learning after this

monograph of 1929 which summarized this research, he wrote: 'I began the study of cerebral function with a definite bias towards such an interpretation of the learning problem (that is, learning as the development of habits and habits as successions of movements). The original programme of research looked toward the tracing of conditioned-reflex arcs throughout the cortex, as the spinal paths of simple reflexes seemed to have been traced through the cord. The experimental findings have never fitted into such a scheme. Rather, they have emphasized the unitary character of every habit, the impossibility of stating any learning as concatenation of reflexes, and the participation of large masses of nervous tissue in the functions rather than the development of restricted conduction-paths.'[23]

From failing to find the neural pathways that had given connectionism its respectability Lashley developed into a fierce critic of S-R theory; his was an unusual case since, as one of his students later expressed it, he began by developing 'just those "Watsonian" notions that, later, he spent most of his professional life refuting' and yet remained a close and loyal friend to Watson, prizing the latter's remark that he was the only thorough-going behaviourist that Watson knew.[24] The research was also important from a methodological point of view, since with its use of the rat, of large groups of animals and of statistical analyses of the results it marked a break with experimental physiology in the European tradition where the typical experiment examined the behaviour of just one or two animals and no quantitative analysis was performed, as in Pavlov's laboratory.

The general growth of interest in the rat is shown graphically in Figure 8.12, which plots the number of papers on rat behaviour published each year from 1910 to 1931.[25] As shown in this graph, there was an accelerating increase during the Twenties: apart from one exceptional year, it took until 1924 before twenty papers were published in a year, yet by 1930 the annual rate had climbed to over sixty. The early part of this rise was due to a burst of papers on motivation beginning around 1923. This was partly an after-effect of the instinct debate, as experimenters' interest turned towards ways of measuring what were becoming known as 'drives'. Such studies included those of Watson's student, Curt Richter, at Johns Hopkins, considerable work on sexual behaviour in rats at Stanford University and a series of experiments at Columbia University which compared the strengths of different drives using the Columbia Obstruction Box.[26]

By 1930 a decreasing number of papers were published on motivational topics and, as shown in Figure 8.12, the overwhelming emphasis was on learning. To a large extent this reflected the accomplishment of Tolman and his students in demonstrating how key issues in learning could be examined by the use of a suitable maze experiment. Tolman dedicated his book of 1932 which reviewed the first ten years of research in the Berkeley laboratory to 'MNA' – *mus norvegicus albinus*, the white rat – and a few years later expressed his conviction that 'everything important in psychology (except such matters as the building of a super-ego, that is everything save such matters as involve society and words) can be investigated in essence through the continued experimental and theoretical analysis of the determiners of rat behavior at a choice-point in the maze'.[27] It would have been difficult in 1920 to find a single psychologist holding such a belief – it was certainly not one that Watson held – but ten years later it was becoming common. And even those who regarded such claims as absurd exaggerations often found it useful, as an aid to that objectivity all agreed was essential in psychology, to formulate their problems in terms of what a rat might do in a particular kind of maze.

The decline and resurgence of strict behaviourism

By 1930 many of the general views that Watson had put forward were widely accepted, but there was little support for his more specific claims, those that characterized 'strict behaviorism', as McDougall had christened the outlook he most detested in American psychology.[1] Watson himself no longer wrote about psychology and those who held views closest to his failed to attract attention.[2] Others who enjoyed more prestige among their colleagues, men like Lashley or Tolman, had moved so far from the theories Watson had advocated that their continued use of the label 'behaviorist' seemed misleading. For them it signified commitment to an entirely objective approach in psychology, for even after Titchener's death in 1927 there were still plenty of psychologists who continued to believe that psychology should also concern itself with subjective states and use introspective methods. But the label did not indicate commitment to the conditioned reflex nor to the rejection of any appeal to internal processes nor to many of the other claims Watson had made after 1913.

Lashley argued that the way to understand the mind was by means of physiological research; he never accepted Watson's suggestion that 'thinking' was based on subvocal speech or some other form of covert response.[3] Tolman's emphasis on the essential goal-directedness of behaviour meant rejecting a basic component of Watsonian behaviourism and, in contrast to Lashley's and to Watson's plea for closer links with physiology, Tolman's approach dropped any pretence that its concepts and variables could be directly related to known physiological processes. His willingness to consider inherited factors in human behaviour and his increasing use of Gestalt terminology placed him still further from Watson. Tolman even began to go back to those issues which twenty years earlier behaviourism had sought to ban from psychology. 'We have been objective and we have refused to introspect. We have assumed that other men likewise were "dumb" and so could not tell us of their conscious experiences. We have sought to build up our psychology as if all the textbooks written up to 1914 (the date of Watson's *Behavior: an Introduction to Comparative Psychology*) had never existed. But after all, we cannot really escape the old questions of *sensation* and *image*, of *feeling* and *emotion*. The good old psychologists in their laboratories, who introspected and filled innumerable pages of their Protokolls with accounts of these processes, were doing something and doing it ably. What, now, in *our* terms was this that they were doing?'[4]

One of the few voices to plead for a return to strict behaviourism was that of Kuo. He became alarmed by the 'conciliatory attitude of the behaviorists towards the traditional psychology' and was horrified by the reception given to the ideas of Gestalt psychology who 'at least as far as their view of learning is concerned, are simply trying to persuade their American colleagues to go back to the Middle ages'. In 1928 Kuo renounced his earlier endorsement of Tolman's 'purposive behaviorism', sharply criticizing the direction in which his Berkeley teacher had since moved and calling for a return to the 'S-R formula directly derived from the basic principles of physics'.[5]

Kuo's attack produced a careful and equally sharp rejoinder from Tolman, summarizing his reasons for using concepts based on inferences from behaviour that were themselves not directly observable in a way that Hobhouse, Jennings or Perry might have done. 'Our doctrine is that behavior (except in the case of the simplest reflexes) is not governed by simple one to one stimulus–response connections. It is governed by more or less complicated sets and patterns of adjustment which get set up within the organism. And in so far as these sets of adjustments cause only those acts to persist and to get learned which end in getting the organism to (or from) specific ends, these sets or adjustments constitute purposes.'[6]

Tolman also defended himself against the accusation that his theories made no reference to physiology. 'I find it an invitation to further investigation. I wonder indeed whether the alternative doctrine of relatively simple one to one stimulus–response connections has now any more acceptable a neurological picture at its service. I doubt if it does. Before Franz and Lashley's results appeared, simple reflex-psychology had a doctrine of specific and insulated sensory–motor paths to fall back on. But now that this simple picture has been taken away, is it, I wonder, any better off as to neurology than is our purposive doctrine?'[7]

Kuo had argued that his 'theory of the prepotent stimulus', which was essentially a conditioned reflex model, could account entirely for the behaviour of a rat in a maze. At the end of his rejoinder Tolman admitted that this was a remote possibility, but one only worth taking seriously when it had been worked out in detail. In reply Kuo simply dismissed Tolman's views as 'really McDougall's animism under disguise' and he never developed the notion of 'prepotent stimuli'.[8] By this time he had returned to China and his position at the University of Shanghai made it difficult to pursue scholarly interests.[9]

For a while no one took on the task that Tolman had suggested to Kuo, that of developing a detailed theory of maze behaviour based on a conditioning model. Furthermore, there was very little experimental work using conditioning procedures, despite the importance Watson had attached to the conditioned reflex, and there were also few experimental studies involving children. In one experiment on 'building likes and dislikes in children' a two-year old boy was given orange juice, which was unexpectedly followed by the sound of a click and a large mouthful of vinegar, and he subsequently showed a distaste for oranges and a tendency to shiver on hearing a clicker.[10] But such Watsonian studies stand out from the journals of this era and only emphasize the absence of any sign elsewhere of Watson's direct influence.

A major problem for anyone who wished to take Watson's advice seriously and concentrate upon the ideas and results generated in Pavlov's laboratory for almost thirty years was that they were effectively inaccessible except in bare outline. The problem was to a large extent overcome in 1927 when the English translation of Pavlov's *Conditioned Reflexes* appeared. This acted as an important stimulus to learning theory in America, at first by encouraging careful comparisons between the procedures employed for the trial-and-error or instrumental studies of Western comparative psychology and those used by Pavlov. Were such differences simply products of an historical accident or did they reflect two different processes of learning? Could the learning displayed by a rat who ran to a goal-box without entering a *cul-de-sac* or who swiftly opened a puzzle box be analysed in the same way as that of a dog salivating before the arrival of food? By now Herbert Spencer's thoughts on learning were rarely cited and never read, but for the first time there was a sustained effort to sort out the muddle he had helped to create when, in the 1870 edition of his *Principles of Psychology*, he had added Alexander Bain's idea of instrumental conditioning to the idea of classical conditioning which he had gained from David Hartley and expressed so much more clearly in the first edition of his book fifteen years earlier.

With the appearance of Pavlov in English there was at first wide agreement that Watson, once again, was wrong and that trial-and-error learning could not be explained in terms of Pavlov's conditioned reflex.[11] In Poland Konorski and Miller had already reached the same conclusion.[12] In 1932 Tolman made clear his view that there existed a variety of forms of learning, of which instrumental and classical conditioning represent two distinct types.[13] The conditioned reflex was regarded as an important, but not a central, concept.

The intellectual atmosphere soon began to change. By the mid-1930s a group of learning theorists centred around Clark Hull at Yale University were becoming a powerful force in American psychology. For these 'neo-behaviorists', as they were later called, and for the 'radical behaviorists' inspired by Skinner who became influential at a later date, one or another model of conditioning has occupied a key theoretical position; what to Tolman and others had seemed to represent different forms of learning have either been treated as secondary or ignored.

There appear to have been two main reasons for the sudden resurgence of a form of strict behaviourism and its subsequent long grip on American psychology. One factor was the powerful institutional position occupied by Hull and his colleagues. Like other forms of research in psychology, animal studies in the 1920s were at best supported by small grants from University coffers. The productivity of Tolman's laboratory at Berkeley was helped considerably by the relatively generous attitude of the University of California towards such research. But even at Berkeley financial support consisted of lump sums towards specific projects which were normally of the order of fifty or a hundred dollars. The amounts often did not seem worth the tedious paperwork that was required to obtain them.[14]

One sign of a change in the financing of research was the launching in 1926 of the 'Behavior Research Fund'. This was based in Chicago and contributions were from private citizens. Although the money was 'to be devoted to research in problems of human behavior', a large part was used to support Lashley's work on brain lesions in rats. With some awkwardness the Foreword to his monograph noted that 'it may puzzle some that the approach to human behaviour should lead through such an apparent by-path, but a frontal attack is not necessarily the most effective . . . the old academic liberty of the Renaissance is the greatest safeguard to successful scientific enterprise'.[15]

The real change to big science came with the Laura Spelman Rockefeller Memorial. In an earlier era the vast sums in this fund might have gone towards a new university or hospital, but instead its administrator, a young man named Beardsley Ruml with a Ph.D. in psychology from Chicago, drew up plans to distribute some forty million dollars over a period of seven years so as to promote research in the social sciences.

One grant from this Rockefeller fund went to support the research on childhood fears that Mary Cover Jones carried out with Watson. Other, larger grants went towards the founding of child study centres at a number of different universities. The main impact on animal psychology came from the relationship between the fund and Yale University. The psychology instructor at Chicago, James Angell, who had acted as graduate supervisor to both Watson and Carr, and later to Ruml, moved in 1920 to a post with the Carnegie Foundation, taking Ruml with him as administrative assistant. A year later Angell became president of Yale, where his achievements included the expansion of psychology and the attraction of outside support for research in the social sciences from various bodies, but pre-eminently from the Rockefeller fund that his former assistant now directed.[16]

A major product of the flow of money into Yale was the Institute of Human Relations. In 1929 Clark Hull was appointed as a member of this institute and, despite its name and Hull's own previous interest in a range of topics in human psychology including hypnosis and aptitude testing, it rapidly became a centre for the development of learning theory based on a conditioning model and for research on maze-running by rats.[17] In the same year Robert Yerkes at last obtained from the National Research Council the massive funding for primate research that he had sought for over fifteen years and he re-entered academic life as a research professor at Yale.

At the best of times members of university psychology departments found it difficult to maintain a steady research programme. Even when some financial support was forthcoming from their university to pay for equipment, materials or other expenses, there was the difficulty of time. Teaching undergraduates in an effective way, supervising graduate students, taking on summer school teaching to supplement modest salaries and making a fair contribution towards the general running of their department and the university, all of these added up to a satisfying, comfortable, but reasonably full, occupation.

In 1876 Daniel Gilman had hoped that the ideal of the university as a centre of research, for which professors would be allowed at least as much time as was expected to be spent on their teaching, would spread across America. Forty years later James Cattell carried out a survey to find out how many members of the American Psychological Association could be counted as professional scientists, by which he meant that they spent at least half of their working hours on research; he found that only sixteen of the two hundred or so members qualified.[18] Presumably Watson and Yerkes were among these sixteen, since, with regard to time if not to money, they had enjoyed unusually favourable conditions for their research at Johns Hopkins and Harvard. Most of their colleagues further west could only maintain a scientific career at the cost of severe pressure on other aspects of their lives.

The opportunities offered by the newly funded research institutes were excellent by any standards. Those in such institutions became even more favoured relative to their fellow psychologists as economic depression dried up the supply of university positions for those with new Ph.Ds and increased the teaching loads of those who already held posts. This increased still further the power to determine the future of psychology of those sitting on committees which distributed research money. An individual might bubble with ideas, might have published a series of excellent papers and served as an inspiring teacher to his graduate students, but, if for reasons other than academic ones, he was denied funding, he would not be able to carry out much experimental work nor support his students, so that they need not look to other fields to find jobs.

Ruml's decision to direct the Rockefeller funds into research came from his belief that social problems were only temporarily assuaged by direct grants to welfare organizations and that in the long run what was needed was 'the production of a body of fact and principle', knowledge that will be 'expected in time to result in substantial social control'.[19] Within psychology, research on animal behaviour was one likely candidate, especially for someone familiar with such work from his graduate student days. It was intellectually attractive, in that the work from Tolman's laboratory plus the confrontation between American comparative psychology, Gestalt theory and Pavlov's research on conditioning had produced a large set of well-defined and basic theoretical questions. Is there a single form of learning? Can problem-solving be reduced to trial-and-error learning? Is reward necessary for learning? How is reward related to drive? Is there a difference between spatial and response learning? How are emotional responses acquired? Is avoidance learning solely an instance of a conditioned reflex? What is the physiological basis of this or that aspect of learning or motivation? Furthermore, these and other such questions appeared to be highly tractable ones, which could well be completely answered by a suitable series of experiments, while the many years of rat research of the kind carried out by Carr's students provided the invaluable background of technique and parametric information needed for such work.

Hull and his co-workers at Yale set about the task of answering such questions in terms of concepts derived from Pavlov and using the experimental approach developed by Tolman. They began to attack Tolman's claims in a concentrated, thorough manner that made his style seem that of the dilettante. Years later he wrote that he had 'liked to think about psychology in ways that proved congenial to me. Since all the sciences, and especially psychology, are still immersed in such tremendous realms of the uncertain and the unknown, the best that any individual scientist, especially any psychologist, can do seems to be to follow his own gleam and his own bent, however inadequate they may be. In fact, I suppose that actually this is what we all do. In the end the only sure criterion is to have fun. And I have had fun.'[20] In contrast, Hull and his colleagues were very serious, both about their science and the need to exploit the powerful position they held at Yale in extending their influence elsewhere.

In order to attract such funding it was, of course, not enough to be studying tractable, intellectually satisfying problems in a serious, scientific manner. The research also needed ideological appeal. This was provided by the environmentalism that, largely due to Kuo and Watson, had become closely associated with behaviourism. Identify the basic principles of learning in a form that is comparable to the laws of mechanics and then, and only then, a solid and comprehensive foundation is obtained, the promise ran, for understanding the complex process whereby an adult human being acquires all those skills, attitudes and feelings that characterize his 'culture'.

It was a bold promise and an unlikely one, since it left out entirely the social dimension to human learning. Even Tolman, with all his enthusiasm for *mus norvegicus albinus*, allowed that the human super-ego and language were not open to profitable approach via maze-running by rats. Since the beginning of the century most psychologists believed it to be self-evident that a large part of what a person learns is from other people, by means of language, imitation or some other process. Yet during this period animal psychology had made singularly little progress in understanding imitation learning, while attempts to study the language learning capacities of apes had only served to emphasize the uniqueness of man in this respect. According to McDougall and Trotter the study of instincts was to provide the basis for social psychology. With the rejection of instinct theory there also went, usually unheeded, rejection of the idea that 'man is throughout a social creature . . . (whose) mind develops entirely out of interactions between him and other human beings, and that an individual not so built up is unthinkable'.[21]

The commitment to the study of the isolated individual, whether rat or man, was by no means confined to the behaviourists. It was part of a common heritage from the experimental tradition of physiology, taken for granted to such an extent that it was rarely commented upon. Thus, for example, the studies of intelligent problem-solving inspired by Gestalt psychology maintain, as recently suggested, an 'idea of the intellectually challenging environment as perfectly described by Daniel Defoe. It is the desert island of Robinson Crusoe – before the arrival of Man Friday. The island is a lonely, hostile environment, full of technological challenge, a world in which Crusoe depends for his survival on his skill in gathering food, finding shelter, conserving energy, avoiding danger.'[22] Applications of the new learning theory to education were to be achieved, as we have seen, by evaluating the role of spacing or of reward in an individual's acquisition of a habit and not by examining some other factor, arguably with at least as great a potential importance, such as the relationship between child and teacher or the parent's role as a model for what a child treats as worth knowing.

Watson's influence on the resurgence of strict behaviourism was indirect. Few of the people involved had ever met him and Hull, for example, did not have a high opinion of Watson's theories.[23] Watson's major contribution was that, more than anyone else of his generation, he had campaigned for the extension of experimental method into areas of human interest that psychology had hitherto hardly touched. The essence of Watsonian behaviourism, shown as much by what he did as by what he wrote, was his belief that experimental evidence is the only means we have for achieving true understanding of some issue and that this kind of science has to be applied directly, even if this means the overthrow of deeply held convictions and fundamental mores of the scientist's own culture. How foolish to legislate on Prohibition when we have no evidence on the effects of alcohol on human behaviour! How irrational to spend large sums on propaganda to reduce the incidence of venereal disease without experimental tests to find out how such propaganda affects sexual behaviour! How absurd to produce innumerable books of advice on bringing up children, but make no scientific effort to find out what factors are really important in human development!

To Watson the 'baby-farm' project for studying under well-controlled conditions the first few years of children of varied racial background was a profoundly sensible idea. His contemporaries were either far from desiring Watson's radical rejection of conventional values or more impressed than he by the enormity of the dangers and problems associated with such ventures. In 1929 a psychologist with general views close to Watson's was suspended from the University of Missouri for approving the circulation among students and their parents of a questionnaire concerning attitudes towards sex that an undergraduate in his department had prepared. A sociology instructor more directly involved in the project was dismissed and the undergraduate, O. Hobart Mowrer, left to continue his studies in the East, eventually ending up at Yale where he became associated with Hull's group at the Institute of Human Relations.[24]

As 'Mr. Behaviourist', Watson created a generally favourable atmosphere for his successors by means of the articles he wrote in the 1920s which startled, enraged and intrigued the readers of the *New Yorker*, *New York Times* or *Harpers Monthly*.[25] Such articles were of considerable help in making subscribers to the 'Behavior Research Fund' or trustees of the Rockefeller Memorial sympathetic towards arguments that investment in behavioural science would produce a technology to solve social problems. A detour around the awe-inspiring barriers blocking the direct assault that Watson wanted was provided by the rat. The relative ease of obtaining clean, unambiguous results from animal experiments was, for example, pointed out by Tolman: 'let it be noted that rats live in cages; they do not go on binges the night before one has planned the experiment'.[26] Presumably because such experiments were rare in the Berkeley laboratory, he did not also add that, as well as preventing them from leaving their cages, conventional ethics allow a human being with suitable status, acceptable motives and a convincing rationale to shock, poison, blind, castrate, drug and surgically interfere with the brain and body of a rat.

Interest in what Watson had to say was not limited to fellow psychologists and readers of up-market New York publications. By a curious turn of fate, the person whose mother, when he was a baby, had acted as an assistant in 1872 in Douglas Spalding's experiments on imprinting and instinct, some fifty years later developed an interest in animal behaviour, although in complete ignorance of the circumstances of his childhood. By then Bertrand Russell was the leading philosopher of the English-speaking world, but also almost penniless. One way for him to earn money was to give lectures in America; another was to write popular books on philosophy. Following his interest in the philosophy of mind, he began to read about recent research in animal psychology; he was provided with an excellent opportunity for getting through a large number of books during a spell in prison for pacifist activities towards the end of the First World War.[27]

On a visit to New York a few years later Russell met his fellow exile from academia, John Watson, who provided hospitality and contacted some of his old friends to see if they could arrange for Russell to lecture at their universities. In 1927 Russell published a brisk and readable book called *Philosophy*, of which a large part was devoted to the basic conceptual issues arising in contemporary physics and contemporary psychology.[28] The aspects of psychology he commented upon were animal research, theories of learning, and behaviourism. The book was generally favourable towards Watson's approach and, though it rejected some of his specific theories – dismissing, for example, Watson's attack on the Law of Effect – it endorsed others with an enthusiasm few psychologists shared. For example, Russell wrote that 'the failure to consider language explicitly has been a cause of much that was bad in traditional philosophy. I think myself that "meaning" can only be understood if we treat language as a bodily habit, which is learnt just as we learn football or bicycling. The only satisfactory way to treat language, to my mind, is to treat it in this way, as Dr Watson does. Indeed, I should regard the theory of language as one of the strongest points in favour of behaviourism.'[29]

For so eminent a philosopher to show such sympathy gave behaviourism considerable prestige. The book was exciting for anyone who had begun to think about philosophical problems and who was also fascinated by the dizzying concepts of relativity and atomic theory and by the prospect of a new science of mind. In September, 1927 McDougall commented on Watson that 'it is the success of his appeal among young students that is the disturbing fact for those who hope much from the splendid development of American universities now going on so rapidly'.[30] In the same month Russell's book was discovered by B. F. Skinner, whose attempts to embark on a career as a writer after leaving college no longer seemed promising; the book inspired him to take up psychology instead.[31] Russell thus played some part in ensuring that, when neo-behaviourism began to falter in the 1950s, Skinner's version of behaviourism was waiting to step into its place.

Notes

The following provides for each section of the book a list of the major sources I used in writing it and in some cases suggestions on further reading on the topic, followed by the notes which are numbered separately for each section. Where a book is cited extensively I have used initials as an abbreviation of its title, as indicated below.

AI: G. J. Romanes *Animal Intelligence* London: Kegan, Paul, Trench & Co., 1882.

AL: F. Lilge *The Abuse of Learning: the Failure of the German University* New York: Macmillan, 1948.

ALI: C. L. Morgan *Animal Life and Intelligence* London: Edward Arnold, 1890.

AP: E. A. Asratyan *I. P. Pavlov; his Life and Work* Moscow: Foreign Language Publishing House, 1953. (Russian original, 1949.)

B: J. B. Watson *Behaviorism* New York: Norton, 1924.

BA: A. Bain *Autobiography* London: Longman Green & Co., 1904.

BICP: J. B. Watson *Behavior: an Introduction to Comparative Psychology* New York: Henry Holt, 1914.

BLO: H. S. Jennings *The Behavior of the Lower Organisms* New York: Macmillan, 1906. (Reprinted in Bloomington, Ind: Indiana University, 1962.)

BM: L. C. Rosenfield *From Beast-machine to Man-machine: the Animal Soul in French Letters from Descartes to la Mettrie* New York: Oxford University, 1940.

BP: B. P. Babkin *Pavlov* Chicago: Chicago University, 1949.

CP: H. Cuny *Ivan Pavlov* (Translation by P. Evans) London: Souvenir Press, 1964. (French original, 1962.)

CR: I. Pavlov *Conditioned Reflexes* (Translation by G. V. Anrep) London: Oxford University, 1927.

CRP: I. Pavlov *Conditioned Reflexes and Psychiatry* (Translation by W. H. Gantt) London: Laurence & Wishart, 1941.

CW: D. Cohen *J. B. Watson: the Founder of Behaviourism* London: Routledge & Kegan Paul, 1979.

DIR: H. Seton-Watson *The Decline of Imperial Russia* London: Methuen, 1964.

DM: R. Descartes *A Discourse on Method* (Translation by J. Veitch) London: Dent, 1912. (French original, 1637.)

EW: A. Bain *The Emotions and the Will* London: Parker & Son, 1859.

FA: N. G. Hale *Freud and the Americans* New York: Oxford University, 1971.

FG: D. W. Forrest *Francis Galton: the Life and Work of a Victorian Genius* London: Elek, 1974.

FP: Y. P. Frolov *Pavlov and his School* London: Kegan, Paul, Trench, Trubner & Co., 1937.

GPHR: V. M. Bechterev *General Principles of Human Reflexology* New York: International Publishers, 1932. (Reprinted New York: Arno, 1973.)

GSH: D. Ross *G. Stanley Hall: the Psychologist as Prophet* Chicago: University of Chicago, 1972.

H: T. H. Huxley *Hume* London: Macmillan, 1878.

HB: E. Nordenskiold *The History of Biology* New York: Tudor, 1928.

HBSC: E. Clarke & C. D. O'Malley *The Human Brain and Spinal Cord* Berkeley: University of California, 1968.

HDN: M. A. B. Brazier 'The historical development of neurophysiology.' In J. Field (Ed) *The Handbook of Physiology, Section 1*: Neurophysiology Vol. 1 Washington, DC: American Physiological Society, 1959.

HEP: E. G. Boring *History of Experimental Psychology* New York: Appleton-Century-Crofts, 1950. (2nd Ed.)

HG: F. Galton *Hereditary Genius* London: Macmillan, 1869.

HI: C. L. Morgan *Habit and Instinct* London: Edward Arnold, 1896.

HM: F. Lange *The History of Materialism* (Translation by E. C. Thomas) London: Truebner, 1880.

HPA: *A History of Psychology in Autobiography* Vols 1, 2 & 3 (Edited by C. Murchison), Worcester, Mass: Clark University, 1930, 1932 & 1936 (Reprinted in New York: Russell & Russell, 1961); Vol. 4 (Edited by E. G. Boring *et al.*) and Vol. 5 (Edited by E. G. Boring & G. Lindzey), New York: Appleton-Century-Crofts, 1952 & 1967; Vol. 6 (Edited by G. Lindzey), Englewood Cliffs, NJ: Prentice Hall, 1974.

HREP: N. Hans *History of Russian Educational Policy: 1701–1917* New York: Russell & Russell, 1931. (Reprinted in 1964.)

HSJ: T. M. Sonneborn 'Herbert Spencer Jennings'.

Biographical Memoirs Vol. 47 Washington, DC: National Academy of Sciences, 1975.

HTHM: J. Priestley *Hartley's Theory of the Human Mind, on the Principle of the Association of Ideas* London: Johnson, 1775.

ICP: C. L. Morgan *An Introduction to Comparative Psychology* London: Scott, 1894.

IHPW: W. Trotter *Instincts of the Herd in Peace and War* London: Ernest Benn, 1916.

ISP: W. McDougall *An Introduction to Social Psychology* London: Methuen, 1908.

JLCL: P. J. Pauly *Jacques Loeb and the Control of Life: Experimental Biology in Germany and America, 1890–1920* Unpublished Ph.D. thesis, Johns Hopkins University, 1980.

JPP: W. James *Principles of Psychology* New York: Holt, 1890.

LCR: I. Pavlov *Lectures on Conditioned Reflexes* (Translation by W. H. Gantt) London: Laurence & Wishart, 1928.

LHAP: W. Wundt *Lectures on Human and Animal Psychology* (English translation by Creighton & Titchener) London: Sonnenschein, 1894. (German original: Leipzig, 1892.)

LK: J. D. Burchfield *Lord Kelvin and the Age of the Earth* London: Macmillan, 1975.

LLD: F. Darwin (Ed.) *The Life and Letters of Charles Darwin* London: Murray, 1887.

LHH: L. Huxley (Ed.) *The Life and Letters of T. H. Huxley* London: Macmillan, 1900. Vols I & II.

LLLG: K. Pearson *The Life, Letters and Labours of Francis Galton* Cambridge: Cambridge University, 1914, 1930.

LLR: E. Romanes (Ed.) *The Life and Letters of George John Romanes* London: Longmans, Green & Co., 1902.

MA: W. Koehler *The Mentality of Apes* (Translation of 2nd German edition by E. Winter) New York: Harcourt, Brace & World, 1925. (Original German, 1917.)

ME: L. T. Hobhouse *Mind in Evolution* London: Macmillan, 1901.

MEA: G. J. Romanes *Mental Evolution in Animals* New York: Appleton & Co, 1884.

MLMA: R. M. Yerkes *The Mental Life of Monkeys and Apes* Behavior Monographs, 3, No. 1 New York: Holt, 1916. (Reprinted in Delmar, New York: Scholars Facsimiles & Reprints, 1979.)

MM: J. O. de la Mettrie *Man a Machine* La Salle, Illinois: Open Court, 1953. (French original, *L'Homme Machine*, 1748.)

NI: F. Galton *Natural Inheritance* London: Macmillan, 1889.

PBAM: E. C. Tolman *Purposive Behavior in Animals and Men* New York: Century, 1932.

PCIC: J. B. Watson and R. Watson *The Psychological Care of Infant and Child* New York: Norton, 1928.

PK: E. Cassirer *The Problem of Knowledge: Philosophy, Science and History since Hegel* New Haven: Yale University, 1950.

PO: V. Bechterev *La Psychologie Objective* Paris: Librairie Felix Alcan, 1913.

PP1: H. Spencer *Principles of Psychology* London: Longman, Brown, Green & Longman, 1855. 1st Ed.

PP2: H. Spencer *Principles of Psychology* London: Longmans, 1870. 2nd Ed.

PSB: W. McDougall *Psychology: the Study of Behaviour* London: Williams & Norgate, 1912.

PSPB: J. B. Watson *Psychology from the Standpoint of a Behaviorist* Philadelphia: Lippincott, 1919.

PSW: I. Pavlov *Selected Works* Moscow: Foreign Publishing House, 1955.

RA: F. Fearing *Reflex Action: a Study in the History of Physiological Psychology* New York: Hafner, 1930.

RCE: G. W. Stocking *Race, Culture and Evolution* New York: Free Press, 1968.

SA: H. Spencer *An Autobiography* London: Williams & Norgate, 1904. Vols. I & II.

SAN: I. Sechenov *Autobiographical Notes* (English translation by K. Haines) Washington, DC: American Institute of Biological Sciences, 1965. (Russian original, 1905.)

SD: R. Hofstadter *Social Darwinism in American Thought* Philadelphia: University of Philadelphia, 1944.

SHE: C. Bibby *T. H. Huxley: Scientist, Humanist and Educator* London: Watts, 1959.

SI: A. Bain *The Senses and the Intellect* London: Parker & Son, 1855.

SP: G. Joncich *The Sane Positivist: a Biography of E. L. Thorndike* Middletown, Conn: Wesleyan University, 1968.

SRC: A. Vucinich *Science in Russian Culture: 1861–1917* Stanford, California: Stanford University, 1970.

SRS: J. Ben-David *The Scientist's Role in Society* Englewood Cliffs, NJ: Prentice-Hall, 1971.

SSW: I. Sechenov *Selected Works* Amsterdam: Bonset, 1968.

TAI: E. L. Thorndike *Animal Intelligence* New York: Macmillan, 1911.

TE: H. Cravens *The Triumph of Evolution: American Scientists and the Heredity–Environment Controversy, 1900–1941* Philadelphia, University of Pennsylvania, 1978.

VB: C. Darwin *The Voyage of the Beagle* London: Dent, 1906. (Original version published in 1839.)

WJ: R. B. Perry *The Thought and Character of William James* Boston: Little, Brown & Co., 1935.

WMcD: W. McDougall 'Autobiographical sketch'. In *HPA* Vol. 1 (1930), pp. 191–224.

YB: P. I. Yakovlev, 'Bechterev.' In M. A. B. Brazier (Ed.) *The Central Nervous System and Behavior* New York: Josiah Macy Foundation, 1959.

Chapter 1: Mental evolution

Charles Darwin and The Descent of Man
Excellent accounts of the development of evolutionary theories during the nineteenth century are provided by Eiseley (1958) and by Himmelfarb (1959). These served as major sources for this section, together with the interesting treatment of the development of Darwin's theories and his views on human evolution in Gruber (1974). A very recent and excellent treatment of Darwinism in general is provided by Oldroyd (1980).

1 See Burckhardt (1977).
2 C. Darwin *The Voyage of the Beagle* (*VB*).
3 *VB*, p. 195.
4 *VB*, p. 206.
5 Gruber (1974).
6 Biographical information is from the two volume autobiography (Wallace, 1903) and the more recent and briefer account in Williams-Ellis (1966). A contemporary tribute deserves quotation: 'Once in a generation a Wallace may be found physically, mentally and morally qualified to wander unscathed through the tropical wilds of America and Asia; to form magnificent collections as he wanders; and withal to think out sagaciously the conclusions suggested by his collections' (Huxley, 1864; p. 35).
7 Jenkin (1867).
8 Wallace (1864).
9 Wallace (1869).

The Spencer–Bain Principle
The general contributions of Spencer and Bain are discussed in any number of texts, but rarely in any detail. Boring (1950; *HEP*) and Hearnshaw (1964) were among the more useful here. Biographical material is from these two sources and from the autobiographies of Bain (*BA*) and of Spencer (*SA*).

There has been some confusion as to whether the Spencer–Bain Principle was originally the idea of Spencer or of Bain. Little doubt about Bain's priority remains after comparing his *The Senses and the Intellect* (*SI*) of 1855 with the first edition, also of 1855, of Spencer's *Principles of Psychology* (*PP1*) and the latter with the second edition (1870) of this work (*PP2*). Spencer was well acquainted with Bain's work long before 1870; in 1859 he wrote a review of Bain's *The Emotions and the Will* (*EW*) and cited Bain in the second edition of his book on topics other than those discussed here.

1 Hull (1973).
2 Flugel (1933).
3 *SI*, Preface.
4 *SI*, p. 289.
5 *SI*, p. 404.
6 *EW*, p. 349.
7 *SA1*, p. 402.
8 *SA2*, p. 171.
9 Haight (1968); Hirshberg (1970).
10 *PP1*, p. 530.
11 *PP1*, p. 543.
12 *PP1*, p. 540.
13 *PP1*, p. 561.
14 *PP1*, p. 465.
15 *PP1*, p. 465.
16 *PP2*, p. 581.
17 *PP1*, p. 574.
18 *SA1*, p. 402.
19 *PP2*, p. 280.
20 *PP2*, pp. 281–6.
21 *PP2*, p. 507.
22 *PP2*, p. 545.
23 *PP2*, p. 545.
24 Himmelfarb (1959).

Douglas Spalding's experiments on instinct
1 Biographical material is from Haldane (1954), which accompanies a reprint of Spalding's first paper, and from Gray (1962; 1968).

2 *SA2*, p. 55.
3 Russell (1967; pp. 15–19); Russell & Russell (1937).

Thomas Huxley and animals as conscious automata
Biographical material is from *The Life and Letters of T. H. Huxley (LLH)*, the biography by Bibby (*SHE*) and Irvine (1955).

1 *SHE*, p. 39.
2 *LLH1*, p. 378.
3 *LLH1*, p. 176.
4 *SHE*, p. 69.
5 *SHE*, p. 70.
6 *SA1*, p. 404.
7 *SHE*, p. 60.
8 Huxley (1874).
9 Huxley (1878, *H*).
10 *H*, p. 184.
11 *H*, p. 191.
12 *H*, p. 192
13 *H*, p. 106.
14 *LLH2*, p. 38.

Chapter 2: Intelligence and instinct

The opening quotation is from the final paragraph of Wallace (1889; p. 478).
Developments in evolutionary theory after Darwin's death are discussed in detail in Romanes (1895). This was the work that Morgan edited and for which he arranged publication following Romanes' death. The concepts of mind held by Romanes, Morgan and other early comparative psychologists have been discussed from a philosophical point of view in an interesting manner by Mackenzie (1977).

The systematic classification of anecdotal evidence by George Romanes
Biographical material is from Hearnshaw (1964) and *The Life and Letters of George John Romanes* (*LLR*). Page references are to *Animal Intelligence* (*AI*) and to *The Life and Letters of Charles Darwin* (*LLD*).

1 The comment was made by Samuel Butler; see Pauly (1982).
2 *LLR*.
3 *AI*, Preface, p. vii.
4 *AI*, p. 26.
5 *AI*, Preface, pp. vii–ix.
6 *AI*, p. 229.
7 *AI*, p. 324.
8 Spalding (1873).
9 *LLH1*, p. 452.
10 *LLD1*, p. 83.
11 *AI*, p. 420.
12 *AI*, p. 353.
13 *AI*, p. 156; for a review of recent evidence on communication in bees, see Gould (1976).

Romanes on mind, instinct and intelligence
Page references are to Romanes' *Mental Evolution in Animals* (*MEA*)
1 *MEA*, p. 15.
2 *MEA*, p. 17.
3 *AI*, p. 6.
4 *MEA*, p. 63.
5 Captain Marryat *Masterman Ready* (1846; p. 273).
6 Marryat (1846; p. 279).
7 *MEA*, p. 159.
8 *MEA*, p. 161.
9 *MEA*, p. 212.
10 *MEA*, pp. 221–9.
11 *MEA*, p. 193.
12 *MEA*, p. 193.
13 *LLR*.
14 *MEA*, p. 12.

Lloyd Morgan and the cinnabar caterpillars
Biographical material on Morgan is from his autobiographical sketch in the second volume of *A History of Psychology in Autobiography (HPA)*, his entry in the *Dictionary of National Biography* and the note by Reynolds (1900). These were supplemented by material in the C. L. Morgan collection in Bristol University Library and by the personal recollections of several of his relatives when I talked with them during 1977 and 1978, as noted below. Page references below are to Morgan's *Animal Life and Intelligence* (*ALI*) and his *Introduction to Comparative Psychology* (*ICP*).

1 *AI*, p. 222.
2 Morgan (1883).
3 Romanes (1895; p. 300).
4 *ALI*, p. 356.
5 *ALI*, p. 399 (Italics as in original).
6 *ALI*, p. 366.
7 *ALI*, p. 422.
8 *ALI*, p. 403.
9 *ALI*, p. 213.
10 *ALI*, p. 444.
11 *LLR*.
12 In these two paragraphs published evidence has been supplemented by information kindly provided by Mr M. C. Morgan, the late Bishop E. R. Morgan, Mrs Mary Denniston and Mrs C. Morgan.
13 *HPA*, p. 249.
14 *ICP*, Preface, p. xi.
15 From the summary Morgan provides in *ICP*, pp. 85–9; 197–206.
16 *ICP*, p. 203.
17 *ICP*, p. 203.
18 *ICP*, p. 89.
19 *ICP*, p. 214. This work on 'distasteful insects' followed that of E. P. Poulton, who found that predators learned to associate 'gaudiness' and 'nastiness', but did not control the early experience of his animals; see Poulton (1890).

20 ICP, p. 215.
21 ICP, pp. 254–8.
22 ICP, pp. 291–4.
23 Morgan (1900).

Morgan on comparative psychology and theories of learning

Morgan's views on Huxley's Belfast paper and on the mind–body problem around this time are also discussed in Morgan (1896*b*).

1 *ALI*, pp. 304; 417; 464.
2 *ALI*, p. 464.
3 *ICP*, p. 4.
4 *ICP*, p. 8.
5 *ICP*, p. 163.
6 *ICP*, p. 186.
7 *ICP*, p. 47.
8 *ICP*, p. 50.
9 *ICP*, p. 29.
10 *ICP*, p. 182.
11 *ICP*, p. 216.
12 *ALI*, p. 415.
13 'Metakinetic' was used by Morgan in *ALI*, pp. 464–70, but the term is not used in the similar argument in *ICP*, pp. 327–36.
14 *ICP*, p. 336.
15 *ICP*, p. 342.
16 Macpherson (1900; p. 10).

Morgan's canon, psychological complexity and instinct

A more detailed analysis of Morgan's treatment of instinct than that provided here is in Richards (1977). References below are to books cited above, plus Morgan's *Habit and Instinct* (*HI*).

1 *ICP*, p. 303.
2 *ICP*, p. 296.
3 *ICP*, p. 370.
4 *ICP*, pp. 376–9.
5 *ICP*, p. 53.
6 *ICP*, p. 59.
7 *ICP*, p. 59.
8 *ICP*, p. 253.
9 *ICP*, p. 244.
10 *ICP*, p. 285.
11 *ICP*, p. 118.
12 *ICP*, p. 114.
13 Tarde (1890); Baldwin (1894).
14 *ALI*, p. 422.
15 *ALI*, p. 138.
16 *HI*, Ch. 2.
17 *HI*, pp. 20–2.
18 *HI*, p. 156.
19 *MEA*, p. 198.
20 *HI*, p. 121.
21 Morgan's conclusions were very similar to those reached by Wallace thirty years earlier in discussing the subject of nestbuilding by birds (Wallace, 1867).
22 *HI*, p. 166.
23 *HI*, p. 171.

Mathematics, heredity and Francis Galton

An excellent account of nineteenth-century debates over geological time is given by Burchfield's *Lord Kelvin and the Age of the Earth* (*LK*). Biographical information on Galton is from Pearson's *Life, Letters and Labours of Francis Galton* (*LLLG*) and from the more recent, and briefer, biography by Forrest (*FG*). Page references to Galton's books are to the second edition of *Hereditary Genius* (*HG*) and to the first edition of *Natural Inheritance* (*NI*).

1 Huxley (1894).
2 *LK*, p. 73.
3 Marchant (1916; pp. 250–1); see also *LK*, p. 78.
4 *LK*, pp. 90–117.
5 *LK*, pp. 93–109.
6 Morgan (1878).
7 The quotation is taken from Bartlett's *Familiar Quotations* (1968) which gives the source as Kelvin's *Popular Lectures and Addresses* (1891–1894), but I have been unable to trace the original quotation.
8 Galton (1879).
9 Mill (1873).
10 *FG*, pp. 1–26.
11 *LLLG1*, pp. 196–207.
12 *FG*, pp. 38–59.
13 Huxley's views on race, class and sex are discussed by Bibby (1959; pp. 25–34).
14 *HG*, p. 12.
15 *HG*, p. 35.
16 *HG*, pp. 338–48.
17 Galton (1908; p. 290); quoted in *FG*, p. 101.
18 Galton (1872); see *FG*, pp. 111–13.
19 *FG*, pp. 114–21.
20 *LLLG*, pp. 156–69.
21 *NI*, p. 80.
22 *FG*, pp. 149–70.
23 Galton (1883; p. 20).
24 Galton (1883; p. 21).
25 *NI*, p. 194.
26 *NI*, p. 155.
27 *NI*, p. 62.
28 Hearnshaw (1964; pp. 66–8); see also Introduction in Pearson (1892).
29 *LK*, p. 21; *LLH2*, pp. 375–9.
30 *LK*, pp. 134–40.
31 E. B. Poulton 'A naturalist's contribution to the discussion upon the age of the earth'. British Association, 1896. Reprinted in Poulton (1908).
32 *LK*, pp. 163–4.

Morgan, Galton and British psychology

1 *HI*, p. 334.

2 *HI*, p. 340.
3 *HI*, p. 340. Italics as in original.
4 Baldwin (1926).
5 Webb (1926; p. 25).
6 Webb (1926; p. 116).
7 From Eiseley (1958).

Summary

1 Wallace (1891; p. 181). This includes a reprint of A. R. Wallace 'The development of human races under the law of natural selection'. *Anthropological Review*, 1864.
2 Wallace (1891; p. 181).

Chapter 3: Experimental psychology and habits

German science and psychology

The account of nineteenth-century German science is based mainly on Ben-David's *The Scientist's Role in Society* (*SRS*) and to a lesser extent on Lilge's *The Abuse of Learning: the Failure of the German University* (*AL*). The development of experimental psychology in German universities is described in a number of standard histories of psychology, of which the classic remains Boring's *History of Experimental Psychology* (*HEP*). Another classic work, Nordenskiold's *History of Biology* (*HB*), describes the impact of evolutionary theory on German biology. This topic is also discussed by Cassirer in *The Problem of Knowledge: Philosophy, Science and History since Hegel* (*PK*). The references to Wundt's *Lectures on Human and Animal Psychology* (*LHAP*) are to the English translation of 1894, which was the first English publication of any of Wundt's work.

1 The professionalization of both humanistic and scientific studies in English universities and its relation to German developments are discussed in Haines (1969).
2 *HEP*, p. 708.
3 H. von Helmholtz, *Popular Lectures on Scientific Subjects*, 1st series; cited in *AL*, p. 61.
4 The standard account of Wundt's intellectual contribution to German psychology is that provided by Boring (*HEP*). In recent years this has been challenged as being totally misleading; see Blumenthal (1975; 1980) and Danzinger (1979). A recent summary account of the immediate background to experimental psychology of a hundred years ago is provided by Littman (1979), while the general conditions favouring the provision of an institutional basis for the new psychology are explored in Ben-David & Collins (1966).
5 This account of Haeckel is mainly from *HB*, pp. 505–27.
6 Cited in *HB*, p. 511.
7 *PK*, p. 177.
8 See *HB*, p. 516.
9 Recapitulation theory was originally suggested by Meckel (1781–1833); see *HB*, p. 516 and Gould (1977).
10 The quotation is from von Baer, as cited in *HB*, p. 519.
11 See *PK*, Ch. 11.
12 *LHAP*, p. 350.
13 *LHAP*, p. 386.
14 *LHAP*, p. 365.

American university reform and Herbert Spencer

A general account of higher education in nineteenth-century America is provided by Hofstadter & Hardy (1953) and Rudolph (1965). Specific developments at Johns Hopkins and Clark Universities are described in Ross's biography of G. Stanley Hall (*GSH*), which provides a valuable and highly readable account of his career. Equally readable is the classic work on the reception in the United States of evolutionary theory and of Spencer's philosophy, Hofstadter's *Social Darwinism in American Thought* (*SD*). Page references are from the 1955 Beacon paperback edition of this work. On Spencer, see also Oldroyd (1980; Chs 15 & 16).

1 Albrecht (1960); cited in Littman (1979; p. 45).
2 *GSH*, pp. 134–43.
3 *GSH*, p. 155.
4 *GSH*, pp. 169–71.
5 *GSH*, Ch. 11.
6 *GSH*, Ch. 15.
7 *SD*, p. 32.
8 Peel (1971); Webb (1926).
9 *SD*, p. 44.
10 Cited by Bibby (1959; p. 24).

William James

This section is based largely on the standard two-volume biography of James by Perry (*WJ*). Page references to James' *Principles of Psychology* (*JPP*) are to the 1901 edition of Vol. 1 and the 1907 edition of Vol. 2.

1 Cited in *WJ1*, p. 474; Ch. 28 contains a valuable discussion of the relationship between James and Spencer.
2 *WJ1*, p. 475.
3 *JPP1*, Ch. 9.
4 *JPP2*, p. 433.
5 *JPP1*, p. 193.
6 *JPP1*, p. 24. James cites Thomas Meynert, an Austrian neuroanatomist, as the main source of these ideas. Meynert appears to have been a key proponent of 'connectionist' ideas, as described in a popular account in German (Meynert, 1874), and in English (Meynert, 1885). I have not been able to consult either of these books.

7 JPP1, p. 112.
8 JPP1, p. 114.
9 JPP1, p. 121.
10 JPP1, p. 122.
11 JPP1, p. 125.
12 JPP2, p. 450.
13 JPP1, p. 33.
14 JPP2, p. 348.
15 JPP2, p. 360.
16 James' changing attitude to automaton theory is described in *WJ*, Ch. 53.
17 *WJ1*, p. 137.
18 *WJ1*, p. 138.
19 See Ross (1972; Ch. 13) for an account of Hall's quarrels with his peers; also *WJ2*, pp. 6–24.
20 *HEP*, pp. 532–40.
21 *HEP*, pp. 528–32.

Edward Thorndike's puzzle boxes and doctoral thesis
Biographical information is from the highly detailed account of Thorndike's life provided by Joncich (*SP*). His thesis work was first reported in Thorndike (1898); this was reprinted in 1911 as the second chapter of his book, *Animal Intelligence* (*TAI*); page references are to the facsimile of the first edition published in Darien, Conn: Hafner, 1970.

1 *SP*, p. 105.
2 Quoted in *SP*, p. 89.
3 *TAI*, p. 123.
4 Quoted in *SP*, p. 146.
5 *TAI*, p. 122.

The Law of Effect and S-R bonds
The sources here are as for the previous section.

1 *SP*, pp. 190; 207.
2 *TAI*, p. 211.
3 Thorndike & Woodworth (1901).
4 Thorndike (1913).
5 *TAI*, p. 240.
6 *TAI*, p. 244.
7 *TAI*, p. 274.
8 Pearson (1892); the metaphor of the brain as a telephone exchange is to be found on p. 42 of the Everyman edition of 1937.
9 Bertrand Russell seems to have been the first to point this out (Russell, 1927; pp. 34–6).
10 Meehl (1950).
11 *TAI*, pp. 253–7.
12 Gardner & Gardner (1969).
13 *SP*, p. 400.
14 Comment by A. H. Maslow, quoted in *SP*, p. 551.
15 *TAI*, pp. 46–8.
16 *TAI*, p. 128.
17 *TAI*, pp. 110–12.
18 *TAI*, p. 119.

Oskar Pfungst and Clever Hans
The most recent English translation of Pfungst's report (New York: Holt, 1965) contains an interesting introduction by R. Rosenthal who discusses the influence of this study on subsequent investigations of non-verbal communication and of telepathy. Very recently a whole symposium was devoted to the topic (Sebeock & Rosenthal, 1981).

1 Watson (1908*a*).

Concluding discussion
1 Maslow (1970; p. 13).

Chapter 4: Reflex action and the nervous system

Rene Descartes and the beast-machine
For all the many commentaries on Descartes' philosophy there has been surprisingly little discussion of his views on animals. One curious, but interesting, paper on the topic is Rodman (1974), which draws heavily on an account of the debates over this topic for a century after Descartes, Rosenfield's *From Beast-machine to Man-machine* (*BM*). A recent discussion of whether Descartes believed animals to be conscious is provided in Cottingham (1978), while an excellent account of the exchanges between Descartes and his contemporaries is to be found in Walker (1983). For an overview of the whole history of human beliefs about the minds and behaviour of our fellow species, see the entry by Singer on the 'History of the study of animal behaviour' in McFarland (1981). Quotations from Descartes' *Discourse on Method* (1637) are from an English translation of 1912 (*DM*).

1 This and the following two quotes are from *DM*, p. 44.
2 *DM*, p. 45.
3 *DM*, p. 46.
4 *DM*, p. 49; despite his belief that knowledge should be available to all, Descartes postponed publication of some of his more controversial writings, as with *The Treatise on Man* noted below.
5 *DM*, p. 58.

The Cartesian reflex
Excellent discussions of the Cartesian reflex and of subsequent physiological research on the topic are to be found in Fearing's *Reflex Action* (*RA*), Clarke and O'Malley's *The Human Brain and Spinal Cord* (*HBSC*) and in Clarke and Dewhurst (1972). Another important seventeenth-century contributor to the idea of reflex action was the Oxford neuroanatomist, Thomas Willis; see Spillane (1981; pp. 53–107).

1 *DM*, p. 43.
2 From Descartes' *Les Passions de l'Ame*; quoted in *HBSC*, p. 332.
3 For the origins of this idea, see the sources listed above.
4 From *Les Passions de l'Ame*; quoted in *HBSC*, p. 331.
5 Notably by Vesalius (1514–1564); see *HBSC*, p. 155.
6 Fulton (1959).
7 An outline sketch of a physiological mechanism for memory is given by Descartes in *The Treatise on Man* (Descartes, 1972; pp. 87–90). Another interesting passage on this topic is quoted in Lashley (1950), but Lashley does not give a precise reference and I have been unable to locate the source.

Julien de la Mettrie's man-machine

A very useful edition of de la Mettrie's *Man a Machine* is the 1953 reprint (*MM*) of a 1912 translation, which also contains the original French version of 1748, the *Eulogy* by Frederick the Great and extracts from de la Mettrie's *The Natural History of the Soul*. An important discussion of de la Mettrie's work is to be found in Vartanian (1960; 1973) and a much older source that was also used here is Lange's *History of Materialism* (*HM*) of 1880.

1 *BM*, Section 2.
2 From the *Eulogy* by Frederick the Great, who is the biographer referred to here; *MM*, p. 6. See also *HM*, pp. 49–91.
3 *MM*, pp. 90–7.
4 *MM*, p. 98.
5 *MM*, p. 103.
6 *MM*, p. 97.
7 *MM*, p. 103.
8 *MM*, p. 117.
9 *MM*, p. 122.
10 *MM*, p. 140.
11 *MM*, p. 148.
12 *MM*, p. 8.
13 *BM*, Section 2, Ch. 3; *HM*, p. 90.
14 Quoted in *HM*, p. 78, which contains a discussion of de la Mettrie's other works.
15 See, for example, the chapter on 'The pursuit of happiness' in Commager (1975), which looks at the theme of happiness in the French Enlightment and its influence on Jefferson, as does Wills (1978; pp. 149–64; 221). A general treatment of Jefferson's beliefs is to be found in Boorstin (1948). The closest link between la Mettrie and Jefferson may have been provided by Pierre Cabanis, famous for his declaration that the brain secretes thought as the liver bile. Cabanis was a disciple of de la Mettrie and also Jefferson's favourite philosopher; see Staum (1974).

David Hartley's Observations

This account is mainly based on the abridged version of his work, Priestley's *Hartley's Theory of the Human Mind* (*HTHM*) of 1775. A recent account of Hartley's life and work is provided by Singer (in preparation; University of Reading).

1 Biographical material is from the *Dictionary of National Biography* (1908).
2 *HTHM*, Preface p. xlix.
3 *HTHM*, Ch. 3, Section 7; pp. 230–48.
4 *HTHM*, Ch. 1, Section 3; pp. 29–37; according to Singer (personal communication) this idea was anticipated by Thomas Willis in *The Souls of Brutes* of 1664.
5 *HTHM*, p. 32.
6 *HTHM*, p. 37.
7 Mill (1873; pp. 68; 123).

The spinal cord and nervous energy

This section is based largely on Fearing's *Reflex Action* (*RA*) and on Clarke and O'Malley's *Human Brain and Spinal Cord* (*HBSC*). A further source was Brazier's 1959 chapter on the 'Historical development of neurophysiology' (*HDN*).

1 *HBSC*, p. 341; *RA*, pp. 74–86.
2 Quoted in *RA*, p. 80.
3 Singer (personal communication) has informed me that the conditioned reflex was set in an evolutionary context in Erasmus Darwin's *Zoonomia* (1794).
4 *HBSC*, pp. 296–303 provides a careful discussion of the relative contributions of Bell and Magendie to this discovery, while an account of the excitement this work provoked is provided in Young (1970; p. 93); see also Cranefield (1973).
5 See, for example, Cohen (1972); an enjoyable way of confirming that Franklin was a remarkable man is to read his autobiography.
6 *HDN*, pp. 14–17.
7 *HDN*, pp. 17–20; 47; *HBSC*, pp. 177–92.
8 The importance of technical developments is stressed in Liddell (1960; Ch. 1).
9 On Mueller, see *HBSC*, pp. 203–6 and *HB*, pp. 382–8.
10 du Bois-Reymond's electrophysiology is described in *HBSC*, pp. 192–203 and in *HDN*, pp. 20–2.
11 *HBSC*, pp. 206–9.
12 On Marshall Hall, see *RA*, pp. 122–45, *HBSC*, pp. 347–51 and the paper on 'Marshall Hall, the grasp reflex and the diastaltic spinal cord' in Jefferson (1960).
13 Jefferson (1960; pp. 94–112) and Young (1970) provide excellent accounts of phrenology.
14 *HBSC*, pp. 483–8.
15 Jefferson (1960; p. 116).

Spontaneous activity and the Berlin physiologists

The main sources used in this section were as in the preceding one, namely, Brazier's 'The historical development of neurophysiology' (*HDN*), Clarke and O'Malley's *Human Brain and Spinal Cord* (*HBSC*) and Fearing's *Reflex Action* (*RA*). The latter gives an extended account of the study of inhibition in Ch. 12. A particularly interesting analysis of nineteenth-century German physiology and its ties with professional medicine can be found in the early chapters of Pauly's thesis on Jacques Loeb (*JLCL*).

1 *HB*, pp. 382–8.
2 See Young (1970; pp. 116–17) on how much Bain probably learned from Mueller.
3 Young (1970; p. 174).
4 Smith (1959; pp. 117–18).
5 The argument that changes in the focus of physiology were the result of changes in medical training in Germany is spelt out in *JLCL*, Ch. 2.
6 *HBSC*, p. 352.
7 *RA*, pp. 162–5.
8 This account is based on the replication of Pflueger's experiment described in Lewes (1860; Vol. 2, pp. 151–272).
9 *RA*, p. 189.
10 *SAN*, p. 68; it is not clear which of Pflueger's experiments Sechenov was able to replicate in the eel.

Ivan Sechenov and inhibition
Details of Sechenov's life are mainly from his *Autobiographical Notes* (*SAN*), while the account of his work, including *Reflexes of the Brain*, is based on the English translation of his *Selected Works* (*SSW*). A general survey of Russian science is provided by the two remarkably scholarly and interesting volumes by Vucinich, of which the second, *Science in Russian Culture: 1861–1917* (*SRC*), was used in the preparation of this and the following chapter.

1 *JLCL*, Ch. 2.
2 *SAN*, p. 37.
3 *SRC*, p. 122; see also Seton-Watson (1967; pp. 167–70).
4 According to *JLCL*, Ch. 2, Ludwig's influence on physiology was enormous and seems to have been considerably underrated in most histories of biology, while *SRC*, p. 127 suggests that he was particularly helpful towards Russian students.
5 *RA*, p. 189.
6 *SAN*, pp. 107–8; this also gives Sechenov's views on why there were not more experiments on inhibition.
7 An English translation of excerpts from the original paper is in *HBSC*, pp. 361–5.

Sechenov's extension of physiology to mental processes
The same sources were used in this section as in the preceding one.

1 *SSW*, p. 315.
2 *SSW*, p. 274.
3 *SSW*, p. 317.
4 *SSW*, p. 335 acknowledges Sechenov's debt to Beneke's summary of the psychological theories of the French school of 'sensualism' ('sensationalism'?), but compare *SAN*, p. 52.
5 *SSW*, p. 335.
6 *SSW*, p. xxiii.
7 *SSW*, pp. xxii–xxiv.
8 *SRC*, pp. 54–65; Seton-Watson (1967; p. 380). However, it has been argued that Count Tolstoy's general contribution to Russian education was much more positive than it was viewed at the time or by most subsequent commentators (Hans, 1931; Ch. 5).
9 Seton-Watson (1967; p. 475).
10 du Bois-Reymond (1874); interesting discussions of the effect of this paper are provided by *HB*, pp. 412–13 and by *HM2*, p. 308.
11 Internal evidence and phrasing strongly suggest that this was written as a rebuttal of du Bois-Reymond, but the version I have seen contains no explicit reference to the *Ignorabimus* lecture.
12 *SSN*, p. 338.
13 *SSN*, p. 350.

Chapter 5: Conditioned reflexes

The following translations into English of Pavlov's work were used in writing this chapter: Anrep's 1927 translation of *Conditioned Reflexes* (*CR*), with page references to the Dover reprint of 1960; Gantt's translations of *Lectures on Conditioned Reflexes* (*LCR*) of 1928 and of *Conditioned Reflexes and Psychiatry* (*CRP*) of 1941; and *Selected Works* (*PSW*), published in Moscow in 1955. The only other collection of Pavlov's papers published in English, *Experimental Psychology and Other Essays* (Pavlov, 1957), is a pirated edition of *Selected Works* that has had four new photographs added and sixteen pages of text deleted.
Far and away the richest and most readable biography of Pavlov is that by Babkin (*BP*); the manuscript for this book (*BPms*)n which is held in the Osler Library, McGill University, contains some material not found in the published version. Other biographies of Pavlov used in preparing this chapter were those by Frolov (*FP*), by Asratyan (*AP*) and by Cuny (*CP*), together with the short biographical sketches in the English translations of Pavlov's work by Gantt (in *LCR* and *CRP*) and by Koshtoyants (in *PSW*).
For a clear and concise summary of Pavlov's theories and empirical contributions to the study of conditioning, see any edition of what has become the standard textbook on animal learning theory, Hilgard and Bower's *Theories of Learning* (e.g.

Hilgard & Bower, 1966; 3rd edition). A comprehensive survey of research on classical conditioning is provided by Mackintosh's *The Psychology of Animal Learning* (Mackintosh, 1974; pp. 1–142). For the non-psychologist an excellent exposition of Pavlov's work and an assessment of it relative to modern developments is provided in a compact, cheap and accessible form in Gray (1979). Background material on scientific, educational and social developments is mainly from Vucinich's *Science in Russian Culture* (*SRC*), Seton-Watson's *The Decline of Imperial Russia* (*DIR*) and Hans' *History of Russian Educational Policy: 1701–1917* (*HREP*).
The opening quotation is from Pavlov's article, 'The conditioned reflex', in the *Big Medical Encyclopedia* (see *PSW*, p. 250).

1 Shaw (1932). Shaw may have been wrong about the extent to which the idea of the conditioned reflex is familiar to children. The claim has been made that, although children of up to the age of ten years know about many other psychological phenomena, conditioning provides one of the exceptions (Mischel, 1979).
2 *LCR*, p. 52.
3 Leake (1959) points out that in Russia 'physiology' is often used in a much wider sense than in the West to include any scientific study of the function of living matter. Thus it can include embryology, genetics and pharmacology, as well as psychology. This is close to its earlier use in Germany at the time of Johannes Mueller.
4 Kuhn (1962).
5 Kline (1955); *SRC*, pp. 273–97.
6 Quoted in Dobzhansky (1955).
7 Kline (1955).
8 *AP*, p. 28.
9 *LCR*, p. 13; *CRP*, p. 170 (footnote); *SRC*, p. 300.
10 Quoted in *CRP*, p. 170.

Pavlov's early career

1 This is one of the few comments Pavlov makes about his youth in his brief autobiography; *PSW*, p. 44.
2 *BP*, p. 337.
3 *SRC*, p. 300; *BP*, pp. 18–21; *PSW*, p. 13.
4 *SRC*, p. 63.
5 *BP*, p. 225.
6 Most of the material on Pavlov's personal life related in Babkin's biography, and to a lesser extent in the other biographies, appears to be derived from two Russian sources that have not been translated into English: a biographical sketch by V. V. Savich dating from 1924 and *Reminiscences* by Pavlov's widow, which was published in *Novi Mir* in 1946; see also *FP*, pp. 264–71.
7 *SRC*, p. 186.
8 Quoted in *SRC*, p. 60.
9 *SRC*, p. 187.
10 *SRC*, p. 203.
11 *BP*, pp. 67–73; *SRC*, p. 303.
12 *CP*, pp. 30–1; *LCR*, p. 16; *BP*, p. 62.
13 *BPms*, Ch. 6 makes it clear that Pavlov enjoyed two further advantages which I have not seen discussed in any published source. One was the political protection that his patron, Prince Oldenburgski, could offer as a progressive, yet very powerful, member of the governing aristocracy. The second was the services of an exceptional technician, E. A. Hanike, who was an eccentric and indispensable member of the laboratory. Hanike built equipment, carried out biochemical assays and handled Pavlov's research budget, just as Sara looked after his domestic one.

How Wolfsohn, Snarsky, Tolochinov, Pavlov and Babkin began to experiment upon conditional reflexes

1 *LCR*, p. 21; *PSW*, pp. 17–19.
2 *PSW*, p. 18.
3 Quoted in *LCR*, p. 21.
4 *BP*, p. 222.
5 *BP*, p. 69; *AP*, p. 60.
6 *LCR*, p. 18; *BP*, pp. 121; 130–4.
7 *LCR*, p. 25.
8 *BP*, pp. 62; 81; 112; *FP*, p. 246.
9 This account of the early studies of conditioning is based on Pavlov's own scattered remarks on the subject, particularly those in the first lectures on conditioning reprinted in *LCR* and *PSW*, and on the accounts provided by Babkin (*BP*, Chs 14 & 27) and by Koshtoyants (*PSW*, pp. 26–9); see also Anokhin (1968). As in all these accounts, the emphasis in the present chapter is on the role of events within his laboratory in changing Pavlov's outlook from a dualist to a materialist position on mind. However, as Philip Pauly has pointed out to me, it seems unlikely to be pure coincidence that Pavlov's ideas came so close to those expresssed by Beer, Bethe & Uexkuell (1899) or by Loeb (1899). At least a portion of those long hours Pavlov spent reading every night must have been devoted to such works.
10 *PSW*, p. 632; also *LCR*, p. 38.
11 *LCR*, p. 39.
12 *BP*, pp. 224–31; see also *BPms*, Ch. 16; pp. 14–16.
13 *LCR*, pp. 49–58.
14 *LCR*, pp. 65–6.
15 *LCR*, pp. 71–2; *CR*, pp. 51–63.
16 This first occurred in the translation of Pavlov's Huxley lecture (Pavlov, 1906).
17 *LCR*, pp. 81–95.
18 *LCR*, p. 95.

Vladimir Bechterev and Objective Psychology

The two books by Bechterev cited here are his *Psychologie Objective* (*PO*), a French edition of 1913 of the original Russian, and his *General Principles of Human Reflexology* (*GPHR*), using the English translation of 1932. In addition to the French version of *La Psychologie Objective*, there was also a German edition, but never an English one. Biographical material on Bechterev is from the sketch by A. Gerver in *GPHR*, from a chapter by P. I. Yakovlev (*YB*) and one by A. L. Schniermann (Schniermann, 1930). Brief sketches are also provided by Vucinich (*SRC*) and by Babkin (*BP*, with additional material in *BPms*, Ch. 9; pp. 23–6).

1 Material on the 1905 revolution is from *DIR*, Pt. 3 and from Walkin (1963).
2 Walkin (1963: p. 197).
3 *SRC*, p. 229.
4 *HREP*, p. 176; but see *AP*, pp. 19; 27.
5 Thus Koshtoyants (1964) admits to his failure to find any evidence of contact between Pavlov and Sechenov, even though Koshtoyants himself is convinced that there ought to have been some personal link.
6 *GPHR*, p. 195.
7 *YB*, p. 194.
8 *YB*, p. 193.
9 A first-hand comparison between the two laboratories is provided by Babkin, *BP*, pp. 75–83.
10 *YB*, p. 190 cites a biography by V. P. Ossipov.
11 *HREP*, p. 200; *SRC*, p. 322.
12 The present account is based on the French version of this paper (Bechterev, 1906).
13 Bechterev (1906; p. 395, my translation).
14 *PO*, pp. 153–5; *GPHR*, pp. 196–8.
15 *PO*, pp. 255; 262; *GPHR*, p. 205.
16 This account of the Pavlov–Bechterev dispute is from Babkin (*BP*, pp. 89–94; with further detail in *BPms*, Ch. 9, pp. 16–23).

Pavlov's later work

1 *CR*, Lectures 9–12; *LCR*, p. 322.
2 *LCR*, pp. 140; 158; *CR*, p. 250.
3 *CR*, pp. 29–30; 295; *LCR*, pp. 185; 341.
4 In experiments with mice to test for the inheritance of conditioned reflexes; see n. 14 below.
5 *CP*, p. 120.
6 *CRP*, Lectures 50 & 55; *CR*, p. 299.
7 Quoted in *CP*, p. 80.
8 *LCR*, p. 279.
9 *GPHR*, pp. 141–2.
10 *GPHR*, pp. 148–9.
11 *YB*, p. 200.
12 *LCR*, pp. 287–93.
13 *LCR*, pp. 364–7; *CR*, pp. 313–18; *FP*, pp. 214–15.
14 Razran (1958); cf. *AP*, pp. 96–7.
15 *CR*, p. 379; Babkin states that Pavlov deliberately made it almost impossible for outsiders to find out the details of research in his laboratory. Thus, referring to the period before the First World War, Babkin writes that 'the reason Pavlov gave for his reluctance to share the results of his research with the Western world was his fear that inexperienced workers would "muddle" such a novel and complex physiological method as conditioned reflexes' (*BPms*, Ch. 11).
16 Konorski (1948).
17 Konorski (1948); for an introduction to Konorski's work, see Halliday (1979).
18 *PSW*, pp. 306–10; 443–4.
19 *CR*, pp. 141–4.
20 An excellent recent introduction to theories of conditioning which is deliberately non-Pavlovian in its approach is Dickinson (1980).
21 See Mecacci (1979). An idea of the interplay between politics, philosophy and psychology in the Soviet Union is provided by McLeish (1975). Many Western psychologists have been more complimentary about Pavlov's theoretical contributions than the present account; one notable example is Gray (1964), which also attaches a great deal more significance to Pavlov's beliefs about individual differences than most other Western commentaries.
22 *PSW*, Pt XII, p. 651.
23 *BP*, p. 118.

Concluding discussion

1 Twitmeyer (1902); see Coon (1982).

Chapter 6: Comparative psychology and the beginning of Behaviourism

The opening quotation is from Watson's *Behavior: an Introduction to Comparative Psychology* (*BICP*) of 1914 (p. 27 in the reprint of 1967).

1 Beer, Bethe & von Uexkuell (1899). From secondary sources of the time I gather that a similar debate took place in France a few years later between Nuel and Claparede (Nuel, 1904), which attracted considerable attention; see Yerkes (1906).

Jacques Loeb, Herbert Jennings and lower organisms

A major source for much of the material in this section was Pauly's thesis on Loeb, *Jacques Loeb and the Control of Life* (*JLCL*). The same author compares Loeb and Jennings in a way that complements the present account in Pauly (1981). Biographical material on Jennings is from the memoir by Sonneborn (*HSJ*). The primary source for the scientific material in this section was Jennings' *The Behavior of the Lower Organisms* (*BLO*); the 1962 reprint of this book contains an interesting introduction on the relationship between Jennings, Loeb and Watson by D. D. Jensen, who also writes about this in Jensen (1970).

1 The argument came to a head at a public meeting where Jennings was able to demonstrate his claims

in dramatic fashion, thus making a name for himself; see *HSJ*, pp. 173–4.

2 *HSJ*, p. 152.
3 *BLO*, p. 237.
4 *BLO*, p. 212.
5 Jennings (1907); cited in *BLO*, p. xi.
6 *BLO*, p. 276.
7 *BLO*, p. 336.
8 Jennings (1907).
9 Loeb (1910); cited in *JLCL*, p. 18.
10 The Baldwin effect, or principle of organic selection, has never been in the mainstream of evolutionary theory, but has been promoted at various times and in various versions since first proposed by Morgan, Poulton and Baldwin. The most recent discussion I have seen is Waddington (1975; pp. 28–31; 88–9). Among recent psychologists Piaget was the most well-known exponent of the principle, having picked it up from Baldwin and then later been gratified by Waddington's exposition.

The laboratory rat and John Watson's early career
Material on the laboratory rat is from papers by Lockard (1968), Miles (1930) and Richter (1968). The account of Watson's life here and in later sections is based on a variety of sources, of which a major one is Cohen's biography, *J. B. Watson, the Founder of Behaviourism* (*CW*). Other sources include Watson's own autobiographical sketch of 1936 in *HPA3*, which needs to be treated with some caution, an unpublished honours thesis by P. O. Welsh (Welsh, 1963) and several individual papers that are cited below, particularly those by Philip Pauly and Lexa Logue.

1 Lockard (1968); Richter (1968).
2 Biographical material on Meyer is mainly from Leif's biography of Meyer (Leif, 1948). His role in introducing the laboratory rat to America is emphasized by Richter (1968) who may exaggerate Meyer's influence (cf. Miles, 1930). In view of the later relationship between Watson and Meyer it is intriguing that the rats Watson had to look after when he first arrived at Chicago were almost certainly ones sent by Meyer to Donaldson; see unpublished letters from Donaldson to Meyer on March 19th and April 29th, 1900 in the Meyer Papers of the Alan Mason Chesney Medical Archives of the Johns Hopkins Medical Institutions.
3 Miles (1930; p. 326).
4 Small (1900).
5 Small (1901).
6 Given the potential interest of rat behaviour it is quite remarkable that Donaldson, despite his training in psychology, concentrated entirely on questions concerning the rate of growth of the body, brain and spinal cord of different strains of rats.
7 This point was brought home to me some years ago by Jose Linaza who, while working at Sussex, incidentally gave some of his rats from an early age unusually extensive handling of the kind that Watson provided and for the same reason, that he too was interested in developmental questions. The behaviour of Linaza's rats was totally unlike that of any of our other animals.
8 Watson (1907*a*).
9 Watson & Carr (1908).
10 See the Robert M. Yerkes Collection, Yale University Library.
11 Watson (1908b).
12 In view of Watson's later views on consciousness two of the most interesting of these reviews are the ones that D. D. Jensen has drawn attention to, one of Jennings and one of Loeb (Watson, 1907*b*); see also the review of Pfungst's report on Clever Hans, referred to in Ch. 3 (Watson, 1908*a*).
13 Carr (1936).
14 Quoted in *CW*, p. 48.

Robert Yerkes' comparative psychology
This account is based on the primary sources listed below. The only overall survey of comparative psychology during this period that goes into any detail is contained in a fascinating recent doctoral thesis on *The Origins of Behaviourism: American Psychology, 1870–1920* (O'Donnell, 1979) which is particularly interesting on the institutional framework that Watson and Yerkes worked within.

1 Washburn (1932).
2 Washburn (1908; p. 120).
3 Kinnaman (1902).
4 Watson (1908*c*).
5 Cole (1907).
6 Yerkes (1932).
7 Yerkes (1907).
8 Yerkes & Dodson (1908).
9 Yerkes & Morgulis (1909).
10 Yerkes (1911).
11 Yerkes (1911; p. 132).
12 Yerkes (1911; p. 239).
13 Yerkes later commented on the importance he attached to the work of Cole and of Hamilton in a brief historical sketch of early comparative psychology (Yerkes, 1917).
14 Hamilton (1911).
15 Hamilton (1914).
16 Hamilton (1911).
17 Coburn & Yerkes (1915); Yerkes & Coburn (1915).
18 Hunter (1952).
19 Hunter (1914).
20 O'Donnell (1979; pp. 502–15).

American psychology at the beginning of the century
Three books were used extensively in preparing this section: Boring's *History of Experimental*

Psychology (*HEP*), Hale's *Freud and the Americans* (*FA*) and Ross's biography of G. Stanley Hall (*GSH*). O'Donnell (1979; Chs 8, 9) is particularly interesting on this period.

1 Groos (1896; 1901); see excerpts in Bruner, Jolly & Sylvan (1976).
2 Thorndike (1904).
3 Baldwin (1895; p. 185).
4 Baldwin has recently been rediscovered, as noted in the following papers, Broughton (1981), Cairns (1980) and Mueller (1976), and the following books, Broughton & Freeman–Moir (1981) and Russell (1978).
5 Quoted in *HEP*, p. 529.
6 Wissler (1901).
7 Sharp (1899); for a recent discussion of the introduction of mental testing into the U.S.A, as well as a brief account of Baldwin's work, see Cairns & Ornstein (1979).
8 Recent reappraisals of the influence of Wundt and Titchener in America are provided by Blumenthal (1979; 1980), Danzinger (1979) and Leahey (1981).
9 Quoted in *HEP*, p. 515.
10 Angell (1936).
11 *FA*, Pts 1 & 2.
12 William James to Mary Calkins (1909) in Perry (1934; Vol. 2, p. 123).
13 Perry (1934; p. 168).
14 James (1976).
15 James (1904).
16 James (1890; Vol. 1, p. 22).
17 Quoted in *GSH*, p. 385.
18 Quoted in *GSH*, p. 385.
19 The biologist is W. M. Wheeler and the quote from a 1917 address comes from Gould (1977; p. 155).
20 Freud (1910).
21 The visit of Freud, Jung and Ferenczi to Clark University is described in *GSH*, Ch. 18 and *FA*, Ch. 1, while details from Freud's point of view are presented in Jones (1955). Quotations from Freud are from this latter source. Freud's intellectual roots were not very distant from those of American psychologists in that his training in research followed the tradition of the Berlin physiologists; Freud regarded Ernst Bruecke as his most influential teacher. This aspect of Freud's theories is emphasized in the excellent introduction to psychoanalysis in Fancher (1973).
22 Jung's view of the American visit is given in Jung (1961), which includes letters he wrote at the time to Emma Jung.

John Watson's behaviourist manifesto

Major sources for this section were Cohen's biography (*CW*) and Pauly (1979), while the major primary source was Watson's *Behavior: an Introduction to Comparative Psychology* (*BICP*) of 1914 (page references are to the 1967 edition which contains an introduction by R. J. Hernstein).

1 Baldwin (1926; Vol. 1, p. 118).
2 For a recent discussion of this work, see Russell (1978).
3 Dunlap (1932).
4 On Baldwin's disgrace, see Pauly (1979) and an unpublished Ph.D. thesis by Mueller (1974).
5 The comment was in fact printed in the sister journal, *Psychological Bulletin*, 1909, 6, 256.
6 Leif (1948).
7 Leif (1948, p. 261).
8 Beach (1961); Hebb (1959).
9 Yerkes & Watson (1911).
10 Watson & Watson (1913).
11 O'Donnell (1979; p. 518).
12 Watson (1913).
13 *BICP*, p. 21.
14 *BICP*, p. 317.
15 *BICP*, p. 334.
16 *BICP*, p. 17.
17 *BICP*, p. 327.
18 *BICP*, pp. 224–7.
19 Quoted in *FA*, p. 161.
20 Watson (1916*a*); Watson's election to the APA presidency is among a number of intriguing events discussed in Samelson (1981).

Concluding discussion

1 Mill (1950; p. 105).

Chapter 7: Apes, problem-solving and purpose

The opening quotation is from Romanes (1882; p. 471)

1 Wallace (1869; Vol. 1, pp. 62–9).
2 Romanes (1882; p. 481).
3 Romanes (1889).
4 Morgan (1894; p. 253).
5 *BICP*, Ch. 9 discusses some of these cases, as does Katz (1937).
6 Katz (1937).
7 Pfungst (1912); cited in Yerkes (1916).
8 Driesch wrote a two-volume work in English, based on his Gifford Lectures of 1907 and 1908, which he describes as modifications of his earlier publications in German and the present account is based on this (Driesch, 1908). I was helped by the discussion of his ideas in an unpublished paper by two Sussex colleagues, G. Webster and B. C. Goodwin, 'The origin of species: a structuralist approach'. Another useful source was Oppenheimer (1967).
9 Driesch (1908; Vol. 1, p. 166).
10 Driesch (1908; Vol. 2, pp. 89–100).
11 Driesch (1908; Vol. 2, p. 6).

12 This 1894 quote from Driesch is given in Oppenheimer (1967; p. 1).
13 Oppenheimer (1967; p. 72).
14 Driesch (1908; Vol. 1, p. 283).

Leonard Hobhouse and articulate ideas
Biographical material is from two books on Hobhouse: Owen (1974) and Hobson & Ginsberg (1931). Quotations are from his major work in psychology, *Mind in Evolution* (*ME*) of 1901.

1 Hobhouse (1896; p. ix).
2 His political views of the time can be gauged from Hobhouse (1904).
3 *ME*, p. 12.
4 *ME*, p. 2.
5 *ME*, p. 128.
6 *ME*, p. 130.
7 *ME*, pp. 220; 223.
8 *ME*, p. 198.
9 *ME*, p. 234.
10 *ME*, p. 258.
12 *ME*, pp. 86–97.
13 *ME*, p. 135 (footnote).

Wolfgang Koehler's tests of chimpanzee intelligence
Unfortunately there appears to be no biographical study of Koehler, who was reticent about his life and never wrote even a sketch of an autobiography. Details of his life here come mainly from two obituary notices, Henle (1968) and Asch (1968), from *HEP* and from conversations I had in Summer, 1976 with Mrs W. Koehler, Carroll Pratt, Frank Geldard and Hans Wallach. I am also indebted to the help given by the late Hans-Lukas Teuber, whose father was Koehler's predecessor on Tenerife and who spent an evening commenting on an early draft of the present account. Finally, Ilse Gaertner of the Free University of Berlin very kindly spent time on my behalf in checking publications of the Koenigliche Preussische Akademie der Wissenschaften for material on the Tenerife station.

Descriptions of Koehler's experiments are based entirely on the accounts given in his *Mentality of Apes* (*MA*). This was first published in German in 1917; page references here are to an English reprint of 1973.
1 von Ehrenfels (1890); this paper and the early history of Gestalt psychology are discussed in Heider (1973).
2 Wertheimer (1912); an account of the early days of the Gestalt movement is included in a recent biographical study of Wertheimer by his son (Wertheimer, 1980).
3 *HEP*, p. 595.
4 Stumpf (1930).
5 For ease of presentation the account here differs from Koehler's in that it keeps to the chronological order in which the tests were given (fortunately Koehler provides dates for most of them) and it attempts to separate methods and results from interpretation and discussion. Some readers may find Koehler's organization, in which the research is presented in a theoretical and non-chronological sequence more satisfactory.
6 *MA*, p. 18.
7 *MA*, p. 65.
8 *MA*, pp. 16–17.
9 *MA*, p. 192.
10 *MA*, p. 19 (footnote).
11 Koehler (1971; p. 198); this quotation is from a paper that provides an extensive and thoughtful discussion on the methods of animal psychology, originally published in 1921.

Insight
1 *MA*, p. 102.
2 One example is Winston (1975); see Boden (1977; pp. 252–67).
3 Koehler (1971; pp. 205–6).
4 *MA*, p. 220.
5 *MA*, p. 61.
6 Koehler (1915).
7 Koehler (1918).
8 *MA*, pp. 305–8.
9 *MA*, p. 272.
10 *MA*, p. 278–82.
11 Ellis (1938; pp. 228–73) contains three papers reporting animal research carried out in Germany during the 1920s which followed on from Koehler's work. These papers describe experiments with ravens, jackdaws, jays and bees. N. R. F. Maier, an American, spent a two-year visit to Berlin in the early 1930s working on problem-solving in rats.
12 Helson (1973); Luchins (1975).
13 Some subsequent studies on problem-solving by chimpanzees which criticize various aspects of Koehler's work are reprinted in Riopelle (1967).
14 *MA*, p. 17.
15 Wallach (1976).
16 Lashley, Chow & Semmes (1951).

Robert Yerkes' studies of apes
A major source for this section was a 1916 monograph by Yerkes, *The Mental Life of Monkeys and Apes* (*MLMA*). A further source was a comprehensive review by Yerkes and his wife, *The Great Apes: a Study of Anthropoid Life* (Yerkes & Yerkes, 1929). Biographical information is from Yerkes' autobiographical sketch of 1932 in *HPA2*, supplemented by an obituary notice (Elliott, 1956).

1 G. M. Haslerud discusses the relationship between Yerkes and Koehler in his introduction to the 1979 reprint of *MLMA*.

2 Yerkes & Yerkes (1929; p. 181). Apparently Haggerty never published a full account of his work with orang-utans, but briefly described it in Haggerty (1913), while his study of imitation is reported in Haggerty (1909).
3 Watson (1908*c*).
4 *MLMA*, p. 91.
5 *MLMA*, p. 68.
6 *MLMA*, p. 87.
7 *MLMA*, p. 131.
8 JBW to RMY, 12 May 1916; letter in the Robert M. Yerkes Collection, Yale University Library. See also the correspondence between RMY and H. S. Jennings in the same collection on May 16th and June 10th, 1916 and April 13th, 1917.
9 See Ch. 8 below; Yerkes (1921).
10 *MLMA*, p. 142.
11 Furness (1916).
12 Yerkes (1925; p. 175).
13 The quote is from Yerkes & Yerkes (1929; p. 307); the full report on this study is Yerkes & Learned (1925).
14 Yerkes (1925; p. 179).
15 The first study of matching by primates carried out in America was by Weinstein (1941); this cites Kohts, but not Hamilton, and points out that the first reported use of the technique was by Itard in eighteenth-century France (Itard, 1972).
16 Sheaffer (1973; pp. 188–91; 501); the research on marriage was reported in Hamilton (1929).
17 Hamilton (1908).
18 Kohts published reports in Russian in 1921 and 1923, but the present account is based on a later summary paper in French (Kohts, 1928).
19 Yerkes (1927; 1928); see summary in Yerkes & Yerkes (1929; Pt V).
20 Yerkes (1932; p. 404).

Concluding discussion

1 Russell (1927; p. 33). This book provides an admirably clear discussion of many of the key issues in psychology raised by animal research in the late 1920s.
2 Koehler (1971; p. 433); on Koehler's departure from Germany, see Henle (1978).
3 Hull (1952).

Chapter 8: Nature and nurture

The opening quotation is from Barzun (1965; p. 218), which gives the reference: J. S. Mill *Principles of Political Economy*, I, p. 390. However, this does not specify which edition and I have been unable to track down the original source.

1 Wilm (1925; pp. 94–119); Jaynes & Woodward (1974).
2 See Ch. 2 for more detail on Romanes' views on instinct.
3 James (1890; Vol. 2, p. 389).
4 James (1890; Vol. 2, p. 393).
5 Morgan (1895); quoted in Herrnstein (1972).
6 Morgan (1896*a*; Ch. 2).
7 Darwin (1872).

William McDougall, Wilfred Trotter and human instincts

Biographical material is mainly from McDougall's own very readable autobiographical sketch (McDougall, 1930 in *HPA*); this is referred to below as *WMcD* and is supplemented by the excellent discussion provided by Hearnshaw (1964; Ch. 12). Discussion of his work is based on McDougall's *Introduction to Social Psychology* (*ISP*) of 1908 (page references are to the 18th edition of 1932) and on his *Psychology: The Study of Behaviour* (*PSB*) of 1912 (page references are to the Home University Library revised edition of 1914). An interesting discussion of McDougall's ideas in the light of recent developments in philosophy and psychology which gives a much more detailed exposition of his theories than the present account is Boden (1972).

1 *WMcD*, p. 208.
2 *ISP*, p. 2.
3 *ISP*, p. 19.
4 See *PSB*, p. 23.
5 *ISP*, p. 29.
6 *PSB*, p. 138.
7 *ISP*, p. 88.
8 *ISP*, p. 327.
9 *ISP*, p. 385.
10 *ISP*, p. 406.
11 *ISP*, p. 410.
12 Greisman (1979). All references to Trotter's work are based on his *Instincts of the Herd in Peace and War* (*IHPW*) of 1916; page references are to the second edition, 1919. An intriguing portrait of Trotter is provided in Jones (1959).
13 *IHPW*, p. 24.
14 *IHPW*, p. 14.
15 *IHPW*, p. 70.
16 *IHPW*, pp. 183–6.
17 *IHPW*, pp. 203–5.
18 *WMcD*, p. 204.
19 *PSB*, p. 20.
20 *ISP*, p. 355.
21 *WMcD*, p. 216.
22 Heidbreder (1973; p. 268).

The Psychological Darwinism of race and instinct

Excellent accounts of the nature–nurture controversy in American science are given in Stocking's *Race, Culture and Evolution* (*RCE*) and Cravens' *The Triumph of Evolution: American Scientists and the Heredity–Environment Controversy: 1900–1941* (*TE*). A particularly thorough and fascinating study of intelligence testing in this era is provided by Samelson (1979). Briefer accounts of some aspects of this controversy, particularly of the Army programme, can be found in Kamin (1974), and in Gould (1981). Cravens (1978; *TE*) includes an account of the instinct controversy which is dealt with in more detail in the following two papers that proved very helpful in preparing this section: Herrnstein (1972) and Krantz & Allen (1967). The classic review of early instinct theories is Bernard's *Instinct: a Study in Social Psychology* (*I*) of 1924.

1 Haeckel (1900; p. 337).
2 Gasman (1971).
3 Attributed to Dr Edgar Berillion in Barzun (1937; p. 174).
4 Material on Boas is from *RCE*, Chs 7 & 8; *TE*, pp. 90–120.
5 *RCE*, p. 175.
6 Boas (1911).
7 *RCE*, pp. 259–64; *TE*, p. 147.
8 Sumner (1906).
9 Samelson (1979; p. 145).
10 Quoted from Samelson (1977) which, with Samelson (1979), was a source for much of the material described here. See also Gould (1981).
11 Dewey (1917).
12 Lippman (1922; 1923).
13 Lippman (1922; p. 297).
14 Lippman (1922; p. 329).
15 Lippman (1923; p. 10).
16 Samelson (1979; p. 136).
17 Pauly (1980; Ch. 10).
18 *TE*, p. 173.
19 *TE*, p. 286.
20 *TE*, p. 239.
21 *TE*, p. 241; Samelson (1979; pp. 128–35).
22 *TE*, pp. 182–4.
23 *I*, pp. 172–220.
24 Dunlap (1919).
25 *I*, p. 172.
26 Gottlieb (1972).
27 Kuo (1921).
26 Whitman (1919); quoted in Kuo (1921).
29 Kuo (1921; p. 657).
30 *I*, Chs 10 & 20.
31 Kuo (1930).

The final part of John Watson's career

A variety of sources were used in preparing this section, some of which are indicated below. The only extended account of Watson's life during this period is Cohen's biography, *J. B. Watson* (*CW*). Among other interesting and slightly more recent sources is Samelson (1981). Quotations from correspondence between Watson and Meyer are from letters in the Meyer Collection, The Alan Mason Chesney Medical Archives, The Johns Hopkins Medical Institutions, Baltimore.

References to Watson's books are from his *Psychology from the Standpoint of a Behaviorist* (*PSPB*) of 1919 (page references to the second edition of 1924); his *Behaviorism* (*B*) of 1924 (references to the University of Chicago reprint of 1958); and to his book with Rosalie Rayner Watson of 1928, *The Psychological Care of Infant and Child* (*PCIC*).

1 Watson (1916*b*).
2 Meyer to Watson, May 29, 1916; unpublished letter in Meyer Collection, The Alan Mason Chesney Archives, Johns Hopkins Medical Center, Baltimore.
3 Meyer to Watson, June 3, 1916; unpublished letter in Meyer Collection.
4 *PSPB*, p. 215.
5 *PSPB*, pp. 21–220; the first published account of this work was Watson & Morgan (1917).
6 Blanton (1917); see also *PSB*, pp. 252–70.
7 Jennings *et al.* (1917).
8 Jennings *et al.* (1917; p. 72).
9 Jennings *et al.* (1917; pp. 75–6).
10 Jennings *et al.* (1917; pp. 77–8).
11 Jennings *et al.* (1917; pp. 92–4).
12 Watson (1936; p. 278).
13 See *CW*, pp. 107–11 and Watson (1936).
14 *PSPB*, p. 383.
15 I am grateful to Curt Richter who in an interview of July 13th, 1976 was very helpful in conveying the atmosphere in Watson's lab.
16 The first and most detailed of the many versions of this study was Watson & Rayner (1920).
17 The subsequent place of the experiment in social science folklore is delightfully given in Harris (1979).
18 Watson (1919*b*).
19 *CW*, Ch. 6.
20 Watson to Meyer, Aug. 13th, 1920; unpublished letter in the Meyer Collection.
21 Unpublished notes by Meyer on letter to Watson, Aug. 1920; in Meyer Collection.
22 Meyer to Goodnow, Presidential Papers, Eisenhower Library, The Johns Hopkins University.
23 *CW*, Ch. 7.
24 Watson (1936; p. 280); see also Buckley (1982).
25 This charge was made in the first virulently anti-Behaviourist book, Roback (1923).
26 Watson to Raymond Pearl, April 8th, 1921. Pearl Collection, American Philosophical Library, Philadelphia.
27 Jones (1924); a more accessible account of this work is the summary in Watson's *Behaviorism*. For Mary Cover Jones' brief recollection of Watson, see Jones (1974).
28 Watson & McDougall (1928).

29 Watson to Jennings, April 3rd, 1924. Jennings Collection, American Philosophical Library, Philadelphia.
30 *B*, p. 103.
31 *B*, p. 102.
32 *B*, p. 104.
33 Watson & McDougall (1928; p. 94); see *CW*, pp. 232–4.
34 Watson & McDougall (1928; p. 102).
35 Lomax (1978; esp. Ch. 4).
36 Jennings *et al.* (1917; p. 81).
37 *PCIC*, p. 73.
38 Rayner Watson (1930); quoted in Logue (in press).
39 *PCIC*, p. 101.
40 See Lomax (1978) on Freudian influences on child care practice.

Rats, mazes and learning theory

Detailed and extensive reviews of the research discussed in this section can be found in Warden, Jenkins & Warner (1935; 1936) and in Munn (1933). Descriptions of work in the Berkeley laboratory are from Tolman's *Purposive Behavior in Animals and Men* (*PBAM*) of 1932; page references are to the second edition. Unfortunately there is no biographical study of Tolman apart from his autobiographical sketch (Tolman, 1952 in *HPA*). Many of Tolman's early papers are reprinted in Tolman (1966).

1 Carr (1936); see also Beach (1974).
2 Carr (1936; p. 79).
3 The relationship between high school expansion and psychology in the state universities is discussed in O'Donnell (1979; Ch. 11).
4 Hunter (1952; p. 172).
5 O'Donnell (1979; Ch. 11).
6 Such mazes and the results they generated are described in many of the psychology textbooks written in America between 1930 and 1960; see, for example, Woodworth & Schlosberg (1954; Ch. 21) or the earlier editions of this text.
7 Watson (1924; p. 16).
8 *BICP*, Ch. 7.
9 Watson (1917).
10 Tolman (1952; p. 326).
11 Perry (1918).
12 Tolman (1952; p. 329).
13 Kuo (1922).
14 *PBAM*, p. 6.
15 *PBAM*, pp. 48–50.
16 *PBAM*, pp. 54–8.
17 *PBAM*, pp. 77–82.
18 *PBAM*, pp. 164–70.
19 Koehler (1926); Koffka (1928).
20 Tinklepaugh (1928); see also *PBAM*, pp. 74–6.
21 Notable examples are Helson (1927) and Maier (1929).
22 Skinner (1979; p. 30).
23 Lashley (1929; p. 14).
24 Hebb, D. O. Introduction to Dover edition (1963) of Lashley (1929; pp. viii–ix).
25 The count of papers on the behaviour of the rat published by American scientists is based on what is claimed to be a complete bibliography up to August, 1932 published in Munn (1933).
26 This provided a technique for quantifying drives that was based on measuring the resistance they would overcome. The idea was suggested by J. J. B. Morgan in 1923, first implemented by F. A. Moss in 1924 and then developed on a large scale by C. J. Warden and his associates at Columbia University where the 'Columbia Obstruction Box' became famous (Munn, 1950; pp. 84–98).
27 Tolman (1938).

The decline and resurgence of strict behaviourism

Two unpublished papers by Franz Samelson (Psychology Dept, Kansas State University, Kans, 66506) were particularly valuable in preparing this section: 'The stalemate of the Twenties' (presented to the 1980 Cheiron meeting) and 'The social sciences in the Twenties' (presented to the 1982 Cheiron meeting). An interesting and much older paper evaluating developments during this era is Harrell & Harrison (1938). The impact of research funding is discussed by S. Cross in a forthcoming Ph.D. thesis (Cross, in preparation).

1 McDougall (1926).
2 M. Meyer, A. P. Weiss, E. A. Singer and J. R. Kantor are among the people frequently classed as strict behaviourists at this time. However, since they made no direct contribution to animal psychology and do not appear to have influenced its development, they are not discussed here. See Harrell & Harrison (1938).
3 Lashley (1923).
4 *PBAM*, p. 234.
5 Kuo (1928).
6 Tolman (1928).
7 Tolman (1928).
8 Kuo (1929).
9 Kuo (1967; Preface); also Gottlieb (1972).
10 Moss (1924).
11 One of the first discussions of this distinction to follow the English translation of Pavlov was Troland (1928). This contrasted 'reflex' and 'retroflex' action; see Ch. 14 and the discussion of this book by Herrnstein (1972).
12 See Ch. 5 above.
13 *PBAM*, Ch. 21.
14 See Samelson's two unpublished papers listed above and footnotes in some of the experimental papers cited in the previous section.
15 Foreword by Herman M. Adler to Lashley (1929; p. xvi).

16 Samelson (unpublished).
17 Hull (1952); May (1971).
18 Cited in O'Donnell (1979; p. 563) from Cattell, 'Our psychological association and research' in Poffenberger (Ed.) *Cattell*, 3, 339.
19 Quoted in Samelson (unpublished).
20 Tolman (1959; p. 152).
21 The quotation is from Trotter's friend, Ernest Jones (Jones, 1959; p. 153).
22 Humphrey (1976; p. 305).
23 Hull (1952).
24 Mowrer (1974).
25 See Cohen (1979; Ch. 10) for a survey of Watson's popular articles.
26 Tolman (1945).
27 Russell (1968; pp. 32–7; 83–6); see also Clark (1975; pp. 339–53).
28 Russell (1927).
29 Russell (1927; p. 46).
30 Watson & McDougall (1928; p. 4).
31 Skinner (1976; pp. 288–92); it is intriguing that Skinner remembers Russell as *endorsing* Watson's attack on the Law of Effect, whereas in fact Russell criticized Watson on this point and argued in favour of the Law in a way quite similar to the one Skinner later adopted.

The page on which a publication is cited in the text or referred to in the Notes is given in italics after its entry in this bibliography.

Bibliography

Albrecht, F. M. (1960) *The New Psychology in America: 1880–1895*. Unpublished doctoral thesis, The Johns Hopkins University, Baltimore. *247*

Angell, J. R. (1936) Autobiography. In C. Murchison (Ed) *A History of Psychology in Autobiography* Vol. 3. Worcester, Mass: Clark University; pp. 1–38. *254*

Anokhin, P. K. (1968) Pavlov and psychology. In B. B. Wolman (ed.) *Historical Roots of Contemporary Psychology*. New York: Harper & Row; pp. 132–140. *251*

Asch, S. E. (1968) Wolfgang Koehler (Obituary notice). *American Journal of Psychology*, **81**, 110–19. *255*

Asratyan, E. A. (1953) *I. P. Pavlov: his Life and Work*. Moscow: Foreign Language Publishing House. (Russian original, 1949.) *251*

Babkin, B. P. (1949) *Pavlov*. Chicago: Chicago University. *251, 252*

Bain, A. (1855) *The Senses and the Intellect*. London: Parker. *8, 244*

Bain, A. (1859) *The Emotions and the Will*. London: Parker. *8, 244*

Bain, A. (1904) *Autobiography*. London: Longman, Green & Co. *244*

Baldwin, J. M. (1895) *Mental Development in the Child and the Race*. New York: Macmillan. *160, 246, 254*

Baldwin, J. M. (1926) *Between Two Wars (1861–1921)*. Boston: Stratford. *247, 254*

Barzun, J. (1965) *Race*. New York: Harper Row. (Original publication, 1937.) *256, 257*

Beach, F. A. (1961) Biographical memoir of K. S. Lashley. *Biographical Memoirs*, **35**, 163–204. Published for the National Academy of Sciences in New York: Columbia University. *254*

Beach, F. A. (1974) Autobiography. In G. Lindzey (Ed) *A History of Psychology in Autobiography* Vol. 6. Englewood Cliffs, NJ: Prentice Hall; pp. 31–58. *258*

Bechterev, V. (1906) La psychologie objective. *Revue Scientifique*, (Revue Rose), Sept., 5th Series, **6**, 353–7; 390–6. *126, 252*

Bechterev, V. M. (1913) *La Psychologie Objective*. Paris: Librairie Felix Alcan. *127, 173, 218, 252*

Bechterev, V. M. (1932) *General Principles of Human Reflexology*. New York: International. (Reprinted in New York: Arno, 1973). *130, 252*

Beer, T., Bethe, A. & von Uexkuell, J. (1899) Vorschlaege zu eine objectivirenden Nomenklatur in der Physiologie des Nervensystems. *Biologisches Zentralblatt*, **19**, 517–21. *136, 251, 252*

Ben-David, J. (1971) *The Scientist's Role in Society*. Englewood Cliffs, NJ: Prentice-Hall. *247*

Ben-David, J. & Collins, R. (1966) Social factors in the origin of a new science: the case of psychology. *American Sociological Review*, **31**, 451–65. *247*

Bernard, L. L. (1924) *Instinct: a Study in Social Psychology*. London: George Allen & Unwin. *216, 257*

Bibby, C. (1959) *T. H. Huxley: Scientist, Humanist and Educator*. London: Watts. *245–7*

Blanton, M. G. (1917) The behavior of the human infant during the first thirty days of life. *Psychological Review*, **24**, 456–83. *220, 257*

Blumenthal, A. L. (1975) A reappraisal of Wilhelm Wundt. *American Psychologist*, **30**, 1081–6. *247*

Blumenthal, A. L. (1979) The founding father we never knew. *Contemporary Psychology*, **24**, 449–53. *254*

Blumenthal, A. L. (1980) Wilhelm Wundt and early American psychology; a clash of cultures. In R. W. Rieber & K. Salzinger (Eds) *Psychology: Theoretical–Historical Perspectives*. New York: Academic; pp. 25–42. *247, 254*

Boas, F. (1911) *The Mind of Primitive Man*. New York: Macmillan. *213, 257*

Boden, M. A. (1972) *Purposive Explanation in Psychology*. Cambridge, Mass: Harvard University. *256*

Boden, M. A. (1977) *Artificial Intelligence and Natural Man*. Hassocks, Sussex: Harvester. *255*

Bois-Reymond, E. du (1874) The limits of our knowledge of nature. *Popular Science Monthly*, **5**, 17–32. *107, 250*

Boorstin, D. J. (1948) *The Lost World of Thomas Jefferson*. Boston: Henry Holt. (Reprinted as Beacon Paperback, 1960). *249*

Boring, E. G. (1929; 1950) *History of Experimental Psychology*. New York: Appleton-Century-Crofts. (2nd ed: 1950) *244, 247, 254*

Brazier, M. A. B. (1959) The historical development of neurophysiology. In J. Field (ed.) *The Handbook of Physiology, Section 1: Neurophysiology* (Vol. 1). Washington, DC: American Physiological Society. *249, 250*

Broughton, J. M. (1981) The Genetic Psychology of James Mark Baldwin. *American Psychologist*, **36**, 396–407. *254*

Broughton, J. M. & Freeman-Moir, D. J. (1981) *The Cognitive-Developmental Psychology of James Mark Baldwin*. Norwood, NJ: Ablex. *254*

Bruner, J. S., Jolly, A. & Sylvan, K. (1976) *Play*. London: Penguin. *254*

Buckley, K. W. (1982) The selling of a psychologist: John Broadus Watson and the application of behavioural techniques to advertising. *Journal of the History of the Behavioral Sciences*, **18**, 207–21. *258*

Burchfield, J. D. (1975) *Lord Kelvin and the Age of the Earth*. London: Macmillan. *246*

Burckhardt, R. W. (1977) *The Spirit of the System: Lamarck and Evolutionary Biology*. Cambridge, Mass: Harvard University. *244*

Cairns, R. B. (1980) Developmental theory before Piaget. *Contemporary Psychology*, **25**, 438–40. *254*

Cairns, R. B. & Ornstein, P. A. (1979) Developmental psychology. In E. Hearst (ed.) *The First Century of Experimental Psychology*. Hillsdale, NJ: Erlbaum; pp. 459–512. *254*

Carr, H. A. (1936) Autobiography. In C. Murchison (ed.) *A History of Psychology in Autobiography* Vol. 3. Worcester, Mass: Clark University; pp. 69–82. *253, 258*

Cassirer, E. (1950) *The Problem of Knowledge: Philosophy, Science and History since Hegel*. New Haven: Yale University. *247*

Clark, R. (1975) *The Life of Bertrand Russell*. London: Jonathan Cape. *259*

Clarke, E. & O'Malley, C. D. (1968) *The Human Brain and Spinal Cord*. Berkeley: University of California. *249, 250*

Clarke, E. & Dewhurst, K. (1972) *An Illustrated History of Brain Function*. Oxford: Sandford. *249*

Coburn, C. A. & Yerkes, R. M. (1915) A study of the behavior of the crow, *Corvus Americanus*, by the multiple-choice method. *Journal of Animal Behavior*, **5**, 75–114. *155, 254*

Cohen, D. (1979) *J. B. Watson: The Founder of Behaviourism*. London: Routledge & Kegan Paul. *253, 254, 257–9*

Cohen, I. B. (1972) *Franklin: Scientist and Statesman*. New York: Scribner. *249*

Cole, R. W. (1907) Concerning the intelligence of raccoons. *Journal of Comparative Neurology & Psychology*, **17**, 211–61. *149, 150, 253*

Commager, H. S. (1975) *Jefferson, Nationalism and the Enlightenment*. New York: Braziller. *249*

Coon, D. J. (1982) Eponymy, obscurity, Twitmyer and Pavlov. *Journal of the History of the Behavioral Sciences*, **18**, 255–62. *252*

Cottingham, J. (1978) A brute to the brutes? Descartes' treatment of animals. *Philosophy*, **53**, 551–9. *248*

Cranefield, P. F. (1973) *The Way In and the Way Out* (History of Medicine Series, No. 41), New York: Futura. *249*

Cravens, H. (1978) *The Triumph of Evolution: American Scientists and the Heredity–Environment Controversy, 1900–1941*. Philadelphia: University of Pennsylvania. *257*

Cross, S. (In preparation) *The Human Factor: Interdisciplinary Enterprise and the Making of Behavioral Science in America, 1918–1949*. Unpublished Ph.D. thesis, The Johns Hopkins University, Baltimore. *258*

Cuny, H. (1964) *Ivan Pavlov* (Translation by P. Evans). London: Souvenir. (French original, 1962.) *251, 252*

Danzinger, K. (1979) The positivist repudiation of Wundt. *Journal of the History of the Behavioral Sciences*, **15**, 205–30. *247, 254*

Darwin, C. (1859) *On the Origin of Species by Means of Natural Selection*. London: Murray. *4, 5, 12, 17, 21, 23, 28, 45, 46, 55, 57, 112*

Darwin, C. (1871) *The Descent of Man and Selection in Relation to Sex*. London: Murray. *2, 5–8, 12, 15, 21, 22, 24, 26, 45, 111, 112*

Darwin, C. (1872) *The Expression of the Emotions in Man and Animal*. London: John Murray. *206, 256*

Darwin, C. (1906) *The Voyage of the Beagle*. London: Dent. (Original version published in 1839.) *244*

Darwin, E. (1794) *Zoonomia*. London: J. Johnson. *249*

Darwin, F. (1887) *The Life and Letters of Charles Darwin*. London: Murray. *245*

Dennis, W. (1948) *Readings in the History of Psychology*. New York: Appleton-Century-Crofts.

Descartes, R. (1912) *A Discourse on Method* (Translation by J. Veitch). London: Dent. (French original, 1637). *85–8, 248, 249*

Descartes, R. (1972) *Treatise on Man* (Translated by T. S. Hall). Cambridge, Mass: Harvard University. *87, 88, 249*

Dewey, J. (1917) The need for social psychology. *Psychological Review*, **24**, 266–77. *215, 257*

Dickinson, A. (1980) *Contemporary Animal Learning Theory*, Cambridge: Cambridge University. *252*

Dobzhansky, T. (1955) The crisis of Soviet biology. In E. J. Simmons (ed.) *Continuity and Change in Russian and Soviet Thought*. Cambridge, Mass: Harvard University; pp. 329–46. *251*

Driesch, H. (1908) *The Science and Philosophy of the Organism* 2 vols. London: Black. *178, 179, 255*

Dunlap, K. (1919) Are there any instincts? *Journal of Abnormal Psychology*, **14**, 307–11. *216, 257*

Dunlap, K. (1932) Autobiography. In C. Murchison (ed.) *A History of Psychology in Autobiography* Vol. 2. Worcester, Mass: Clark University; pp. 35–61. *254*

Ehrenfels, C. von (1890) Ueber Gestaltqualitaeten. *Vierteljahresschrift fuer Wissenschaftliche Philosophie*, **14**, 249–92. *184, 255*

Eiseley, L. (1958) *Darwin's Century*. New York: Doubleday. *244, 247*

Elliott, R. M. (1956) R. M. Yerkes (Obituary notice). *American Journal of Psychology*, **69**, 487–94. *256*

Ellis, W. D. (1938) *A Source Book of Gestalt Psychology* London: Routledge & Kegan Paul. *256*

Fancher, R. E. (1973) *Psychoanalytic Theory: the Development of Freud's Thought*. New York: Norton. *254*

Fearing, F. (1930) *Reflex Action: a Study in the History of Physiological Psychology*, New York: Hafner. *249, 250*

Flugel, J. C. (1933) *A Hundred Years of Psychology*. London: Duckworth. *244*

Forrest, D. W. (1974) *Francis Galton: the Life and Work of a Victorian Genius*. London: Elek. *246*

Freud, S. (1910) *Three Contributions to Sexual Theory* (Translated by A. A. Brill) *Journal of Nervous & Mental Diseases*, Monograph Series, No. 7. (German original, 1905; English translation by J. Strachey of the 8th German edition in *The Complete Psychological Works of Sigmund Freud* Vol. 7. London: Hogarth, 1953.) *164, 254*

Frolov, Y. P. (1937) *Pavlov and his School*. London: Paul, Trench, Trubner & Co. *251*

Fulton, J. F. (1959) Historical reflections on the backgrounds of neurophysiology. In C. McBrooks & P. C. Cranefield (eds.) *The Historical Development of Physiological Thought*. New York: Hafner; pp. 109–36. *249*

Furness, W. H. (1916) Observations on the mentality of chimpanzees and orang-utans. *Proceedings of the American Philosophical Society*, **55**, 281–90. *200, 256*

Galton, F. (1869) *Hereditary Genius*, London: Macmillan. *46, 246*

Galton, F. (1872) Statistical enquiries into the efficacy of prayer. *Fortnightly Review*, **12**, 125–35. *46, 246*

Galton, F. (1879) Psychometric experiments. *Brain*, **2**, 149–62. *246*

Galton, F. (1883) *Inquiries into Human Faculty and its Development*. London: Macmillan. *47, 246*

Galton, F. (1889) *Natural Inheritance*. London: Macmillan. *48, 246*

Galton, F. (1908) *Memories of My Life*, London: Methuen. *246*

Gardner, R. A. & Gardner, B. T. (1969) Teaching sign language to a chimpanzee. *Science*, **165**, 664–72. *248*

Gasman, D. (1971) *The Scientific Origins of National Socialism: Social Darwinism in Ernst Haeckel and the German Monist League*, London: MacDonald. *257*

Gottlieb, G. (1972) Zing Yang Kuo. *Journal of Comparative & Physiological Psychology*, **80**, 1–10. *257, 259*

Gould, J. L. (1976) The dance-language controversy. *Quarterly Review of Biology*, **51**, 211–44. *245*

Gould, S. J. (1977) *Ontogeny and Phylogeny*. Cambridge, Mass: Harvard University. *247, 254*

Gould, S. J. (1981) *The Mismeasure of Man*, Cambridge, Mass: Harvard University. *257*

Gray, J. A. (1964) *Pavlov's Typology*. London: Pergamon. *252*

Gray, J. A. (1979) *Pavlov*. London: Fontana Paperback. *251*

Gray, P. H. (1962) D. A.Spalding: the first experimental behaviorist. *Journal of General Psychology*, **67**, 299–307. *245*

Gray, P. H. (1968) Prerequisite to an analysis of behaviorism. *Journal of the History of the Behavioral Sciences*, **4**, 365–77. *245*

Greisman, H. C. (1979) Herd instinct and the foundations of Biosociology. *Journal of the History of the Behavioral Sciences*, **15**, 357–69. *257*

Groos, K. (1898) *The Play of Animals*. London: Chapman & Hall. (German original, 1896). *254*

Groos, K. (1901) *The Play of Man*. New York: Appleton. *254*

Gruber, H. (1974) *Darwin on Man*. London: Wildwood House. *244*

Haeckel, E. (1900) *The Riddle of the Universe at the Close of the Nineteenth Century* (Translation by J. McCabe of *Die Weltraetsel*), London: Watts. *212, 257*

Haggerty, M. E. (1909) Imitation in monkeys. *Journal of Comparative Neurology & Psychology*, **19**, 337–455. *196, 256*

Haggerty, M. E. (1913) Plumbing the minds of apes. *McClures Magazine (NY)*, **41**, (August) 151–4. *196, 256*

Haight, G. S. (1968) *George Eliot*, Oxford: Clarendon. *244*

Haines, G. (1969) *Essays on the German Influence upon English Education and Science*. Hamden, Conn: Connecticut College. *247*

Haldane, J. B. S. (1954) Introducing Douglas Spalding *British Journal of Animal Behaviour*, **2**, 1–11. *245*

Hale, N. G. (1971) *Freud and the Americans*. New York: Oxford University. *254*

Hall, G. S. (1904) *Adolescence: Its Psychology and its Relations to Physiology, Antropology, Sociology, Sex, Crime, Religion and Education* (2 vols). New York: Appleton. *160, 163*

Halliday, M. S. (1979) Jerzy Konorski and Western psychology. In A. Dickinson & R. A. Boakes (eds.) *Mechanisms of Learning and Motivation*. Hillsdale, NJ: Erlbaum; pp. 1–18. *252*

Hamilton, G. V. (1908) An experimental study of an unusual type of reaction in a dog. *Journal of Comparative Neurology & Psychology*, **18**, 329–41. *201, 256*

Hamilton, G. V. (1911) A study of trial and error reactions in mammals. *Journal of Animal Behavior*, **1**, 33–66. *153–5, 254*

Hamilton, G. V. (1914) A study of sexual tendencies in monkeys and baboons. *Journal of Animal Behavior*, **4**, 295–318. *154, 254*

Hamilton, G. V. (1929) *A Research in Marriage*. New York: Boni. 201, *256*

Hans, N. (1931) *A History of Russian Educational Policy: 1701–1917*. New York: Russell & Russell. (Reprinted in 1964). *250, 251*

Harrell, W. & Harrison, R. (1938) The rise and fall of behaviorism. *Journal of General Psychology*, **18**, 367–421. *258, 259*

Harris, B. (1979) Whatever happened to Little Albert? *American Psychologist*, **34**, 151–60. *258*

Hearnshaw, L. (1964) *A Short History of British Psychology: 1840–1940*. London: Methuen. *244, 245, 247, 256*

Hebb, D. O. (1959) K. S. Lashley (Obituary notice). *American Journal of Psychology*, **72**, 142–50. *254*

Heidbreder, E. (1973) William McDougall and social psychology. In M. Henle, J. Jaynes & J. J. Sullivan (eds.) *Historical Conceptions of Psychology*. New York: Springer; pp. 267–75. *211, 257*

Heider, F. (1973) Gestalt theory: early history and reminiscences. In M. Henle, J. Jaynes & J. J. Sullivan (eds.) *Historical Conceptions of Psychology*. New York: Springer; pp. 63–73. *255*

Helson, H. (1927) Insight in the white rat. *Journal of Experimental Psychology*, **10**, 378–96. *258*

Helson, H. (1973) Why did their predecessors fail and the Gestalt psychologists succeed? In M. Henle, J. Jaynes & J. J. Sullivan (eds.) *Historical Conceptions of Psychology*. New York: Springer; pp. 74–82. *256*

Henle, M. (1968) Wolfgang Koehler (Obituary notice). *American Philosophical Society: Year Book*, p. 139–45. *255*

Henle, M. (1978) One man against the Nazis – Wolfgang Koehler. *American Psychologist*, **33**, 939–44. *256*

Herrnstein, R. J. (1972) Nature as nurture: behaviorism and the instinct doctrine. *Behaviorism*, **1**, 233–52. *257, 259*

Hilgard, E. R. & Bower, G. H. (1966) *Theories of Learning* (3rd edition). New York: Appleton-Century-Crofts. *251*

Himmelfarb, G. (1959) *Darwin and the Darwinian Revolution*. New York: Doubleday. *244*

Hirshberg, E. W. (1970) *George Henry Lewes*. New York: Twayne. *244*

Hobhouse, L. T. (1896) *The Theory of Knowledge*. London: Methuen. *179, 255*

Hobhouse, L. T. (1901) *Mind in Evolution*. London: Macmillan. *180, 182, 230, 255*

Hobhouse, L. T. (1904) *Democracy and Reaction*. London: Fisher Unwin. (Reprinted in Hassocks, Sussex: Harvester Press, 1972.) *255*

Hobson, J. A. & Ginsberg, M. (1931) *L. T. Hobhouse: his Life and Work*. London: George Allen & Unwin. *255*

Hofstadter, R. (1944) *Social Darwinism in American Thought*. Philadelphia: University of Pennsylvania. *62, 247*

Hofstadter, R. & Hardy, C. D. (1953) *The Development and Scope of Higher Education in the United States*. New York: Columbia University. *247*

Hull, C. L. (1952) Autobiography. In E. G. Boring *et al.* (eds.) *A History of Psychology in Autobiography* Vol. 4. New York: Russell & Russell; pp. 143–62. *256, 259*

Hull, D. L. (1973) *Darwin and his Critics*. Cambridge, Mass: Harvard University. *244*

Humphrey, N. K. (1976) The social function of intellect. In P. P. G. Bateson & R. A. Hinde (eds.) *Growing Points in Ethology*. Cambridge: Cambridge University; pp. 303–17. *259*

Hunter, W. S. (1914) The delayed reaction in animals and children. *Behavior Monographs*, **2**, No. 6. Published by Williams & Wilkins. (A large part of this monograph is reprinted in Dennis, 1948). 156–8, *254*

Hunter, W. S. (1952) Autobiography. In E. G. Boring *et al.* (eds.) *A History of Psychology in Autobiography* Vol. 4. New York: Russell & Russell; pp. 163–88. *254, 258*

Huxley, L. (1900) *The Life and Letters of T. H. Huxley* Vols. I & II. London: Macmillan. *245*

Huxley, T. H. (1863) *Evidence as to Man's Place in Nature*, London: Williams & Norgate, *5, 244*

Huxley, T. H. (1874) On the hypothesis that animals are automata. *Nature*, **10**, 362. (Reprinted in Huxley, 1893). *19, 20, 245*

Huxley, T. H. (1878) *Hume*. London: Macmillan. *20, 245*

Huxley, T. H. (1893) *Methods and Results (Collected Essays Vol. 1)*. London: Macmillan.

Huxley, T. H. (1894) Geological reform. In *Discourses Biological and Geological (Collected Essays Vol. 8)*. London: Macmillan; pp. 305–39. (Original publication, 1869). *44, 246*

Irvine, W. (1955) *Apes, Angels and Victorians: a Joint Biography of Darwin and Huxley*. London: Weidenfeld & Nicolson. *245*

Itard, J. (1972) *The Wild Boy of Aveyron*. London: NLB. *256*

James, W. (1890) *Principles of Psychology*. New York: Holt. *63–8, 82, 206, 247, 248, 254, 256*

James, W. (1904) Does consciousness exist? *Journal of Philosophy, Psychology & Scientific Methods*, **1**, 477–91. (Reprinted in James, 1976.) *163, 254*

James, W. (1976) *Essays in Radical Empiricism*. Cambridge, Mass: Harvard University. (Originally published in 1912 as a collection edited by R. B. Perry). *163, 254*

Jaynes, J. & Woodward, W. (1974) In the shadow of the Enlightenment. *Journal of the History of the Behavioral Sciences*, **10**, 3–15; 144–59. *256*

Jefferson, G. (1960) *Selected Papers*. London: Pitman Medical. *250*

Jenkin, F. (1867) The origin of species *North British Review*, **46**, 277–318. (Reprinted in Hull, 1973). *5, 244*

Jennings, H. S. (1906) *The Behavior of the Lower Organisms* New York: Macmillan. (Reprinted in Bloomington, Ind: Indiana University, 1962). *140–2, 253*

Jennings, H. S. (1907) The interpretation of the behavior of the lower organisms. *Science*, **27**, 698–710. *253*

Jennings, H. S. *et al* (1917) *Suggestions of Modern Science concerning Education*. New York: Macmillan. *220, 257*

Jensen, D. D. (1970) Polythetic biopsychology: an alternative to behaviorism. In J. H. Reynierse (ed.) *Current Issues in Animal Learning*. Lincoln, Neb: University of Nebraska; pp. 1–31. *253*

Joncich, G. (1968) *The Sane Positivist: a Biography of E. L. Thorndike*. Middletown, Conn: Wesleyan University. *248*

Jones, E. (1955) *Sigmund Freud: Life and Work* Vol. 2. London: Hogarth. *254*

Jones, E. (1959) *Free Associations*. London: Hogarth. *257, 259*

Jones, M. C. (1924) A laboratory study of fear: the case of Peter. *Pedagogical Seminary*, **31**, 308–15. *225, 259*

Jones, M. C. (1974) Albert, Peter and John B. Watson. *American Psychologist*, **29**, 581–3. *258*

Jung, C. G. (1961) *Memories, Dreams and Reflections* (Translated by R. & C. Winston), New York: Random House. *254*

Kamin, L. J. (1974) *The Science and Politics of IQ*. Hillsdale, NJ: Erlbaum. *257*

Katz, D. (1937) *Animals and Man: Studies in Comparative Psychology*. London: Longman. *255*

Kinnaman, A. J. (1902) The mental life of two Macacca rhesus monkeys in captivity. *American Journal of Psychology*, **13**, 98–148; 173–218. *149, 253*

Kline, G. L. (1955) Darwinism and the Russian Orthodox Church. In E. J. Simmons (ed.) *Continuity and Change in Russian and Soviet Thought*. Cambridge, Mass: Harvard University; pp. 307–28. *251*

Koehler, W. (1915) Optische untersuchungen am Schimpansen und am Haushuhn. *Abhandlungen der Koeniglich Preussischen Akademie der Wissenschaften*, (Berlin) physikalisch-mathematisch Klasse (Whole no. 3). *193, 255*

Koehler, W. (1918) Nachweis einfacher Strukturfunctionen beim Schimpansen und beim Haushuhn. *Abhandlungen der Koeniglich Preussischen Akademie der Wissenschaften*, (Berlin) physikalisch-mathematisch Klasse (Whole no. 2). (Parts of this paper are reprinted in Ellis, 1938; pp. 217–27). *194, 255*

Koehler, W. (1925) *The Mentality of Apes* (Translation by E. Winter of 2nd German edition). New York: Harcourt, Brace & World. (Original German edition, 1917). *185–95, 203, 233, 255, 256*

Koehler, W. (1926) *Gestalt Psychology*. New York: Appleton. *258*

Koehler, W. (1971) *Selected Papers* (edited by M. Henle). New York: Liveright. *255, 256*

Koffka, K. (1928) *The Growth of Mind* (2nd edition). New York: Harcourt Brace. *258*

Kohts, N. (1928) Recherches sur l'intelligence du chimpanze par la methode de 'choix d'apres modele'. *Journal de Psychologie Normale et Pathologique*, **25**, 245–75. *201, 256*

Konorski, J. (1948) *Conditioned Reflexes and Neuron Organization*. Cambridge: Cambridge University. *252*

Koshtoyants Kh. S. (1964) *Essays on the History of Physiology in Russia*. Washington, DC: American Institute of Physiological Sciences. (Russian original, 1946). *252*

Krantz, D. L. & Allen, D. (1967) The rise and fall of McDougall's instinct doctrine. *Journal of the History of the Behavioral Sciences*, **3**, 326–38. *257*

Kuhn, T. S. (1962) *The Structure of Scientific Revolutions*. Chicago: University of Chicago. *251*

Kuo, Z. Y. (1921) Giving up instincts in psychology. *Journal of Philosophy*, **18**, 645–64. *217, 257*

Kuo, Z. Y. (1922) The nature of unsuccessful acts and their order of elimination in animal learning. *Journal of Comparative Psychology*, **2**, 1–28. *231, 258*

Kuo, Z. Y. (1928) The fundamental error of the concept of purpose and the trial and error fallacy. *Psychological Review*, **35**, 414–33. *236, 259*

Kuo, Z. Y. (1929) Purposive behavior and the prepotent stimulus. *Psychological Review*, **36**, 547–50. *236, 259*

Kuo, Z. Y. (1930) The genesis of the cat's response to the rat. *Journal of Comparative Psychology*, **11**, 1–35. *218, 257*

Kuo, Z. Y. (1967) *The Dynamics of Behavior Development*. New York: Random House. *259*

Lange, F. (1880) *The History of Materialism* (Translation by E. C. Thomas) London: Truebner. *249, 250*

Lashley, K. S. (1923) The behavioristic interpretation of consciousness. *Psychological Review*, **30**, 237–72; 329–53. *236, 259*

Lashley, K. S. (1929) *Brain Mechanisms and Intelligence* Chicago: University of Chicago. (Reprinted in New York: Dover, 1963). *234, 235, 258, 259*

Lashley, K. S. (1950) In search of the engram. *Symposia of the Society of Experimental Biology*, **4**, 454–82. *249*

Lashley, K., Chow, K. L. & Semmes, J. (1951) An examination of the electrical field theory of cerebral integration. *Psychological Review*, **58**, 125–36. *256*

Leahey, T. H. (1981) The mistaken mirror; on Wundt's and Titchener's psychologies. *Journal of the History of the Behavioral Sciences*, **17**, 273–82. *254*

Leake, C. D. (1959) Danilevsky, Wedensky and Ukhtomsky. In M. A. B. Brazier (Ed) *The Central Nervous System and Behavior*. New York: Josiah Macy Foundation; pp. 151–62. *251*

Leif, A. (1948) *The Commonsense Psychiatry of Adolph Meyer*. New York: McGraw Hill. *253, 254*

Lewes, G. H. (1860) *The Physiology of Common Life*. London: Blackwood. *10, 112, 250*

Liddell, E. G. T. (1960) *The Discovery of Reflexes*. Oxford: Clarendon. *249*

Lilge, F. (1948) *The Abuse of Learning: the Failure of the German University*. New York: Macmillan. *247*

Lippman, W. (1922) The mental age of Americans. *New Republic*, **32**, 213–15; also 246–8; 257–98; 328–30. *215, 257*

Lippman, W. (1923) A future for the tests. *New Republic*, **33**, 9–10; see also 116–20; 145–6. *215, 257*

Littman, R. A. (1979) Social and intellectual origins of experimental psychology. In E. Hearst (ed.) *The First Century of Experimental Psychology*. Hillsdale, NJ: Erlbaum; pp. 39–86. *247*

Loeb, J. (1900) *Comparative Physiology of the Brain and Comparative Psychology*. New York: Putnam. *139, 252*

Loeb, J. (1910) Die Bedeutung der Tropismen fuer Psychologie. *Rapports et Comptes Rendues, VIme Congres Internationale de Psychologie*, Geneve: Librairie Kuendig; pp. 281–306 (English translation, The significance of tropisms for psychology, in *Popular Science Monthly*, 1911, **79**, 105–25). *253*

Lockard, R. B. (1968) The albino rat. *American Psychologist*, **23**, 734–42. *253*

Logue, A. (in press) The growth of behaviorism. In C. Buxton (ed.) *Points of View in the Modern History of Behaviorism*. New York: Academic. *258*

Lomax, E. M. R. (1978) *Science and Patterns of Child Care*. San Francisco: Freeman. *258*

Luchins, A. S. (1975) The place of Gestalt theory in American psychology. In S. Ertle, L. Kemmler & K. M. Stadler (eds.) *Gestalttheorie in der Modernen Psychologie.* Darmstadt: Steinkopf; pp. 21–44. *256*

McDougall, W. (1908) *An Introduction to Social Psychology.* London: Methuen. *207, 208, 211, 216, 256, 257*

McDougall, W. (1912) *Psychology: the Study of Behaviour.* London: Williams & Norgate. *256, 257*

McDougall, W. (1926) Men or robots? In C. Murchison (ed.) *Psychologies of 1925.* Worcester, Mass: Clark University; pp. 273–305. *236, 259*

McDougall, W. (1930) Autobiographical sketch. In C. Murchison (ed.) *A History of Psychology in Autobiography* Vol. 1. Worcester, Mass: Clark University; pp. 191–224. *256*

McFarland, D. (1981) *The Oxford Companion to Animal Behaviour* Oxford: Oxford University. *248*

McLeish, J. (1975) *Soviet Psychology.* London: Methuen. *252*

Maier, N. R. F. (1929) Reasoning in white rats. *Comparative Psychology Monographs,* **6**, No. 29. *258*

Mackenzie, B. D. (1977) *Behaviourism and the Limits of Scientific Method.* London: Routledge & Kegan Paul. *245*

Mackintosh, N. J. (1974) *The Psychology of Animal Learning.* London: Academic. *251*

Macpherson, H. (1900) *Herbert Spencer.* London: Chapman & Hall. *246*

Marchant, J. (1916) *Alfred Russell Wallace; Letters and Reminiscences* Vol. 1. London: Cassell. *246*

Marryat, Captain (1846) *Masterman Ready.* London: Dent. *30, 31, 245*

Maslow, A. H. (1970) *Motivation and Personality* (2nd edition). New York: Harper. *248*

May, M. A. (1971) A retrospective view of the Institute of Human Relations at Yale. *Behavior Science Notes,* **6**, 141–72. *259*

Mecacci, L. (1979) *Brain and History* (Translated by H. A. Buchtel). New York: Brunner/Mazel. *252*

Meehl, P. E. (1950) On the circularity of the Law of Effect. *Psychological Bulletin,* **47**, 52–75. *248*

Mettrie, J. de la (1953) *Man a Machine* La Salle, Illinois: Open Court. (French original, 1748). *90–2, 249*

Meynert, T. (1874) *Zur Mechanik des Gehirnbaues.* Vienna. *248*

Meynert, T. (1885) *Psychiatry.* New York. *248*

Miles, W. R. (1930) On the history of research with rats and mazes: a collection of notes. *Journal of General Psychology,* **3**, 324–37. *253*

Mill, J. S. (1843) *A System of Logic.* London: Longman. *8, 10, 53*

Mill, J. S. (1873) *Autobiography.* London: Longman. *246, 249*

Mill, J. S. (1950) *On Bentham and Coleridge.* London: Chatto & Windus. (Original publication as separate essays, 1838 and 1840 respectively). *174, 254*

Mischel, W. (1979) On the interface of cognition and personality. *American Psychologist,* **34**, 740–54. *251*

Morgan, C. L. (1878) Geological time. *Geological Magazine (Series* **2**), **5**, 154–62; 199–207. *45, 246*

Morgan, C. L. (1883) Letter *Nature,* **27**, 313. *32, 245*

Morgan, C. L. (1890) *Animal Life and Intelligence.* London: Edward Arnold. *32, 37, 39, 42, 245*

Morgan, C. L. (1894) *An Introduction to Comparative Psychology* London: Scott. *37–42, 51, 245, 255*

Morgan, C. L. (1895) Some definitions of instinct. *Natural Sciences,* **7**, 321–9. *205, 256*

Morgan, C. L. (1896*a*) *Habit and Instinct.* London: Edward Arnold. *42–4, 49–51, 68, 246, 247*

Morgan, C. L. (1896*b*) Animal automatism and consciousness. *The Monist,* **7**, 1–17. *246*

Morgan, C. L. (1900) *Animal Behaviour.* London: Edward Arnold. *246*

Moss, F. A. (1924) Note on building likes and dislikes in children. *Journal of Experimental Psychology,* **7**, 475–8. *237, 259*

Mowrer, O. H. (1974) Autobiography. In G. Lindzey (ed.) *A History of Psychology in Autobiography* Vol. 6. Englewood Cliffs, NJ: Prentice Hall; pp. 327–64. *259*

Mueller, R. H. (1974) *The American Era of James Mark Baldwin.* Unpublished Ph.D. thesis, University of New Hampshire, Durham. *254*

Mueller, R. H. (1976) A chapter in the relationship between psychology and sociology in America: James Mark Baldwin. *Journal of the History of the Behavioral Sciences,* **12**, 240–53. *254*

Munn, N. L. (1933) *An Introduction to Animal Psychology: the Behavior of the Rat.* Boston: Houghton Mifflin. *258*

Munn, N. L. (1950) *Handbook of Psychological Research on the Rat.* Boston: Houghton Mifflin. *258*

Murchison, C. (1930, 1931 & 1934) *A History of Psychology in Autobiography* (Vols. 1, 2 & 3). Worcester, Mass: Clark University. (Reprinted in New York: Russell & Russell, 1961)

Nordenskiold, E. (1928) *The History of Biology.* New York: Tudor. *247, 250*

Nuel, J. P. (1904) *La Vision.* Paris: Bibliotheque Nationale. *253*

O'Donnell, J. M. (1979) *The Origins of Behaviorism: American Psychology, 1870–1920.* Unpublished Ph.D. thesis, University of Pennsylvania, Philadelphia. (University Microfilm International). *253, 254, 258, 259*

Oldroyd, D. R. *Darwinian Impacts.* Milton Keynes: Open University. *244, 247*

Oppenheimer, J. M. (1967) *Essays in the History of Embryology and Biology.* Cambridge, Mass: MIT. *255*

Owen, J. E. (1974) *L. T. Hobhouse, Sociologist.* London: Nelson. *255*

Pauly, P. (1979) Psychology at Hopkins. *The Johns Hopkins Magazine,* **30**, (December) 36–41. *254*

Pauly, P. J. (1980) *Jacques Loeb and the Control of Life: Experimental Biology in Germany and America, 1890–1920.* Unpublished Ph.D. thesis, The Johns Hopkins University, Baltimore. *250, 253, 257*

Pauly, P. J. (1981) The Loeb-Jennings debate and the

science of animal behavior. *Journal of the History of the Behavioral Sciences*, **17**, 504–15. *253*

Pauly, P. J. (1982) Samuel Butler and his Victorian critics. *Victorian Studies*, **25**, 161–80. *245*

Pavlov, I. (1897) *Lectures on the Work of the Digestive Glands* St. Petersburg: Kushneroff. (English translation by W. H. Thompson, London: Charles Griffin, 1902; 2nd ed., 1910). *120*

Pavlov, I. (1906) Scientific study of the so-called psychical processes in the higher animals. *The Lancet*, **2**, 911–15. *123, 252*

Pavlov, I. (1927) *Conditioned Reflexes* (Translation by G. V. Anrep). London: Oxford University. *131, 203, 237, 250, 252*

Pavlov, I. (1928) *Lectures on Conditioned Reflexes* (Translation by W. H. Gantt). London: Lawrence & Wishart. *250–2*

Pavlov, I. (1941) *Conditioned Reflexes and Psychiatry* (Translation by W. H. Gantt). London: Lawrence & Wishart. *250, 252*

Pavlov, I. (1955) *Selected Works*. Moscow: Foreign Publishing House. *250–2*

Pavlov, I. (1957) *Experimental Psychology and Other Essays*. New York: Philosophical Library. *250*

Pearson, K. (1892) *The Grammar of Science*. London: J. M. Dent & Sons. *75, 247, 248*

Pearson, K. (1914; 1930) *The Life, Letters and Labours of Francis Galton*. Cambridge: Cambridge University. *246*

Peel, J. D. Y. (1971) *Herbert Spencer: the Evolution of a Sociologist*. London: Heinemann. *247*

Perry, R. B. (1918) Docility and purposiveness. *Psychological Review*, **25**, 1–25. *230, 258*

Perry, R. B. (1935) *The Thought and Character of William James*. Boston: Little, Brown & Co. *247, 248, 254*

Pfungst, O. (1912) Zur Psychologie der Affen. *Bericht ueber den V Kongress fuer Experimentelle Psychologie*, pp. 200–5. *177, 255*

Pfungst, O. (1965) *Clever Hans: the Horse of Mr. von Osten*. New York: Holt. (German original, 1908). *78–81, 248*

Poulton, E. P. (1890) *The Colours of Animals*. London: Kegan Paul, Trench & Co. *246*

Poulton, E. P. (1908) *Essays on Evolution*. Oxford: Clarendon. *247*

Preyer, W. (1882) *Die Seele des Kindes*. Leipzig: Fernan. (English translation, *The Mind of the Child*. New York: Appleton, 1888–9). *56, 61*

Priestley, J. (1775) *Hartley's Theory of the Human Mind, on the Principle of the Association of Ideas*. London: Johnson. *92–4, 106, 249*

Razran, G. (1958) Pavlov and Lamarck. *Science*, **128**, 758–60. *252*

Reynolds, S. H. (1900) Note on Prof. C. L. Morgan. *Proceedings of the Bristol Naturalists Society*, **9**, Pt. 1, p. 1. *245*

Richards, R. J. (1977) Lloyd Morgan's theory of instinct; from Darwinism to neo-Darwinism. *Journal of the History of the Behavioral Sciences*, **13**, 12–32. *246*

Richter, C. P. (1968) Experiences of a reluctant rat-catcher: the common Norway rat, friend or foe? *Proceedings of the American Philosophical Society*, **112**, 403–15. *253*

Riopelle, A. J. (1967) *Animal Problem Solving*. London: Penguin. *256*

Roback, A. A. (1923) *Behaviorism and Psychology*. Cambridge, Mass: University Bookstore. *258*

Rodman, J. (1974) The dolphin papers. *North American Review*, Spring issue, 12–25. *248*

Romanes, E. (1902) *The Life and Letters of George John Romanes*. London: Longman, Green & Co. *245*

Romanes, G. J. (1882) *Animal Intelligence*. London: Kegan, Paul, Trench & Co. *25–8, 31–3, 177, 245, 255*

Romanes, G. J. (1884) *Mental Evolution in Animals*. New York: Appleton. *27–31, 33, 245*

Romanes, G. J. (1889) On the mental faculties of *anthropithecus calvus*. *Nature*, **40**, 160–2. *177, 255*

Romanes, G. J. (1895) *Darwin and After Darwin*. London: Longman, Green & Co. *245*

Rosenfield, L. C. (1940) *From Beast-Machine to Man-Machine: the Animal Soul in French Letters from Descartes to la Mettrie*. New York: Oxford University. (Revised edition, 1968). *248, 249*

Ross, D. (1972) *G. Stanley Hall: the Psychologist as Prophet*. Chicago: University of Chicago. *247, 248, 254*

Rudolph, F. (1965) *The American College and University: a History*. New York: Knopf. *247*

Russell, B. (1927) *An Outline of Philosophy*. London: Allen & Unwin. (Published in the U.S.A. under the title, *Philosophy*.) *202, 240, 248, 256, 259*

Russell, B. (1967) *The Autobiography of Bertrand Russell* Vol. 1. London: George Allen & Unwin *245*

Russell, B. (1968) *The Autobiography of Bertrand Russell* Vol. 2. (1914–1944). London: George Allen & Unwin. *259*

Russell, B. & Russell, P. (1937) *The Amberley Papers: The Letters and Diaries of Lord and Lady Amberley*. London: Hogarth Press. *245*

Russell, J. (1978) *The Acquisition of Knowledge*. London: Macmillan. *254*

Samelson, F. (1977) World War I intelligence testing and the development of psychology. *Journal of the History of the Behavioral Sciences*, **13**, 274–82. *257*

Samelson, F. (1979) Putting psychology on the map: ideology and intelligence testing. In A. R. Buss (ed.) *Psychology in Social Context*. New York: Irvington; pp. 103–68. *257*

Samelson, F. (1981) Struggle for scientific authority: the reception of Watson's behaviorism. *Journal of the History of the Behavioral Sciences*, **17**, 399–425. *254, 257*

Schniermann, A. L. (1930) Bechterev's reflexological school. In C. Murchison (ed.) *Psychologies of 1930*. Worcester, Mass: Clark University; pp. 221–42. *252*

Sebeock, T. A. & Rosenthal, R. (1981) *The Clever Hans Phenomenon: Communication with Horses, Whales, Apes and People*. New York: N.Y. Academy of Sciences. *248*

Sechenov, I. (1965) *Autobiographical Notes* (Translation by K. Haines). Washington, DC: American Institute of Biological Sciences. (Russian original, 1905.) *250*

Sechenov, I. (1968) *Selected Works.* Amsterdam: Bonset. *106–8, 124, 250*
Seton-Watson, H. (1964) *The Decline of Imperial Russia.* London: Methuen. *251*
Seton-Watson, H. (1967) *The Russian Empire: 1801–1917.* Oxford: Oxford University. *250*
Sharp, S. E. (1899) Individual psychology. *American Journal of Psychology*, **10**, 329–91. *254*
Shaw, G. B. (1932) *The Adventures of the Black Girl in Search of God.* London: Constable. *110, 251*
Scheaffer, L. (1973) *O'Neill: Son and Artist.* London: Elek. *256*
Singer, B. (In preparation) David Hartley, 1705–1757. Psychology Dept, Reading University. *249*
Skinner, B. F. (1976) *Particulars of My Life.* London: Jonathan Cape. *259*
Skinner, B. F. (1979) *The Shaping of a Behaviorist.* New York: Knopf. *258*
Small, W. S. (1900) An experimental study of the mental processes of the white rat. I. *American Journal of Psychology*, **11**, 133–64. *144, 253*
Small, W. S. (1901) An experimental study of the mental processes of the white rat. II. *American Journal of Psychology*, **12**, 206–39. *144, 253*
Smith, H. W. (1959) The biology of consciousness. In C. McBrooks & P. F. Cranefield (eds.) *The Historical Development of Physiological Thought.* New York: Hafner; pp. 109–36. *250*
Sonneborn, T. M. (1975) Herbert Spencer Jennings *Biographical Memoirs, Vol. 47.* Washington, DC: National Academy of Sciences; pp. 143–219. *253*
Spalding, D. A. (1873) Instinct; with original observations on young animals *Macmillans Magasine*, **27**, 282–93 (Reprinted in Haldane, 1954). *14, 30, 245*
Spencer, H. (1855) *Principles of Psychology* 1st edn. London: Longman. *2, 10–12, 28, 64, 244*
Spencer, H. (1870) *Principles of Psychology* 2nd edn. London: Longman. *2, 12, 13, 28, 64, 237, 244*
Spencer, H. (1904) *An Autobiography* Vols. I & II. London: Williams & Norgate. *244, 245*
Spillane, J. D. (1981) *The Doctrine of the Nerves.* Oxford: Oxford University, *249*
Staum, M. S. (1974) Cabanis and the science of man. *Journal of the History of the Behavioral Sciences*, **10**, 135–43. *249*
Stocking, G. W. (1968) *Race, Culture and Evolution.* New York: Free Press. *257*
Stumpf, C. (1930) Autobiography. In C. Murchison (ed.) *A History of Psychology in Autobiography* Vol. 1. Worcester, Mass: Clark University; pp. 389–481. *255*
Sumner, W. G. (1906) *Folkways.* Boston: Ginn. *214, 257*
Tarde, G. (1890) *Les Lois de l'imitation.* Paris: Alcan. (English translation by E. C. Parsons, New York: Holt, 1903). *246*
Thorndike, E. L. (1898) *Animal Intelligence: an Experimental Study of the Associative Processes in Animals* Monograph Supplement No. 8, *Psychological Review.* 68–72, *248*
Thorndike, E. L. (1904) Review of Hall's *Adolescence. Educational Review*, **28**, 217–27. *160, 254*
Thorndike, E. L. (1911) *Animal Intelligence.* New York: Macmillan. (Reprinted in Darien, Conn: Hafner, 1970). *74–6, 248*
Thorndike, E. L. (1913) Educational diagnosis. *Science*, **37**, 142. *248*
Thorndike, E. L. & Woodworth, R. S. (1901) The influence of improvement in one mental function upon the efficiency of other functions. *Psychological Review*, **8**, 247–61. *73, 248*
Tinklepaugh, O. L. (1928) An experimental study of representative factors in monkeys. *Journal of Comparative Psychology*, **8**, 197–236. *233, 234, 258*
Tolman, E. C. (1928) Purposive behavior. *Psychological Review*, **35**, 524–30. *236, 237, 258*
Tolman, E. C. (1932) *Purposive Behavior in Animals and Men.* New York: Century. *231–4, 258, 259*
Tolman, E. C. (1938) The determiners of behavior at a choice point. *Psychological Review*, **45**, 1–41. *236, 258*
Tolman, E. C. (1945) A stimulus-expectancy need-cathexis psychology. *Science*, **101**, 160–6. *240, 259*
Tolman, E. C. (1952) Autobiography. In E. G. Boring *et al* (eds.) *A History of Psychology in Autobiography* Vol. 4. Worcester, Mass: Clark University; pp. 323–39. *258*
Tolman, E. C. (1959) Principles of purposive behaviorism. In S. Koch (ed.) *Psychology: a Study of a Science* Vol. 2. New York: McGraw Hill; pp. 92–157. *259*
Tolman, E. C. (1966) *Behavior and Psychological Man.* Berkeley: California. *258*
Troland, L. T. (1928) *The Fundamentals of Human Motivation.* New York: van Nostrand. *259*
Trotter, W. (1916) *Instincts of the Herd in Peace and War.* London: Ernest Benn. *209, 210, 257*
Twitmyer, E. B. (1902) *A Study of the Knee-Jerk* Philadelphia. (Reprinted in the *Journal of Experimental Psychology*, 1974, **103**, 1047–66.) *133, 134, 252*
Vartanian, A. (1960) *La Mettrie's L'Homme Machine: a Study in the Origins of an Idea.* Princeton, NJ: Princeton University. *249*
Vartanian, A. (1973) Man-machine from the Greeks to the computer. In P. P. Wiener (ed.) *Dictionary of the History of Ideas* Vol. 3. New York: Scribners, pp. 131–46. *249*
Vucinich, A. (1970) *Science in Russian Culture: 1861–1917.* Stanford, Calif: Stanford University. *250–2*
Waddington, C. H. (1975) *The Evolution of an Evolutionist.* Edinburgh: Edinburgh University. *253*
Walker, S. (1983) *Animal Thought.* London: Routledge & Kegan Paul. *248*
Walkin, J. (1963) *The Rise of Democracy in Pre-Revolutionary Russia.* London: Thames & Hudson. (New York: Praeger). *252*
Wallace, A. R. (1864) The origin of human races and the antiquity of man deduced from 'the theory of natural selection' *Anthropological Review*, **2**, 158–87. (Reprinted in Wallace, 1870). *5, 244*
Wallace, A. R. (1867) The philosophy of birds' nests.

Intellectual Observer, July issue. (Reprinted in Wallace, 1870). *6, 246*

Wallace, A. R. (1869*a*) Geological climates and the origin of species. *Quarterly Review*, **126**, 359–94. (Reprinted in Wallace, 1870). *6, 244*

Wallace, A. R. (1869*b*) *The Malay Archipelago.* London: Macmillan. *176, 255*

Wallace, A. R. (1870) *Natural Selection.* London: Macmillan.

Wallace, A. R. (1889) *Darwinism.* London: Macmillan. *23, 245*

Wallace, A. R. (1891) *Natural Selection and Tropical Nature*, 2nd edition. London: Macmillan. *247*

Wallace, A. R. (1903) *My Life.* London: Chapman & Hall. *244*

Wallach, H. (1976) Empiricist was a dirty word. *Swarthmore College Bulletin* (Alumni issue, April), pp. 1–5. *256*

Warden, C. J., Jenkins, T. N. & Warner, L. H. (1935; 1936) *Comparative Psychology* Vols. 1 & 3. New York: Ronald Press. *258*

Washburn, M. F. (1908) *The Animal Mind.* New York: Macmillan. *148, 149, 253*

Washburn, M. F. (1932) Some recollections. In C. Murchison (ed.) *A History of Psychology in Autobiography* Vol. 2. Worcester, Mass: Clark University; pp. 333–58. *253*

Watson, J. B. (1903) *Animal Education.* Chicago: University of Chicago. *145, 146*

Watson, J. B. (1907*a*) Kinaesthetic and organic sensations: their role in the reactions of the white rat to the maze. *Psychological Monographs*, **8**, No. 33. *146, 253*

Watson, J. B. (1907*b*) Reviews of books by H. S. Jennings and by J. Loeb. *Psychological Bulletin*, **4**, 288–91; 291–3. *253*

Watson, J. B. (1908*a*) Review of Pfungst's *Das Pferd des Herrn von Osten. Journal of Comparative Neurology & Psychology*, **18**, 329–31. *81, 248, 253*

Watson, J. B. (1908*b*) The behavior of noddy and sooty terns. *Carnegie Publications*, **103**, (Washington) 187–255. *147, 148, 253*

Watson, J. B. (1908*c*) Imitation in monkeys. *Psychological Bulletin*, **5**, 169–79. *149, 253, 256*

Watson, J. B. (1913) Psychology as the behaviorist views it. *Psychological Review*, **20**, 158–77. *137, 169–71, 228, 254*

Watson, J. B. (1914) *Behavior: an Introduction to Comparative Psychology.* New York: Henry Holt. (Reprinted, 1967). *136, 171, 172, 229, 234, 236, 252, 254, 255*

Watson, J. B. (1916*a*) The place of the conditioned reflex in psychology. *Psychological Review*, **23**, 89–116. *173, 254*

Watson, J. B. (1916*b*) Behavior and the concept of mental disease. *Journal of Philosophy, Psychology & Scientific Method*, **13**, 587–96. *218, 257*

Watson, J. B. (1917) The effects of delayed feeding upon learning. *Psychology*, **1**, 51–60. *230, 258*

Watson, J. B. (1919*a*) *Psychology from the Standpoint of a Behaviorist.* Philadelphia: Lippincott. *222, 225, 257*

Watson, J. B. (1919*b*) A schematic outline of the emotions. *Psychological Review*, **26**, 165–96. *223, 258*

Watson, J. B. (1924) *Behaviorism.* New York: Norton. *225–7, 229, 257, 258*

Watson, J. B. (1936) Autobiography. In C. Murchison (ed.) *A History of Psychology in Autobiography* Vol. 3. Worcester, Mass: Clark University; pp. 271–81. *253, 257, 258*

Watson, J. B. & Carr, H. A. (1908) Orientation of the white rat. *Journal of Comparative Neurology & Psychology*, **18**, 27–44. *147, 253*

Watson, J. B. & McDougall, W. (1928) *The Battle of Behaviorism.* London: Kegan Paul, Trench, Trubner. *226, 258, 259*

Watson, J. B. & Morgan, J. J. B. (1917) Emotional reactions and psychological experimentation. *American Journal of Psychology*, **11**, 163–77. *219, 257*

Watson, J. B. & Rayner, R. (1920) Conditioned emotional reactions. *Journal of Experimental Psychology*, **3**, 1–14. *222, 258*

Watson, J. B. & Watson, M. I. (1913) A study of the response of rodents to monochromatic light. *Journal of Animal Behavior*, **3**, 1–14. *169, 254*

Watson, J. B. & Watson, R. R. (1928) *The Psychological Care of Infant and Child.* New York: Norton. *227, 228, 257, 258*

Watson, R. R. (1930) I am the mother of a behaviorist's sons. *Parents Magazine*, Dec. pp. 16–18; 67. *227, 258*

Webb, B. (1926) *My Apprenticeship.* London: Longman, Green & Co. *247*

Weinstein, B. (1941) Matching-from-sample by rhesus monkeys and by children. *Journal of Comparative Psychology*, **31**, 195–213.

Welsh, P. O. (1963) *The Brave New World of John B. Watson.* Unpublished honors thesis, Harvard University, Cambridge, Mass. *253*

Wertheimer, Max (1912) Experimentelle studien ueber das Sehen von Bewegung. *Zeitschrift fuer Psychologie*, **61**, 162–227. *184, 255*

Wertheimer, Michael (1980) Max Wertheimer, Gestalt prophet. *Gestalt Theory*, **2**, 3–17. *255*

Whitman, C. O. (1919) *The Behavior of Pigeons.* Washington, DC: Carnegie Institute, Publication No. 257; pp. 1–161. *217, 257*

Whytt, R. (1751) *An Essay on the Vital and Other Involuntary Motions of Animals* Edinburgh. *94*

Williams-Ellis, A. (1966) *Darwin's Moon.* London: Blackie. *244*

Wills, G. (1978) *Inventing America: Jefferson's Declaration of Independence.* New York: Doubleday. *249*

Wilm, E. C. (1925) *The Theories of Instinct.* New Haven: Yale University. *256*

Winston, P. H. (1975) *The Psychology of Computer Vision.* New York: McGraw Hill. *255*

Wissler, C. (1901) The correlation of mental and physical tests. *Psychological Review Monographs*, **3**, No. 16. *161, 254*

Woodworth, R. S. & Schlosberg, H. (1954) *Experimental Psychology* (Revised edition). New York: Holt, Rinehart & Winston. *258*

Wundt, W. (1894) *Lectures on Human and Animal Psychology* (Translation by Creighton and Titchener) London: Sonnenschein. (German original of 2nd edition, Leipzig, 1892). *51, 53, 58, 68, 247*

Yakovlev, P. I. (1959) Bechterev. In M. A. B. Brazier (ed.) *The Central Nervous System and Behavior.* New York: Josiah Macy Foundation. pp. 187–210. *252*

Yerkes, R. M. (1906) Objective nomenclature, comparative psychology and animal behavior. *Journal of Comparative Neurology & Psychology*, **16**, 380–9. *253*

Yerkes, R. M. (1907) *The Dancing Mouse.* New York: Macmillan. *151, 254*

Yerkes, R. M. (1911) *Introduction to Psychology.* New York: Henry Holt. *153, 254*

Yerkes, R. M. (1916) *The Mental Life of Monkeys and Apes* Behavior Monographs, **3**, No. 1. New York: Holt. (Reprinted in Delmar, New York: Scholars Facsimiles & Reprints, 1979). *196–9, 255, 256*

Yerkes, R. M. (1917) Methods of exhibiting reactive tendencies characteristic of ontogenetic and phylogenetic states. *Journal of Animal Behavior*, **7**, 11–28. *254*

Yerkes, R. M. (1921) *Psychological Examining in the United States Army.* Washington, DC: Memoirs of the National Academy of Science, No. 15. (Excerpts reprinted in Dennis, 1948.) *256*

Yerkes, R. M. (1925) *Almost Human.* London: Jonathan Cape. *256*

Yerkes, R. M. (1928) The mind of a gorilla; Pts. 1 & 2. *Genetic Psychology Monographs. 201, 256*

Yerkes, R. M. (1929) The mind of a gorilla; Pt. 3. *Comparative Psychology Monographs, 5,* (Baltimore) 1–92. *201, 256*

Yerkes, R. M. (1932) Psychobiologist. In C. Murchison (ed.) *A History of Psychology in Autobiography* Vol. 2. Worcester, Mass: Clark University; pp. 381–407. *253, 256*

Yerkes, R. M. & Coburn, C. A. (1915) A study of the behavior of the pig, *Sus scrofa*, by the multiple-choice method. *Journal of Animal Behavior*, **5**, 185–225. *155, 156, 196, 254*

Yerkes, R. M. & Dodson, J. D. (1908) The relation of strength of stimulus to rapidity of habit formation. *Journal of Comparative Neurology & Psychology*, **18**, 459–82. *151, 152, 254*

Yerkes, R. M. & Learned, B. W. (1925) *Chimpanzee Intelligence and its Vocal Expression.* Baltimore: Williams & Wilkins. *256*

Yerkes, R. M. & Morgulis, S. (1909) The method of Pavlov in animal psychology. *Psychological Bulletin*, **6**, 257–73. *152, 153, 254*

Yerkes, R. M. & Watson, J. B. (1911) Methods of studying vision in animals. *Behavior Monographs*, **1**, No. 2. *169, 254*

Yerkes, R. M. & Yerkes, A. W. (1929) *The Great Apes: a Study of Anthropoid Life.* New Haven: Yale University. *256*

Young, R. M. (1970) *Mind, Brain and Adaptation in the 19th Century.* Oxford: Clarendon. *249, 250*

The names of people referred to in the text are given in CAPITALS.

Name and subject index